*Created and Directed by Hans Höfer*

**INSIGHT GUIDES**

# amazonWILDLIFe

Produced by Hans-Ulrich Bernard

Photography by Luiz Claudio Marigo, Michael
and Patricia Fogden and others

Editorial Director: Geoffrey Eu

HOUGHTON MIFFLIN COMPANY

APA PUBLICATIONS

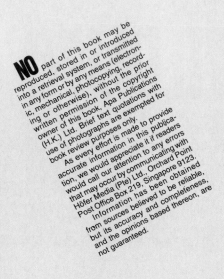
# AMAZON WILDLIFE

*First Edition (Revised)*
© 1994 APA PUBLICATIONS (HK) LTD
*All Rights Reserved*
Printed in Singapore by Höfer Press Pte Ltd

Distributed in the United States by:
**Houghton Mifflin Company**
222 Berkeley Street
Boston, Massachusetts 02116-3764
ISBN: 0-395-66426-8

Distributed in Canada by:
**Thomas Allen & Son**
390 Steelcase Road East
Markham, Ontario L3R 1G2
ISBN: 0-395-66426-8

Distributed in the UK & Ireland by:
**GeoCenter International UK Ltd**
The Viables Center, Harrow Way
Basingstoke, Hampshire RG22 4BJ
ISBN: 9-62421-146-9

Worldwide distribution enquiries:
**Höfer Communications Pte Ltd**
38 Joo Koon Road
Singapore 2262
ISBN: 9-62421-146-9

# ABOUT THIS BOOK

**a**mazon *Wildlife* is the fourth guidebook from APA Publications devoted to the nature and nature reserves of the world. As the home to the largest tropical rain forest and one of the last major expanses of natural habitat in the world, the Amazon has been in recent years the focus of attention of environmentalists and nature lovers throughout the world.

Fortunately, the gloomy picture of massive nature destruction, as widely depicted by the press, is not entirely correct. More than 80% of the region's forest is intact, and becoming increasingly accessible to nature tourism. *Amazon Wildlife* is the first book to not only provide a general description of the region's nature, but also to give precise, hitherto unavailable information on where to find and how to approach South America's tropical ecosystems.

## The Right Staff

*Amazon Wildlife* was conceived by **Hans-Ulrich Bernard**, who also edited *Insight Guide: Southeast Asia Wildlife*. Bernard, together with APA founder **Hans Höfer** (who is a firm believer in the nature series, having previously published guides to *Indian Wildlife* and *East African Wildlife*) and Editorial Director **Geoffrey Eu**, guided the book through the editorial process.

In his other life, Bernard works as a scientist at the Institute of Molecular and Cell Biology at the National University of Singapore. Besides his research work as a virologist, he is an ardent advocate of environmental protection. He is also a keen field ornithologist, having traveled extensively around the national parks of the world in search of fine-feathered friends.

For this book, Bernard was able to gather a team of outstanding professional experts on the tropics of South America. He contributed several chapters himself, among them the pieces on mammals, insects, traveling in the region, boat excursions from Manaus, and national parks close to Rio.

Several contributors are scientists at the University of Vienna, whose biologists have undertaken a number of expeditions to Brazil and Peru.

**Wilfried Morawetz**, professor at the Institute of Botany, wrote about the region's different ecosystems, as defined by their plant cover. Colleague **Walter Hödl**, professor of herpetology at Vienna's Institute for Zoology, highlights the amazing diversity of Amazonia's frogs. He also describes Manaus, the ecology of floating meadows, and nature observation in the Guianas.

**Martin Henzl** penned the piece on reptiles, and also provided some beautiful photographs of reptiles, amphibians, and insects.

**Peter Krügel** described a typical plant family of the neotropics, the bromeliads, and the ecological niches created by these plants.

**Manfred Aichinger** and **Margarethe Roithmair** spent four years in a small village in the Amazon of Peru and their concise account of and fascinating daily life in this village helps to shed some light on native cultures and customs.

**William Overal** from the Goeldi museum in Belém sacrificed valuable research time to author chapters on the Amazon River, Indians and Caboclos, and the island of Marajó. Overal and Brazilian colleague **João Batista da Silva**, a specialist on orchids, expand on some particularly interesting species of

*Bernard*  *Morawetz*  *Hödl*  *Quintela*

Amazonian flora.

**Martin Kelsey**, a professional ornithologist, is in charge of the Latin America program of the International Council for Bird Preservation in Cambridge, England. Kelsey has spent several years studying the avifauna of Colombia. He provided the excellent chapter on birds, and also wrote about the impact of the illegal trade in parrots. Kelsey and his wife, **Claudia Camacho**, a former employee of the Colombian tourist association, wrote the chapters on Colombia's national parks, which also extend to several areas beyond the Amazon.

**George Monbiot**, nature journalist from Oxford, has written and broadcast widely about the Amazon's ecological problems, such as those detailed in his book *Amazon Watershed*. Here, he is responsible for the chapters on fishes, the Manaus fish market, the role of modern Man in the Amazon, conservation strategies, and the village of Tomé Acú.

**Klaus Jaffe**, professor of biology at the Simon Bolivar University of Caracas, contributed the chapter on the numerous national parks of Venezuela.

**William L. Murphy**, author of *A Birder's Guide to Trinidad and Tobago* and operator of ecotours to the region, wrote the article on Trinidad.

**Nigel Dunstone** from the University of Durham, U.K., has led several expeditions to the region and is responsible for describing the national parks of Ecuador and Peru.

**Carlos E. Quintela**, who works in the Latin America program of The Nature Conservancy, USA, is on leave and serving as an advisor and executive director in the national park department of the government of Bolivia in La Paz. He wrote about the national parks of Bolivia, the Pantanal in Brazil and also gives an account of the problems of Brazil's Atlantic forest.

**Jeanne Mortimer**, presently in charge of the Worldwide Fund for Nature program on the protection of marine turtles, is based in Kuala Lumpur, Malaysia. She had previously worked toward the protection of the Amazon's river turtles and describes her experiences at the Rio Trombetas. Mortimer is also a contributor to *Southeast Asia Wildlife*.

## Picture Perfect

Most of the authors have contributed some photography to this book. However, this book owes much of its visual appeal and excellence to the outstanding work of **Luiz Claudio Marigo** from Rio de Janeiro, a wonderful talent in the difficult field of nature photography, and to **Michael** and **Patricia Fogden**. The Fogdens are based in Scotland but spend much of their time in Latin America, where they have built up an impeccable reputation as the region's leading nature photographers. Marigo and the Fogdens have spent 15 years in the neotropics creating an unrivaled photographic collection. We are pleased to be able to display their formidable skills in the pages of this book.

—APA Publications

*Kelsey*

*Camacho*

*Dunstone*

*Monbiot*

*Marigo*

# CONTENTS

## Features & People

# Features & Places

## Maps

# TRAVEL TIPS

# NATURE'S GREATEST SHOW

The word "Amazon" stands for the world's mightiest river as well as for the surrounding biogeographic region, the tropical forests of the Amazon river system. These forests grow in the largest area on earth dominated by moist tropical climate and evolved our planet's greatest diversity of plant and animal species. The Amazon is one of the largest remaining contiguous tracts of nature on earth, the only large one in the tropics. More than 80% of these forests are still intact, in spite of localized heavy destruction and the justified fear of further losses.

The introductory chapters of this book try to describe the region's ecosystems, the most conspicuous groups of plants and animals, and Man's interaction with the nature of the Amazon. Subsequent chapters deal with particular geographic areas, with national parks, or, where parks have not yet been gazetted or developed, with nature explorations from the cities of the region. Included is the description of some nature reserves outside the Amazon biogeographical region, which are close to the large cities, and the gateways to the Amazon.

Information and illustrations were gathered by professional biologists and two of the best nature photographers that work in South America. The result is the first comprehensive guidebook to the Amazon for the nature tourist and a documentation of the beauty of the region in this time of increasing environmental concern and conservation movements. With the growing interest in the protection of tropical nature, ecotourism may become an important source of income for the people of the Amazon, and one which is likely to make sustainable use of natural resources.

**Preceding pages**: argumentative jaguars; iguana out on a limb; a Striped-backed Tree Frog in leafy repose; Flagellum Spider carrying a cluster of young; mini-monster: The Leaf Katydid defends itself with spines. **Left**, Venezuela's Angel Falls.

The vast lowland basin that surrounds the Amazon river and its tributaries, Amazonia, as it is often referred to, has a size of 3.7 million square km. The Andes in the west, the Guayana shield in the north, the Brazilian shield in the south and the Atlantic in the east form its geographical borders. The area has a homogenous moist and warm climate, a prerequesit for the growth of tropical rain forest. This climate changes at Amazonia's geographical borders: to the west to the cold mountane climate of the Andes, to the north and south to the dry climate of the *llanos* of Venezuela and the *cerrado* and *chaco* of Bolivia and Brazil.

The extreme diversity of plant and animal life in tropical rain forests may stem from the lack of stress from drought or frost – major selective forces in all other ecosystems on earth. In moist tropical climate, temperatures hover in the lowland around 28° C throughout the year, with daily maxima and minima between 32 and 23° C. This climate lacks the extreme heat of the subtropics, and colder temperatures occur only in the mountains. The annual precipitation exceeds normally 2,000 mm, which is about three times more than in most parts of Europe or North America. Much of this precipitation falls throughout the year in form of showers, that most often develop in the afternoon. Prolonged rainfall may occur during the rainy season, which lasts for three or four months, in most parts of the region from November to March.

Moist tropical climate is also found elsewhere in South America. Some of these areas along the Caribbean coast and in the Guayanas are more or less contiguous with Amazonia and hold many common plants and animals. Another region on the Pacific slope of the Andes, is efficiently separated from the Amazon by this mountain chain. Consequently, it evolved a fairly different flora and fauna, which is related to that of Central America. To the south, the dry *cerrado* of central Brazil is an efficient border to the distribution of Amazonian animals and plants, but moist coastal and river corridors have permitted some migration to the Atlantic rain forest, an area of tropical forest in southeastern Brazil, 2,500 km to the south of the equator.

One of the most important factors that influenced fauna and flora of the Amazon was South America's geographic isolation over a long geological period: about 100 million years ago, South America was not connected with North America but linked to Africa, Antarctica and Australia to form the supercontinent, Gondwanaland. Subsequently, Gondwanaland broke apart and the four continents drifted apart on separate tectonic plates. For several ten million years, through most of the tertiary, South America existed as an island, without contact with other continents. Many families of animals and plants that became extinct elsewhere were preserved on this island continent. Examples among the mammals are the edentates, which only occur in the tropical America, or the marsupials, which, however, survived in even larger diversity on another isolated continent, Australia.

South America fused with North America only a few million years ago, in the pliocene, by formation of the isthmus of Panama. Plants and animals could now be exchanged between the two continents. Today, some of the South American species like the Nine-banded Armadillo have spread several thousand kilometers into the northern subcontinent, and ungulates and cats, which arrived from the north, are widespread in South America.

In spite of continental drift, South America remained throughout the tertiary close to the equator with its moist and warm climate. However, the climate varied in coincidence with the Ice Age, when it became drier rather than cooler. During this period, much of the Amazon was covered by a savanna-like vegetation, and the rain forest recessed to some small pockets. These areas served as refuges for the rain forest fauna and flora and as centers of evolution of new species. This process of temporary isolation may have contributed to the biological richness of the Amazon, and areas of particular pronounced biodiversity are seen as remnants of previous moist refuges.

**Upland rain forest in Brazil's Pará state.**

# THE AMAZON RIVER

The Amazon River contributes almost one fifth of the total annual amount of freshwater discharged into the oceans of the world. It has a water flow five times that of the Congo, and 12 times that of the Mississippi. It discharges into the Atlantic within 24 hours as much water as the Thames carries past London in one year. Ocean-going freighters can travel 3,720 km inland from the Atlantic to Iquitos, Peru, where they are nearer to the Pacific Ocean than the Atlantic. This makes the Amazon the longest navigable natural waterway in the world. As the force of the river pushes far out into the ocean, freshwater can be recarded more than 100 km off the coast of South America.

The Amazon River is 6,470 km long and has a total water flow of 160,000 to 200,000 cubic meters per second, varying seasonally. In the dry season, it flows at an average velocity of 2.6 km per hour. This is remarkable since the main river drops only 70 m in the 3,100 m between Peruvian frontier and its mouth.

The true source of the Amazon was discovered in 1953 to be a small stream, the Huarco, rising near the summit of Cerro Huagra in the Andes mountains of Peru. The Huarco becomes the Rio Toto, then the Rio Santiago, then the Rio Apurimac, followed by the Rio Ene, then the Rio Tambo, which flows into a main tributary of the Amazon, the Rio Ucayali which is still in Peru. The Amazon flows across Peru to Brazil where it is known as the Rio Solimões until it reaches the confluence with the large tributary, the Rio Negro which is near Manaus. The last 1,600 km from Manaus to the mouth is called the Rio Amazonas. In the 990 km from the source of the river to Atalaya, on the Rio Tambo, it drops 4,450 m in altitude, but from Atalaya to the Atlantic, it drops only another 194 m.

The word *Apurimac* in the Indian language means "Great Speaker" because of the roar of its rapids. Due to the sudden drop in altitude, the Rio Apurimac portion of the Amazon is full of falls and rapids.

**A small tributary of the Amazon winds its way through lowland forest.**

The steep Andean mountain slopes of the first 990 km of the river are completely different from the lowlands in their aspect, climate, human culture, and in every other feature. However, most people associate the Amazon with the lowland area bordered by tropical rain forest.

The official source of the Amazon on Mount Huagra is a small spring about 15 cm wide seeping out of the spongy high altitude grassland. The small stream formed from this spring, the Huarco, is soon joined by others filled by the melting snow. The clear water of the first kilometer of the Huarco soon turns a murky yellow with mine tailing and then joins the Rio Toro. As the river drops down to the Rio Apurimac, the mountain slopes are covered with a moist, rich cloud forest. The sediments on the bed of the lower Amazon are up to 2,000 m deep in places. After leaving the Andes, the present-day Amazon River flows over the tertiary lake bed until below Tefé in Brazil. The last part of it flows through quaternary sediments.

The upper part of the river has the cold Andean climate; around the source, snow falls frequently. As one descends, the climate becomes progressively warmer, more humid and tropical. The region of cloud forest is moist and cool, but is below the frost level. The rainfall in the lowlands near to the Andes at Iquitos in Peru is high, about 2,600 mm. In the central part of Amazonia the rainfall is considerably less – 1,770 mm at Manaus, but it increases again near the coast to 2,277 mm around Belém.

One of the most important features of the Amazon river system is the different types of water which occur within the basin. This can readily be seen near Manaus, where the Amazon and the Rio Negro flow together. The Amazon has a muddy brown color and is full of silt and alluvial matter. In local terminology, this is called a white water river. The Rio Negro is the color of strong tea and has very little silt, and is an example of a black water river. Where these two big rivers converge at Manaus, the river water is divided into two colors for 15 and 25 km downstream until they mix together.

Apart from their different appearances,

the two water types differ in many important ways that have a profound effect on the flora and fauna of the region, and, as a result, on the settlements of Man also. The white water rivers are only slightly acidic or not at all, with an average ph of 6.5 to 7.4. They have much less humic matter, about 14.1 mg/liter in the Rio Solimões near Manaus. In contrast, the Rio Negro has a ph of 4.6 to 5.2, making it very acidic, and has 26.6 mg/liter of humic matter. The water properties cause different plant species to grow in areas flooded by the different water types. For examples, the famous Amazon water-lily, *Victoria amazonica* grows only in white water areas. It cannot grow in the acidic black water.

cut into one bank and build up silt on the opposite margin. They are characterized by a large number of oxbow lakes made by former channels of the river that have become isolated.

On the meandering rivers one encounters many shortcuts dug through the flooded forest for small boats to use in the flood season to avoid the necessity of sailing around the long loops of the river. This, and the fact that all the numerous sandy beaches are flooded and can be traveled over in boats, makes flood-season traveling much quicker than dry season navigation. There are also numerous lateral channels near the main river. These are called *paranás*. The *paranás* are

In addition to the black and white water types in the Amazon river system, a few rivers have clear water. The biggest clear water river is the Rio Tapajós, which flows north into the Amazon 800 km above its mouth. The Rio Tapajós has only 2.26 mg/liter of humic matter.

The Amazon River itself cuts a reasonably straight line from its beginning in Peru to its mouth, being interrupted by only a few islands such as the large Careiro Island near Manaus. Nevertheless, many of its tributaries, such as the Juruá, are meandering snake-like rivers that have changed their course many times. These winding rivers constantly

heavily settled side ducts of the main stream, and are much used as shortcuts.

Seasonally different rainfalls and the very slight relief of the Amazon bring about regular drastic but predictable changes of the river level: in Manaus, high water at the end of the rainy season in March towers 15 m over the low water marks at the end of the dry season in October.

The forests that are inundated for several months by the rising water are called *igapó* along black water rivers and *várzea* around white water rivers. Migration between the river bed and these flooded forests is an important annual rhythm for many species of aquatic life. For many fishes, the river bed is poor in food, and the flooding of the forest makes access to special ecological niches like fruit dropping from forest trees. For species like manatees, the rise of the water opens access to the floating meadows, which had shrunken to small puddles during the dry season and permits grazing on a rich diversity of water plants.

For the nature tourist who travels by boat, high water permits the access to wildlife observation in the flooded forest that is now accessible by the convenience of a boat ride. The wildlife is concentrated at the treetops, which are not only accessible but even on eye level. The end of the rainy season is also the

wave up to five meters in height whose roar can be heard many kilometers away. Although it is a predictable phenomenon and the local people are usually prepared for it, they still make to high river banks.

Geological research has revealed a very peculiar history of the Amazon. Some 250 million years ago, what is today the Amazon River Basin was an oceanic gulf that opened westward to the Pacific Ocean. At that time, all rivers originated on two Archean massifs, the Guayan and the Brazilian shields and ran towards this western gulf. Then about 100 million years ago, this gulf was cut off by the uplifted Andes mountain chain. A huge inland sea grew, which, of course, contained

season of optimal food supply for many land animals. And species like monkeys are much more agile during this period than other times of the year. For nature observation by boat, such as most tours that start in Manaus, the time from the end of the rainy season, namely from March onward, is therefore optimal.

Close to the coast, a peculiar phenomenon of the Amazon is the powerful tidal bore, the *Pororoca*, which occurs twice per year during the highest tides of the Atlantic. It is a

**Left**, flooded forest along the Rio Negro. **Above**, floodplain on Marajó Island, at the mouth of the Amazon.

many of the original oceanic fauna. These animals could survive only if they evolved to tolerate freshwater, that eventually replaced the original salt water. These geological events are the reason for today's presence in the Amazon river of some species – like stingrays, which elsewhere only occur in seawater.

Slowly, up to 4,000 m deep sediments filled this inland sea and created today's Amazon basin. Eventually, about 60 million years ago, the sea broke through its eastern escarpment, and the Amazon river system eroded into these sediments. A river which would now drain into the Atlantic was born.

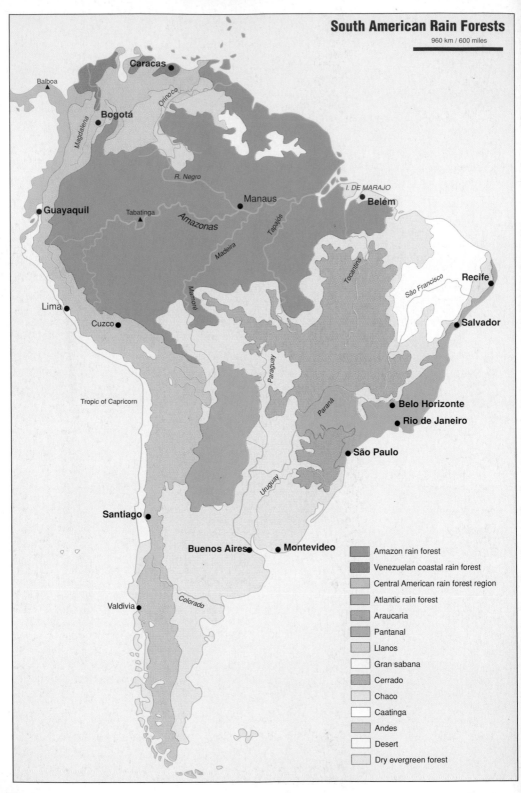

# South American Rain Forests

960 km / 600 miles

Balboa ▲

Caracas ●

*Orinoco*

Bogotá ●

*Magdalena*

*R. Negro*

Manaus ●

*I. DE MARAJO*

Belém ●

Guayaquil ●

Tabatinga ▲

*Amazonas*

*Tapajós*

*Madeira*

*Tocantins*

*São Francisco*

Recife ●

Lima ●

Cuzco ●

*Mamoré*

Salvador ●

*Paraguay*

Tropic of Capricorn

*Paraná*

Belo Horizonte ●

Rio de Janeiro ●

São Paulo ●

*Uruguay*

Santiago ●

Buenos Aires ●

Montevideo ●

Valdivia ●

*Colorado*

| | |
|---|---|
| ▮ | Amazon rain forest |
| ▮ | Venezuelan coastal rain forest |
| ▮ | Central American rain forest region |
| ▮ | Atlantic rain forest |
| ▮ | Araucaria |
| ▮ | Pantanal |
| ▮ | Llanos |
| ▯ | Gran sabana |
| ▮ | Cerrado |
| ▮ | Chaco |
| ▯ | Caatinga |
| ▮ | Andes |
| ▯ | Desert |
| ▮ | Dry evergreen forest |

Amazonia contains the world's largest closed tropical forest system with an estimated size of about 3.7 million square km, roughly the size of the Indian subcontinent.

In strict botanical terminology only the lowlands and some isolated mountain ranges are covered by vegetation belonging to the Amazonian plant kingdom, but adjacent or even disjunct vegetation types in South America like the Atlantic coastal forest of southeastern Brazil show strong floristic or structural relationships.

Species diversity is high and at least 20% of all existing Angiosperms are found in Amazonia. The age of some plant families (up to 120 million years) and several migration and separation events of the tropical forests during earth's history permitted this tremendously diverse evolution. Migration and separation was triggered on the one side by the frequent flooding of the Amazon basin by the sea or by droughts during the Northern Ice Age.

Species diversity survived the climatic disasters only in "refuge areas" of mountain regions, invading and speciating newly into the Amazon basin when conditions became again favorable. Therefore, the present floristic composition of most areas is relatively young.

The most stable habitats possibly are located in the Brazilian coastal mountain ranges, the Central Brazilian shield and the Guayana highlands, regions where many ancient plant families occur in great abundance (like *Winteraceae, Monimiaceae* spp.). Plant families which are typical for the neotropics (neotropical endemics) are the cacti and bromeliads. Both families are thought to originate in rain forest climate, but at present most species are concentrated in dry or mountain climates.

Several phenomena and life forms of Amazonia are also found in the rain forests of Africa and Asia but are unknown in the temperate regions. One of them is the extremely high diversity of plant life. Rain forests are mostly composed of many different tree species (up to 250 species per hectare) in contrast to the uniform fir, maple or beech forests of the northern temperate hemisphere.

Trees are mostly evergreen and broadleafed and highly sensitive to longer durations of coldness or drought. The active shedding of old and nonfunctional branches occurs frequently in tropical trees and old leaves are also shed continuously throughout the year. Subsequently, leaf growth may occur abruptly by flushing: young leaves sprout quickly and

hang down as red, yellow, purple or even blue bundles of soft foliage; only after weeks they turn green and move into their normal position.

The structural diversity of flowers is incredible, going from minute and hardly visible green flowers to enormously big and fleshy ones. The high frequency of bright red flowers (almost lacking in non-tropical regions) is related to bird pollination. Unusual in temperate climate, pollination also occurs by beetles, bats or even mammals. These plant-animal interactions are often highly complicated and evolved over millions of years.

**Preceding pages:** forest undergrowth: which of the leaves is an insect? **Above**, Euterpe palms.

Another example of these interactions are plants like Cecropia and Acacia which supply peculiar ant species with living space and food (edible bodies; foliar nectaries). In return, the plant is protected by the ants which kill or drive away presumable predators. Some of these ants even kill neighboring plants by bites or "chemical weapons". Anybody who has tried to climb a tree inhabited by ants will surely remember their effective defense mechanisms.

Typical for tropical rain forests are lianas and epiphytes. Lianas can build up an unresolvable network of knots and ramifications, many of them strangling each other or the supporting tree. Epiphytes are herbs and

tually and replaces it by developing its own trunk. It is generally believed by botanists that the plant types of the temperate zones are derived from tropical rain forest plants representing impoverished and specialized branches of this complex vegetation.

**Vegetation Types:** Ninety percent of the Amazonian lowlands are covered by the *"terra firme"* forest, a name ("firm ground") that refers to the fact that these forests are never flooded during the rainy season. On poorer soils and in drier areas one finds patches of savanna that are small in comparison to the huge forested area, but sometimes still as big as Uruguay or half the size of Germany.

shrubs that live in the crowns of other trees to get more light. They represent a special life form which is functionally independent of the supporting tree and not at all parasitic. They had to evolve mechanisms to store water and get nutrients in the top of the forest, without connection to the soil.

Oddities are species that begin their life as a liana to shed their roots afterwards and continue as an epiphyte while some other shrubs or trees transform after a certain time into lianas. Strangling figs use yet another strategy. They begin as small epiphytic sprout on a high branch of a tree, develop into a liana that strangles the supporting tree even-

An idiosyncrasy of the Amazon are the flooded forests (*várzea* and *igapó*) which cover two percent of the region. Towards the Atlantic, the seasonally flooded forest is replaced by tidally influenced vegetation, the river mangrove. Structure and floristics of the lowland forest do not change very much up to 900 m on the Andean slopes. The forest of smaller stature at higher levels (up to 2,500–3,000 m) is called mountain rain forest, and is replaced at even higher levels by elfin forests. The timber line is formed by a low scrub which turns into *"paramo"* – vegetation in the moister areas (north of 8° S) and into the *"puna"* in the southern drier parts.

**Terra firme:** Specialists distinguish different forest types, which are never subject to overflooding, but there is still no satisfactory system of well-defined units. Most of the *terra firme* forests are found on deep, well-drained latosols, the typical reddish brown, loam-like soil of the tropics, which can be free of rocks and with a depth of more than 50 m. Also significantly different, are the forests on white sands. Brazilians also distinguish a "heavy dense forest" (*mata pesada*) in contrast to the rarer and lighter "liana forest" (*mata do cipó*).

A primary *terra firme* forest (i.e. a virgin forest that is not disturbed by Man), has usually a closed canopy 30 to 40 m in height,

Leaves of different species are very similar in shapes and textures, most probably due to an adaption to the constant and peculiar climate. They are mostly medium-sized elliptic, with a smooth surface and a long pronounced drip tip, which is thought to facilitate the run-off rain water.

The ground layer is dark, and is reached by only two to four percent of the full sunlight. Consequently, there are few herbs and bushes and one can easily penetrate without using a bush knife. The few herbal plants are reminiscent of plants that can be cultivated at home: their adaption to poor light, poor soil and other unfavorable conditions allow them to grow even in dimly lit living rooms of

with interspersed emergents that many reach up to 50 m. Most of the trees found here are evergreen with relatively thin stems. They branch only at the top and form a closed canopy high above the ground. A cross section of the trunk is irregularly shaped, since the tree extends often into buttress or stilt roots. Palms of all sizes and types determine the character of a *terra firme* forest, while giant trees with enormous stems and crowns are rare.

**Left**, primary forest is not a dense jungle and is surprising easy to walk through. **Above**, lower mountain forest is often shrouded in mist.

houses in temperate climates.

Most of the other plants at the ground level are seedlings from trees. Nevertheless, there are a few shrubs (e.g. *Psychotria*: from the Coffee family or *Rinorea* from the violet family) which are regular woody members of the undergrowth. Although hardly any other plant flowers there, some trees bear their flowers along the trunks (*cauliflory*) or even at the roots (*rhizoflory*).

A special case is seen in some species of *Duguetia* which send long stolons from the trunk along the ground bearing flowers at the end, sometimes in a distance of more than six meters from the mother tree.

For a visitor who is not a botanical expert, identification of the thousands and thousands of plant species is impossible. In fact, a well-guided observation training is necessary to be able to observe the true diversity of a rain forest. Even well-trained specialists find it sometimes difficult to identify more than 10 different flowering or fruiting tree species per day.

The visitor who is unable to identify the exact species may enjoy to concentrate on the structures of the rain forest: the great variety of lianas and of diverse palms which can be stemless, bushy, climbing, or trees of impressive height; buttresses and stilt roots, spiny trunks or unusual bark formations;

need many thousand years to restore them, the denuded earth may remain for years even without grass or herbs and represents a red Amazonian desert.

If the destroyed areas are not too large and some of the biomass remained intact, a secondary forest can develop. Along the road or within cultivated land, one finds dense patches of low forest which are sometimes shown to visitors as "Jungle". These areas are recuperating, they are indeed formerly logged forests (in Brazil's "*capoeira*"). Such forests contain a flora different from and poorer than the virgin forest, which are made up of thin-stemmed lianas and aggressively growing trees and shrubs.

epiphytes like orchids, bromeliads and *Peperomias* high up in the canopy. Scanning of the ground will reveal climbing aroids or *Marcgraviaceae*, and an enormous diversity of fallen flowers and fruits.

**Secondary Growth:** Logged and burnt tropical forests can hardly recuperate. Almost all nutrients of a tropical rain forest are fixed in living plants and animals. When released from dead and rotting organisms they are immediately absorbed by the surrounding plants and returned into the life cycle. The soil is extremely poor and a humus layer is lacking. With the destruction of the forest, the nutrients are gone forever and it would

Characteristics plants of these habitats are the Cecropias of the fig family (*Moraceae*), with large silvery palmate leaves that grow in clusters at the end of the branches. Cecropias are inhabited by millions of small aggressive ants. In the western Amazon the Cecropias may be replaced by the somewhat similar Balsa wood trees (*Ochroma*).

The giant herb *Phenakospermum guyanensis* from the banana family (*Musaceae*) is reminiscent of the travelers tree from Madagascar and is indeed related to it. Its banana-like foliage and Strelitzia-like inflorescence can reach more than 10 m in height.

Shrubs with opposite leaves and an bright orange latex belong to the genus *Vismia* (*Clusiaceae*). Unbranched slender trees with a dense cluster of large terminal pinnatified leaves are young individuals of *Jacaranda copaia* (*Bignoniaceae*). Once grown to larger height, they change growth and leaf form, and become a typical member of the mature Amazonian forest.

**Mountain Rain Forests:** From a height of about 900 m onwards the structure of the lowland forest changes gradually: the mean height of the trees becomes lower, the crowns more spherical and much more branched. Trees don't grow straight but crooked and are less buttressed. The shrub layer is better

the crowns but even at eye level. Due to the permanent moisture by fog and the low evaporation rate, small epiphytes like mosses, lichens and ferns cover stems and rocks. Small thin-leaved and transparent filmy ferns (*Hymenophyllaceae*) inhabit slippery rocks and earth holes. In valleys, along creeks and waterfalls – a rich herbaceous flora has developed with beautiful *Gesneriads*, *Zingibers* and *Marants*, growing under the shelter of large tree ferns.

This forest includes "living fossils" like the conifer *Podocarpus* with leaf-like needles, the little known genus *Mollinedia*, a shrub with small greenish grey flowers, or treelet of the genus *Hedyosmum*, known from

developed than at lower altitudes. Species diversity decreases and species composition changes with increasing height. Especially *Melastoms* which are recognized by their opposite leaves with three to five parallel nerves and by showy red, blue or white flowers dominate the forest.

Epiphytes are crowding the branches and trunks of the trees and include striking individuals of hanging cacti, orchids, aroids and bromeliads, which grow not only high up in

**Left, tree ferns are common in the cloud forest. Above, *Espletia* plants in the high plateau of Ecuador.**

fossil records from a time more than a hundred million years ago.

At higher levels, namely between 1,500 to 2,800 m, the forest can turn into elfin forest, a special type of mountain rain forest with enchanting structures. The low trees are even more gnarled and inclined, branching close to the ground and held together by lianas and nests of epiphytes. The ground is composed of an up to two meters thick layer of rotten wood and leaves, roots, and branches bent down from the trees. One walks on a soft and elastic carpet-like surface as long as one doesn't fall into a hole only covered by a thin layer of organic matter. Any noise is muffled by the ground and the thick moss covering the trunks.

Red and fleshy flowers (*Fuchsia*, *Ericaceae*) shine out of the thicket and are visited and pollinated by hummingbirds, which are common at this height. Lichens, ferns and thousands of small *Peperomias*, orchids, aroids and bromeliads cling to rocks and trees, or grow on the ground. Another plant characteristic of this forest is a steel-blue fern with simple-shaped leaves.

**Paramo:** At a height of 2,500 to 3,000 m the mountain rain forests turn into a closed shrub woodland which eventually gives way to the *paramos*, one of the most typical high altitude vegetation of the neotropics. This veg-

and only once with a single huge inflorescence (similar to many *Agave* species), to die afterwards.

At even higher altitudes, regular frost is common and snow may cover in the morning tropical plants like wild passion flowers (*Passiflora*), mountain-bamboo (*Chusquea*) or members of the numerous species of *Melastoms* and *Ericaceae*. That typical for high tropical altitudes is *Weinmannia* (*Cunoniaceae*), a shrub or tree with tiny white flowers and small simply pinnatified leaves. A special feature are the *Polylepis* forests (*Rosaceae*) occurring at more than 4,000 m in height on large boulders.

An interesting feature found here is the

etation goes up to 4,800 m, and ends at the zone of permanent snow. *Paramos* are open grass- and shrublands with small forest pockets at humid and sheltered sites. They can be recognized best by large rosule plants with silvery and hairy leaves, which belong to many different families.

The most frequent and best known genera are *Espeletia* (*Compositae*) and *Puya* (*Bromeliaceae*). *Espeletia* can vary from a stemless rosule plant to a rosule tree covered by the old foliage (reminiscent of a monk: *frailejón* in Spanish). *Puya* is best known because of its hapaxanty, i.e. the plant flowers after many years of vegetational growth

occurrence of many plant genera. These which are well known from temperate or alpine areas like Plantains (*Plantago*), Barberry (*Berberis*), Valerian (*Valeriana*), Violet (*Viola*) or the many members of the *Compsitae* (sunflower family).

In the drier regions, further South (Peru) the *paramo* is replaced by the *puna* vegetation. This word describes open land with tuft grasses, (thorny) shrubs or extensive cacti associations, which may be covered with several *Tillandsia* species (bromeliads). *Azorella*, also found in some *paramos*, forms large green cushions with an extremely hard surface that are built up by a single plant with countless branches that end in minute hard leaves and flowers. The inside of a more or less spherical cushion is filled with dead leaves.

**Flooded Forests of Lowland Amazonia:** Typical for the Amazon river system is a forest type which is highly adapted to survive long periods of submergence. Following the rainy season in the Andes, the water level of some rivers may rise up to 15 m every year (e.g. the Rio Solimões at Manaus) and the water covers large areas of forests up to eight months.

In comparison to the surrounding *terra firme* forests, these flooded forests hold fewer large trees, but those that occur frequently

during the high water which covers even adult trees completely. To prevent soaking, the bark is covered, in many species, by thick cork tissues, and the leaves are protected by water repelling cuticles. As soon as a part of the crown is submerged, the metabolism is slowed down to a minimum, but leaves are not shed and they can function normally as soon as the water levels fall. Part of the tree that emerge out of the flood may then flower or fruit during this season. A crucial point for survival is the oxygen supply that is partly ensured by aerial roots. It is also suspected that the roots are capable of switching to an anaerobic (i.e. without oxygen) metabolism if necessary.

have well-developed buttress or stilt roots. Herbs are almost lacking and the shrub layer is only composed by saphings of larger trees. The epiphytic flora is richer than in the *terra firme*.

The flooded forest is called *igapó* in the nutrient poor black water systems. *Igapó* is poorer in species composition than the flooded forest along nutrient rich white water rivers that is called *várzea*. Trees in both habitats share the same problem to stay alive

**Left**, flooded forest in Rio Negro. **Above**, after the rainy season, the flooded forest is accessible by dugout.

Trees of the flooded forest evolved several special mechanisms for seed dispersal. One of the method of dispersals takes advantage of the fishes that enter the forest during the flood. The ripe fruits of some species are swallowed entirely by fish and the undamaged seeds are deposited after passing through the digestive tract. Some fruits float easily and are dispersed by currents. Seedlings have to survive in the first years mostly under water, but usually grow quickly during the dry periods.

The flora of the flooded forest is much poorer than in the *terra firme*. Some species are confined to one of the two types of flooded forest, like *Pseudobombax munguba* to the *várzea* or the palm *Leopoldina pulchra* to the *igapó*, others like *Annona hypoglauca* occur in both types. The species composition is often better known than that of the *terra firme* forest, since plant collecting can be done comfortably from the boat during high water.

Quite different to the *várzeas* and *igapós* are the "mangroves". Mangroves grow along the coast or in river mouths in sea- or brackish water that is influenced by tides. With approximately 10 typical species, the mangroves of the Amazon are much poorer developed than in Asia where they probably originated. Mangroves exhibit clear

zonations: an outer belt is dominated by *Rhizophora*, a species with a bark rich in tannin which is often used for tanning. The inner belt holds species like *Avicennia* and *Laguncularia*.

Mangrove plants have similar adaptions to their extreme habitat. The salt of the sea water can be filtered by the roots or can be excreated by special glands at the leaves. Stilt roots (*Rhizophora*) or pneumatophores (*Avicennia*, *Laguncularia*) provide the necessary oxygen. Seeds germinate usually on the mother trees and drop as fully developed plantules into the mud, rooting there quickly (vivipary).

**Savannas:** The *terra firme* forest is often interrupted by areas of open shrub- and woodland or grass savannas. Like the forests, these open habitats can be further subgrouped.

The "savannas of Humaitá" represent the largest open areas in the central Amazon. They evolved probably on old river beds which did not offer an appropriate soil for forests. The poorly permeable latosol allowed only grasses and a limited number of forest species to establish. Furthermore, savanna plants like *Curatella americana*, *Xylopia aromatica* or *Palicourea rigida* may have either become newly established on these poor soils or may have perpetuated as a relict of the widespread savannas which covered almost the whole continent during dry periods of the Pleistocene, one to two million years ago. Another example of forest species tolerant to these conditions is *Physocalymma scaberrimum*. Most of the tree species have leaves with silicate incrustations, that lead to a rough surface and brittle structure.

Completely different are the *campinas* or Amazonian *caatingas*, small groups of dense growth of trees and shrub in the midst of open white sand areas. Signs of early human settlements in *campinas* and the low degree of endemism of plants lead to the conclusion that these open areas are possibly of human origin.

The small and gnarled trunks of the trees are densely covered with orchids, ferns and bromeliads. There is no special *campinas* flora: the plant inhabitants are mostly similar to those of the next forests on white sand, which are sometimes connected with the *campinas* by a transitional zone, the *campinarana*.

The savannas of Colombia and Venezuela are called *llanos*. They are usually represented by grasslands interrupted with small forest patches or gallery forests, frequently with Mauritia palms. Their origin is uncertain, nevertheless, their extension certainly has been influenced by Man. Significantly different are the coastal savannas of the Guianas, which are rich in species and often swampy.

**Above**, palm flowers in the Peruvian Amazon. **Right**, leaves of Cecropia trees are easily recognizable.

The Bromeliads are a specialty of the tropics of the New World, and its members leap to the eye in all ecosystems. About 2,500 bromeliad species are known from tropical and warm temperate parts of the Americas, only *Pitcairnia feliciana* is recorded from western Africa (Guinea). Yet *Bromeliaceae* is not an important source of plant materials for human use. Several bromeliads yield edible fruits but only the pineapple (*Ananas comosus*) supports a large commercial industry with a production of about seven million tons world-

inflated leaf bases) are termed tank bromeliads. The arrangement of tanks varies: in some species, water accumulates in a few or many separate leaf axils, yet in others the central leaves form a large tank surrounded by a few separated tanks provided by other leaf axils.

The tank bromeliads in certain areas constitutes a major feature of the total landscape. Epiphytic tank bromeliads (called *parásitos* in Latin America) usually occur within or below the forest canopy and so far has no proofs that they absorb nutrients by their roots from their

wide. All bromeliads are herbaceous perennials and most reproduce both by seeds and vegetatively. Pollination is by by hummingbirds, bats, and insects as bees, moths, and butterflies in all known cases.

Specific habitats of bromeliads range from sea level to over 4,000 m, from the Peruvian desert to rain forests of eastern Brazil. The three subfamilies: *Pitcairnioideae* are terrestrial, some of the *Bromelioideae*, and most of the *Tillandsioideae* are epiphytic (grow upon other plants or even on telephone wires as *Tillandsia usneoides*, "Spanish moss").

Bromeliads which collect water in their leaf axils (due to tightly-fitting, overlapping and

hosts. Bromeliad leaves bear special epidermal cellular structures (trichomes) that absorb water and minerals from the leaf surface. During and after rains, water dripping from trees is enriched with minerals leached from tree leaves. The tanks also tap pollen, dead leaves, twigs, and seeds falling from the trees which break down in the tanks to form a nutritive soup available to bromeliads and their inhabitants. The total water capacity varies with size and species, from a few milliters to several liters (up to 45 liters stated for a *Vriesia imperialis*, a native to Brazil).

Many organisms use tank bromeliads for various purposes: ants, birds and bats eat the leaves, flower stalks, nectar, fruit, seeds or pollen; for others the bromeliad provides concealment from humidity or prey, and amphibians seek out impounding bromeliads not only to keep moist and avoid heat but also for breeding.

Centipedes, scorpions, roaches, ants, snakes, salamanders, and lizards frequently occupy the older leaf axils outside the central water-collecting portion of the rosette which can no longer hold water, instead they create a terrarium-like environment, with decomposed in deep shade, but these surely are end points of a continuum. Nearly all of the major groups of fresh water invertebrates have been reported from bromeliad tanks including worms, snails, crabs and the aquatic insects such as dragonflies, stonesflies, caddisflies, bugs, beetles, and flies. The latter seem to have exploited those habitats most successfully with members of over 20 families reported to date. Larvae of over 200 mosquito species have been found in water in bromeliad tanks, and their numbers usually dominate most commu-

organic material that resembles peat, is well suited to support animals requiring a moist but well-aerated substrate.

Bromeliad tanks provide relatively stable habitats for aquatic animals because of the long life span of the plant and the retention of water throughout the year. About 500 species of aquatic organisms are now recorded from this special ecosystem. There seem to be two extreme types of food chains, one based on algae in sun exposed tanks, and one on detritus

**Blooming bromeliads: *Aechmea fasciata* in the Brazilian rain forest (left) and *Bromelia guzmania* from the cloud forest in Ecuador (right).**

nities. Adult female mosquitos of some bromeliad breeding species (*Aedes, Culex, Anopheles*) are capable of transmitting certain infectious diseases as yellow fever, filariasis, encephalitis, and malaria. Thus, the term "bromeliadmalaria" has been coined, although the bromeliad is neither disease, carrier nor host, only the host of the carrier. In parts of South America, malaria has abated significantly by reducing the local bromeliad population.

While bromeliads usually represent only a small component of the complex tropical forest community, they often play a significant role in the formations and their tanks represent focal points for much biological activity.

In the Brazilian forests, legends has it, there exist fierce carnivorous plants capable of swallowing a man whole. Perhaps, but in botanical expeditions to Amazonia, one has yet to encounter anything more terrifying than small insect-eaters comparable to our pitcher plants or sundews.

There are, however, many interesting plants, and one can observe the equally interesting native uses to which many of them are put. Some plants yield food, some medicines, some are made into weapons, others

**Orchids**: For most part of the Amazon, orchids have taken to the trees in order to compete for light in the forest. This mode of existence is distinct from parasitism because the orchid does not harm the host tree on which it is perched. Plants living on trees or other vegetation are called epiphytes. Groups that have adopted this piggy-back life style in the Amazon include ferns, mosses, lichens, cacti, bromeliads (relatives of the pineapple), orchids, and many other flowering plants.

While becoming an epiphyte means more

into tools, and there are those like the hallucinogenic plants that play a part in the culture of folklore of the area. Others like the rubber tree and rosewood, are vital to the economy.

Some plants that originated in the Amazon, like the Brazil nut, have remained commercial specialties of the region. Others have spread throughout the tropics of the world – some can be cultivated much better in other parts of the world, like the rubber tree, or that they started as ornamental plants with a successful march throughout the tropics of the world. The number of such plants useful to Man is legion, and it is quite difficult and somewhat arbitrary to select a small sampling.

light and probably less competition, certain ecological adaptations are required: the aerial habitat is hotter and dryer than conditions on the forest floor, and orchids have special vellum roots for the absorption of water and nutrients. Succulent leaves and pseudobulbs are adaptations for storing water, and the frequent association between orchids and ants probably helps both to protect the orchid from destructive insects and to enrich the humus soil collected around the roots. In some cases, the ants themselves carry this soil from the forest floor, while in other cases this soil is deposited by rain. Symbiosis between orchids and fungi is another adaptation for living on a reduced budget.

Germination of orchid seeds in nature appears to always depend on an association with ectomycorrhizal fungi which provide almost all of the necessary nutrients, since the seed is very small and has no endosperm as a food reserve. Orchid flowers are highly specialized for pollination by insects, and the mechanisms by which pollen is transported from one flower to another are often quite complex and bizarre, involving diverse types of mimicry. The orchid bees (sub-

stachya concreta, Rodriguezia lanceolata, and Schomburgkia gloriosa, all of which can be found throughout the Amazon in diverse types of habitats. Other species are more restricted in their occurrence, both geographically and ecologically: among these orchids are Acacallis firmbriata and Gongora quinquenervis, both of which are found only in river flood forest. A. fimbriata is rare, occurring in low densities in Pará, Amazonas, Rondônia, and Amapá, and the destruction of its habitat could lead to its extinction.

family Euglossinae) are known to visit orchids which mimic the scent of female bees and which exploit the sexual ardor of their male visitors. Most orchid seeds are wind distributed and the very few lucky one lodge on tree branches where they can develop.

About 500 species of orchids have been recorded for the Amazon Basin. Many of these are widely distributed, such as Catasetum macrocarpum, Encyclia fragans, Epidendrum nocturnum, Oncidium cebolleta, O. nanum, Orleanesia amazonica, Poly-

Eulophia alta, a terrestrial species growing in open areas, is found widely in the Amazon and has become a roadside flower.

Species like Zigosepalum labiosum, Koellensteimia graminea, Paphinia cristata, Pleurothallis sp., and Maxillaria sp. show a preference for humid, shady habitats. Catasetum longifolium, although widely distributed, is known only from one type of habitat: the crowns of Morichi palm trees (Mauritia sp.). Acacallis cyanea grows in floodplain forest in Pará and Amazonas, but rarely in the open, sandy soil campinas where so many orchids come down to ground level. Most orchid species with distinct habitat

**Left and above, a selection of wild orchid species from the Amazon lowland forest.**

preferences are found in these *campinas*, in flooded forests, or in the Amazon uplands or Andean foothills.

Some species can be considered endemic to certain Amazon localities. *Brassavola fasciculata*, for example, is known only from one collection made in 1975 on the upper Amazon. *Cattleya eldorado*, likewise, is restricted to white sand *campinas* and black water floor forests on the Rio Negro and Rio Uatuma near Manaus. *C. araguaiensis* occurs only on the banks of the Rio Araguaia, and *Catasetum pulchrum*, as far as is known, grows only in restricted localities in Pará. *Catasetum pileatum* is restricted to the upper Rio Negro region.

tropics and subtropics of both hemispheres. Due to the singular beauty of its flowers, the orchid family is well-known, not only to botanical specialists but also to amateur horticulturalists and collectors whose knowledge is often impressive.

Within the Amazon, orchids are still the subject of intensive study due to the extensive geographic areas awaiting systematic exploration and the difficulty with which these areas are reached. These same two factors, extensive areas and difficult access, are also saving graces for the preservation of the Amazon orchids. Two factors threaten these species at present: the destruction of habitats which are sometimes restricted, and

There are cases of apparently restricted or disjunct geographical distributions among Amazon orchid species that, in reality, reflect a lack of systematic plant exploration in the region. This is probably the case of *Catsetum multifidum* which was collected only in Rondônia before being found at the Serra do Carajás, some 1,000 km away.

The orchid family is among the most evolved of the monocots, showing an impressive range of floral characters that vary from microscopic single flowers to enormous inflorescences. This is the most diversified family of flowering plants, with over 25,000 species mostly concentrated in the

the indiscriminate collection of orchids for commercial purposes.

Many large-scale development, agricultural, highway, and colonization projects undertaken in the Amazon since the 1970s have been responsible for the deforesting of huge tracts and have endangered several orchid species. The simple expansion of the urban area of Manaus seems to have extirpated *Cattleya eldorado* from its former habitat. It is possible that several species of orchids, especially micro-orchids that are not sought after by collectors, have disappeared or will disappear from the face of the earth before even becoming known to botanists.

Many regions previously considered unsuitable for agriculture or colonization are now being included in Amazon development projects. The sand from *campinas* is often used in the construction of roads and houses when development encroaches on these patchy environments. This happened near the Tucuruí dam site on the lower Rio Tocantins to several orchid habitats. The Palha *campina* in Vigia, near to Belém, also succumbed, and this former orchid paradise is today completely disturbed and its orchid flora on the way to extinction.

Only the preservation of whole plant and animal communities in national parks, biological reserves, and the like can insure that

woven with the rubber industry. The great rubber boom at the turn of the century led to the fabulous excess of the city of Manaus and to prosperity throughout the whole region. But when the rubber plant, *Hevea brasiliensis* (*Euphorbiaceae*), was exported to Asia and grew better there, the rubber boom in Amazonia ended. Yet today, rubber still ranks as the major industry of Amazonia, and one of the most important exports. Most of the rubber grown is wild rubber, and its collection is a time-consuming job. It is extracted in the dry season, so the same men can work as rubber gatherers and Brazil nut harvesters.

Their routine is interesting in this age of automation. The gatherer begins his circuit

the joy of seeing orchids in their native habitat will not be lost to future generations. But how are these reserves to be created when we know so little about where orchids are found and how they propagate themselves? Surely it is difficult to preserve without more complete knowledge, but a complete regional survey for orchids is time consuming, while habitat destruction is frighteningly fast and terribly irreversible.

*Seringueira*, **Rubber Tree:** The wealth of the Amazon region has been inextricably inter-

**Left** and **above**, colorful orchid species from the elfin forest of the Andes.

about 5 a.m. along a trail devised to include 100 to 120 rubber trees. He makes a diagonal incision on the trunk of each tree with a special sharp knife, and attaches a small tin below this slit to collect the latex. On a large tree, he may make several slits and collect several tins of latex. This first round takes four to five hours. At midday he begins the circuit again, this time to collect the fresh milky latex from each tin. He must hurry with this round in order to get all the latex back before it congeals. By three or four o'clock, he is ready to process his latex into a large ball of crude rubber for sale. To do this, he builds a smoky fire and erects a large wooden spindle over the smoke. He then pours a little latex onto this spindle and rotates it to form a smoke-darkened ball, gradually adding his day's collection until the ball is about 20 to 40 kg in weight. Then he cuts or slips it off the spindle, and it is ready to sell or trade. The balls are bought by traders (very cheaply, and usually by barter), who ship them to the cities, where they are exported to the south of Brazil.

There have been attempts at establishing rubber plantations in Amazonia, but have not yet proven very successful. Fordlândia, for example, on the Rio Tapajós, was a failure because the locality had unsuitable soil and topography. Another venture from Malaysia, the rosewood oil used in making perfumes.

About 50 factories exist in Amazonia for the extraction of this oil. Many move from place to place, in even the remotest areas. To extract rosewood oil, first a factory with a still is set up in an area of virgin forest. Field men select trees above the legal limit of 30 cm in diameter (about 15 years old). These trees are then felled, cut into transportable logs and taken to the factory, where they are fed into a large chipper, and the resultant chips are carried in wheelbarrows to the stills. It takes two and a half hours for 1,100 pounds of chips to be distilled. Each ton of wood yields 17 to 20 pounds of oil. Nearly all the oil is exported. Seventy percent goes to

also failed due to an attack of the fungus *Microcylus ulei*. Some fungus-resistant strains have been produced in Brazil now, and the plantations are yielding more rubber. Since the quantity of plantation rubber is higher than that of wild rubber, and the labor of collecting it is much less, these plantations should be beneficial to the economy.

**Pau Rosa, Rosewood:** These two species of the *Lauraceae* family, *Aniba duckei Kosterm.* and *A. rosaeodora Ducke*, comprise one of the most valuable forest crops in Amazonia. Although other lauraceous trees have a similar scent, these two are the commercial sources of linalool, the major ingredient of

the United States, 14% to Great Britain, 3% to France, and 10% remains in Brazil for its own cosmetic industry and for later exportation in refined forms.

The rosewood was once a very common tree, but exploitation is making it more scarce. There are laws controlling its use, but in areas near factories, trees soon become scarce. The factories, therefore, must move often in order to keep up with the supply.

Research is now being carried out to find a practical way to use the linalool in the leaves so that entire trees need not be destroyed. Some planting of young trees is also being done, but meanwhile the pleasure of finding a sweet-scented rosewood tree while collecting in the forest is growing rarer.

**Guaraná:** Guaraná *(Paulinia cupana var. sorbilis)* from the *Sapindaceae* family has been cultivated for centuries in Amazonia, and is still an important crop. Humboldt and Bonpland first collected it in cultivation in 1800 from the upper Rio Negro, where the Indians used it as a stimulant and as a medicinal plant. When planted in the shade it takes the form of a liana, and in the open it is a shrub. Guaraná bears fruit three years after planting, but the yield increases greatly after five years. The fruit is usually ground into a powder to make a drink by the natives, although there are other uses as well.

in gourds; the factories add sugar and put in sterile bottles.

The Indians at Maués first remove the red outer pericarp of the fruit and the white aril, leaving the black seed. They dry these seeds rapidly without fermentation on a large straw platter and toast them in a primitive sort of oven. Then they place them in a sack and pound them vigorously to separate the seeds from the seed coats, or integument. The separated seeds are transferred to a wooden pestle and ground to fine powder. This is left out one night to be moistened by the dew, and the following day water and cassava flour are added to form a paste.

The Indians have a primitive form of art

In folk medicine, the drink is an important stimulant, comparable to the African cola bean. It is cited as being useful for heart, liver, kidneys, and as an aphrodisiac and analgesic. Today it is bottled in an aerated form and sold as "Guaraná", a popular soda of the region. The preparation of today's commercial drink is basically the same as the primitive, though there is a reliance on machinery rather than hands to do the job. Also, the Indians mix a stronger brew and keep it

**Far left**, rubber tapping, Amazon style. **Left**, latex trickles to its collecton point. **Above**, preparing a latex ball for transportation.

attained by making models out of Guaraná paste, which are then baked and hardened. In modern times, this has been updated by the Brazilians, who use the paste to make models of animals, which they sell as souvenirs.

*Castanha-do-Pará*, **Brazil nut:** Although Brazil nuts, *Bertholletia excelsa (Lecythidacease)*, are popular the world over, few people are aware that they grow like the segments of an orange, arranged 12 to 24 within a large wood cup *(pyxidium)*. This cup is the fruit of the Brazil nut tree, which grows to 30 m tall and begins bearing when it is 10 to 15 years old. One adult tree may produce 500 kg of nuts each harvest. The fruit of the

Brazil nut tree matures in January, about 14 months after the large, attractive yellow flowers appeared. Harvesting Brazil nuts is a hazardous job. The fruiting season is at the height of the rainy season, so workers have to walk and camp in the heavy rains. In addition, the fruits are heavy, trees are tall, and harvesting is done on the ground after the fruits have fallen, so it is not unusual to meet a worker who has been injured by a falling fruit. Workers gather the fruits, chop off the top third with machetes and shake out the nuts. These are then washed in a nearby stream, dried, and shipped.

The annual commercial yield of Brazil nuts is about 50,000 metric tons, a large per-

*Carapanuaba* has value, over and above its function as a mosquito hatchery. It is a highly resistant wood and is not attacked by termites. The lamellated trunk is actually made up of long, thin, twisted buttresses which radiate from a central core. Each buttress is itself divided into irregular, plant like sections which may be no more than an inch thick.

The resistance of the wood and its unusual shapes dictate its uses – it is a common source in Amazonia and the Guianas for paddles and axe handles, which can be produced with little labor from such thin wood. One frequently encounters a *carapanuaba* tree in the forest with a paddle-shaped piece of buttress missing where someone has taken

centage of which comes from the area around the Rio Purus. Most of the nuts come from wild trees in the forest, but there are also a few experimental plantations that contributes to the annual crop. Although the vast majority of the crop is exported, the natives also use the nuts both for food and oil. The nuts are very rich in oil (60 to 70%), like olive oil, can be burnt for light or used in cooking. The cups, too, are useful. One tribe even used them as containers for their arrow poison.

*Carapanuaba*, Paddlewood: The Amazonian name for this tree comes from the mosquitoes (*carapanã*) that breed and live in the dark damp recesses of its fluted trunk.

the wood to make a paddle.

*Cipó d'água*, water vine: This is a particularly valuable plant to expeditions in the remote forest areas, as it yields drinking water. The water vine is a large liana common on high, non-flooded ground away from rivers. When a piece of the stem, two to three feet long, is cut and held up, a good quantity of fresh, clear water runs out freely. The liquid also contains some minerals. The natives contend that the bulk has medicinal value, but this has not been confirmed. The *cipó d'água* plants belong to the family *Dilleniaceae,* and can be of several species such as *Doliocarpus rolandri*, and *D. coriaceus.*

Water can also be drunk from plants of some other families (e.g., from the surface roots of a species of Cecropia, a common tree in Amazonia), but there is a certain danger involved unless positive identification is made. Some poisonous vines yield a small quantity of water too.

*Escada-de-jabuti*, **monkey ladder vine:** The name *escada-de-jabuti*, or "turtle ladder", as it is called in Brazil, is given to a number of species of *Bauhinia* (*Leguminosae*, *Caesalpinioideae*) with an unusual type of stem growth. It is a common plant in all the tropical forests of South America and various areas have different vernacular names for it. In Venezuela it is called *bejuco de*

attract the interest of botanists making their first trip to tropical South America.

Tropical America is home to numerous ornamental plants and park trees, that started from here assisted by Man to a successful spread around the world. In fact, a visitor to parks in other tropical parts of the world, e.g. in Southeast Asia, may identify in local parks and gardens more plants of American than of endemic origin. Examples are the yellow flowering *Allamanda*, various species of *Plumeria* or *Frangipani*, the raintree *Samanea saman*, the palm *Royastenia*, or *Heliconias*, often considered equal in beauty to orchids. No plant, however has been as successful as the water-hyacinth, although

*cadena*, or "chain liana", and in Guyana it is known as "monkey ladder". In fact, these names all refer to the irregular growth of the liana trunk in a ribbon-like "S" or scalariform pattern. The leaf is shaped rather like a cloven hoof, from which is derived another of its Venezuelan names, *pato de venado*, or "deer's foot". The liana climbs to the top of the tallest trees and flowers in the crown of the forest. The unusual stem of *escada-de-jabuti* is always one of the growth forms to

**Left**, the odd-looking Guaraná fruit is used in the manufacture of soft drinks. **Above**, gathering Brazil nuts in the forest.

the kind of spread it took was certainly not on the mind of the exporting horticulturists.

*Aguapé-purua*, **Water-Hyacinth:** Unfortunately, the water-hyacinth's beautiful blossom, which lasts only one day and wilts after sunset, is not its most notable characteristic. The plants over-efficient system of growth and propagation has made it a problem in waterways throughout the tropical world.

A mother-plant, once established, quickly sends stolons branching in all directions, and on their tips new plants rapidly form to send out stolons of their own. The stolons usually break away or decay when the young plants are established, but often they remain connected. The results after only a few generations – which is only a matter of months – is a huge mat of interconnected plants, its petioles swollen into spherical bladders which keep it afloat. In addition, the water-hyacinth propagates rapidly by seed.

The water-hyacinth, *Eichornia crassipes* (*Pontederiaceae*), originated in Amazonia, has never been a serious problem because the balance of nature keeps it under control. However, it has found its way, either by accident or as the result of importation as an ornament, to other parts of the world where it *has* been a serious problem, due to a population explosion of monumental proportions. It has clogged drainage trenches in Guyana,

widely used killer is *2,4-d*; however, it needs frequent reapplication. There are experiments in biological control as well. In Louisiana, a small moth caterpillar which feeds on the plant has been introduced. Although this has proved effective, it is not enough. Meanwhile, water-hyacinths keep appearing in new areas, where they continue reproducing at their phenomenal rate.

Just as no perfect control has been found, no large-scale use for the water-hyacinth has yet been devised. Its low protein content – 0.85% of its total weight, compared to 16.25% for spinach – makes it useless as food or forage. In India, it has been used as mulch for young tea plants and has been found to have

caused concern in Queensland, Australia, and has become a real concern in several regions of Louisiana (after having come to New Orleans as a horticultural exhibit in 1884). It was reported in the Nile for the first time in 1958, and six years later impeded over 1,000 miles of the water and dam system. In Suriname, within two years after the Brokopondo Dam was opened, water-hyacinths were reported to be covering 53% of the surface of the large lake behind it.

A plant whose seed can remain dormant for 20 years and then give rise to 65,000 new plants in a single season is obviously a challenge to weed control experts. The most

value as a plant fertilizer. But the demand for tea plant mulch hardly approaches the supply of water-hyacinths, and therefore that supply multiples geometrical every day.

**Crops**: Most crops, that grow close to the equator, are today pantropically distributed. Most of them originated, however, only in one continent or even in a very restricted area. Examples are the coconut, which once was endemic to Southeast Asia, or the oil palm, which evolved in West Africa. Many of these widespread crops are of Amazonic origin, like cacao, or papaya, or manioc. Manioc (*Manihot utilisima*), also called yuca or cassava, is the staple carbohydrate crop of

Amazonia, a starchy root crop. Bitter manioc is poisonous unless prepared correctly to remove the cyanide poisons. To do this, the natives soak the roots, then grind and roast them to prepare a flous which is eaten with all meals. Sweet manioc, called *macexeira* in Brazil, is not poisonous. It is boiled like potato and eaten. Another advantage is that manioc grows well in poor soil, and recently has been used to produce alcohol for powering motor vehicles. Manioc is of particular importance since all other cereal crops, even including rice, do not grow well in Amazonia.

Fruit trees are grown mainly in house gardens and around small farms. The most common fruits of Amazonia include: bana-

grown extensively in Amazonia – especially in Rondônia. Coffee is mainly grown in southern Brazil. *Robusta* coffee is being grown locally within Amazonia and yields a poorer quality coffee than *Arabica*.

Jute is an extremely important crop of the floodplain. The plant is grown beside rivers. Upon harvesting, the cane is soaked and then beaten to detach the fibers. The fiber is then sent to factories in Manaus for processing.

The shallow, sloping banks and beaches of the floodplain constitutes a fertile, useful natural area for cultivation of quick growing crops. This area is mostly used to grow bananas, corn and more recently soy. The soil is naturally re-fertilized each year by the

nas, limes, papayas, mangoes, cashew nuts, passion fruits, bread fruit, jackfruit and many lesser known local fruits seen in the markets in Iquitos, Manaus and Belém. Large fruit orchards are non-existent, although cashew is now being cultivated in large quantities in the Salinópolis region of Pará near the estuary.

Cacao or the cocao bean is a native crop of Amazonia. The main source of cocoa was Bahia in eastern Brazil; however, now it is

**Plants from the Amazon are commonly found throughout the tropics. Far left, water-hyacinth and left, papaya tree and its delicious fruit. Above, palm flowers and fruits.**

floods. Sugarcane is found in many house-gardens. Along the *Transamazon Highway*, it is being grown commercially to produce alcohol as a motor fuel.

A relatively recent introduction to Amazonia is black pepper. This is grown mainly by Japanese farmers near Belém and Manaus. This crop requires fertilizer, but because of its high selling price, it is economical even with the additional cost of fertilizer. Fertilization is also facilitated by using chicken manure from chickens raised by the pepper farmers. Some areas can no longer grow pepper because they have been attacked by the disease, *Fusarium*.

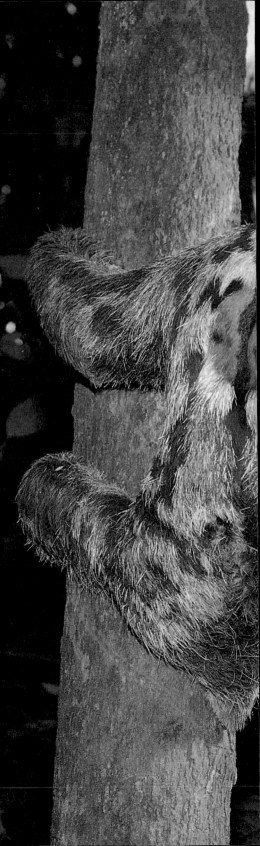

# MAMMALS

Observations of mammals and birds constitute for most people a particularly important part of their nature experience. Birds fascinate because they often have bright colors and inspiring vocalizations and since they are fairly easy to find and to identify. In contrast, most mammals are of subdued coloration, fairly silent, and shy or nocturnal in their activities. But we are attracted to mammals because appearance and behavior are comprehensible to us and make us feel the fairly close relationship. In South American rain forests, monkeys are frequently seen and will satisfy some of this sort of human curiosity. But many other mammals will most likely not be found, although they have high priority on the wish list of the nature observer, such as the jaguar, the largest predator of South America, or anteaters, which look so different from a *bona fide* mammal. In addition to their naturally secretive life-style, many of these mammals have become rare and shy under the extreme hunting pressure of the last five centuries.

To avoid disappointment, it is important to approach the Amazon with the attitude that nature comes as a profusion of plant, insect and bird life. Mammals are an occasional bonus rather than the principle objective of nature observation. On a typical trip of one or a few weeks duration, it is realistic to expect to see about a dozen mammals species, mostly monkeys and the two species of dolphins, also bats, of course, and with some luck sloths, coatis, and otters.

It is one of the best approaches to understand South American mammalian fauna, to ask how these animals relate to similar forms in other parts of the world, and how these relationships came about in time and space. If one orders South America's mammals according to their similarity with animals found elsewhere, one could firstly define a category of animals which are very different from those found anywhere else or in most parts of the world, namely the edentates and the marsupials, Secondly, there are mammals, which have have relatives on other continents, but differ very much from them. Examples are the New World monkeys or certain rodents. A third group of mammals differs only slightly from related forms in North America or even Asia, for example many cats or ungulates.

It is believed that the marsupials and edentates were present in South America, when the southern supercontinent Gondwanaland broke up into South America, Africa, Antarctica and Australia, about 100 million years ago. Protected from the competition with other forms of mammalian life that evolved on other continents, these two orders thrived in geographic isolation for several ten million years, and evolved unusual forms like the Giant Ground Sloth or predatory marsupials, which are extinct today. On other continents, without isolative protection, the edentates could not keep up with the success of higher mammals, and survived therefore as an idiosyncrasy of the South American fauna. The marsupials were somewhat more lucky than the edentates, since they reached from South America an area of Gondwanaland, that later broke off to become Australia. Protected completely from the competition by higher mammals, they proliferated on this island continent – until recently, when Man arrived. Under these circumstances, they survived in Australia in a wide variety of life forms, while competition eliminated all marsupials in South America except the opossums.

It is easy to explain the origin of the third group, those mammals that have related species or races in North America, such as jaguar and puma. They reached South America from the north a few million years ago, when the two landmasses became connected at the isthmus of Panama. Unclear, however, is the exact geographic origin and the means of arrival of intermediately divergent mammals such as the New World monkeys and rodents like the pig-sized capybara. On the one side, primates and rodents evolved only subsequent to the separation of South America and were initially not present on this continent, on the other side, they are far too diverse in South

America from their relatives elsewhere to have arrived from the north only a few million years ago.

Whatever the answer to such questions may be, the combination of evolution in isolation or immigration resulted in today's fauna of Amazonia of 300 to 400 mammalian species. In fact, the exact number depends on the inclusion or exclusion of montane and savanna species, and on the opinion about the species or subspecies status of some animals. This chapter, however will neglect species which occur in South America outside Amazonia in the alpine regions of the Andes, or in the rain forests to the west of the Andes, or in the Atlantic

evolving even hyena-like and sabre-toothed, cat-like animals. Many of these forms became extinct upon the arrival of placental mammals across the Panamanian landbridge, in the Pliocene. Nevertheless, 81 marsupial species exist today in the New World, and about a dozen of them occurs in Amazonia.

America's modern marsupials nearly all belong to the opossum family. They normally have a somewhat rat- and shrew-like appearance, with head-body-tail lengths ranging from 12 to 110 cm. Although of similar appearance, they fill quite diverse ecological niches, and a variety of tree- or ground-dwelling species can coexist in the same geographical range. Examples are the West-

forests of southeastern Brazil.

**Marsupials:** Marsupials, or pouched mammals, are considered primitive on grounds that they split away from the common stock of the placental mammals at a very early time of mammalian evolution, more than 100 million years ago, and that they normally became extinct when confronted with the competition by higher mammals. They probably evolved in a part of the earth which is now North America, and spread from here through South America and the connected and ice-free Antarctica to Australia. Just like in Australia today, they flourished in South America and filled many ecological niches,

ern and the Bare-tailed Woolly Opossum, two of many species with the prehensile tail of specialized tree dwellers. There are also several species of Mouse Opossums, and the Water Opossum. This latter one is the only marsupial on earth that lives in the water. It has adopted with the help of its webbed hind-feet an otter-like life-style and diet.

Interestingly, the insectivores, primitive placental mammals like shrews, hedgehogs and moles, have a complementary distribution to the marsupials. They occur worldwide but not in Australia and most of South America. While they have passed the Panamanian landbridge in the Pliocene, they have only reached the slopes of the Andes, and no species occurs in the Amazon or Orinoco river basin.

**Edentates**: The edentates are comprised of three quite diverse looking groups, the anteaters, the sloths and the armadillos, and 16 of the 29 living species occur in the Amazon. The edentates are confined to the tropical America, but have relatives, the pangolines, in Africa and Asia. In earlier geological periods, many more species of edentates inhabited South America. Their most impressive representative was the Elephant-sized Giant Ground Sloth, which still inhabited the continent as a contemporary of Man, and occurs in many Indian legends.

Edentates means "toothless," but only the anteaters have no teeth at all. Their snout is extended to form a rigid trunk. They have a long tongue, which can be pushed far out, 60 cm in case of the Giant Anteater. Ants are trapped between backward pointing spines by rapid movements of the tongue, more than a hundred times per minute, and then they are masticated by pressure against horny papillae of the mouth. This anatomy doesn't completely protect anteaters from bites, and they avoid aggressive types like leafcutting or army ants.

The Giant Anteater, a beautiful animal with a diagonal black and white shoulder stripe and a bushy tail, can exceed a com-

While outcompeted or preyed upon by modern mammals, the group still holds strongly on some niches in today's ecosystems. Important for their competitiveness is their low metabolic rate, which allows them to specialize in nutrient poor diets, namely ants in case of the anteaters and armadillos, and a wide range of leaves in case of the sloths. This low metabolic rate protects armadillos from overheating in their subterrenean life-style.

**Left**, not a mouse, but a Mouse Opossum. **Above**, the smallest of the three anteater species – a Silky Anteater.

bined body and tail length of two meters, and weigh up to 40 kg. It occurs in forest and savanna alike, and is specialized to a life on the ground. In contrast, the Tamandua or Lesser Anteater is specialized to forage in trees. It reaches only half the size of the Giant Anteater, and is adjusted to its arboreal life through the possession of a prehensile tail. The same adaption exists in the Pigmy or Silky Anteater, which has a size of less than 50 cm. It is rarely observed due to its nocturnal life-style. While anteaters are not hunted for food, pet trade and trophy hunt has reduced their numbers in many areas drastically and they face regional extinction.

Sloths are divided into two-toed and three-toed sloths. These names are slightly misguiding, since both subgroups have three toes on their feet: the count rather refers to the number of their fingers. They have round heads with flat faces, and their long arms and legs are extended into long curved claws, which are used to anchor their body while climbing upside down along branches. The sometimes greenish color of their long fur is due to the growth of blue green algae. Color and slow movements camouflage them from the detection by predators.

In many forests the long-haired two-toed sloths of the genus *choloepus* are the most abundant larger mammal. In contrast to the

attacked, they either crouch to the ground or roll up into a ball to protect their unarmored belly. Most species are medium-sized, but the Giant Armadillo can reach a combined head-body-tail length of 1.5 m, and a huge extinct species, had a three meter long body shell used by aboriginal Indians as roof for huts or burial sites. The Giant Armadillo, which is heavily hunted for food, is considered endangered. But most of the other four species in the Amazon, like the Nine-banded Armadillo, which has spread all the way into the southern United States, are faring well, although being considered for human consumption. Another species, the Yellow Armadillo, can be found further south in the

choosy monkeys, both species of this animal feed on a variety of leaves, which are only digestable over weeks in their large stomach and intestine system.

The brown-dappled three-toed sloths, of which there are three species, are also common throughout Amazonia, though more frequently encountered at the edges of the forest and on riverbanks. They feed exclusively on the leaves and fruits of the Ymbahuba tree, which explains why they have never survived very long in captivity.

The armadillos evolved an armor of bony plates, that makes them at first glance look more like a reptile than a mammal. When

Pantanal. Armadillos are often observed passing the road when one is driving at night, or dead on the street side, when their attempts were futile.

**Bats**: Nocturnal life-style and similar general appearance discourage most people to make detailed observations of these interesting mammals. Bats are the only mammals capable of active flight, and this ability has facilitated their distribution throughout the world. There are nearly 1,000 bat species known worldwide and about 100 of them occur in Amazonia. Taxonomists have put bats into 19 different families. Seven of them are restricted to the New World, and three more occur both in the New and the Old World. Missing from the New World is unfortunately the attractive group of flying foxes and fruit bats, wich is so typical for the tropics of Asia and Africa. Their niche is filled by the Spear-nosed Bats, which, however, are smaller than flying foxes.

Similar in their flapping nightly activities, bats have nevertheless become anatomically quite diverse during their evolution into a variety of ecological niches. The prototype bat is an insect eater catching flying insects with the teeth or with the tail membrane. They use echolocation to navigate at night, and to localize their prey. Some bat species have become carnivoric and prey on small

by the sleeping mammal, and laps the small trickle of blood. However, it rarely bothers people sleeping in the open air.

**Primates**: Two families of monkeys occur in South America's rain forests, namely the small marmosets and tamarins, and the larger capuchin monkeys. New World monkeys only live in trees and are quite different from their relatives in Africa and tropical Asia. Advanced primates like great apes and gibbons, or primitive forms like lemurs don't exist in America. A long and, in case of the capuchins, often prehensile tail and a broad nose with nostrils pointing sideways rather than downwards are typical for New World species. There are 51 monkey species in

mammals, even on other bats, on lizards and on frogs. A specialist, the Fishing or Bulldog Bat, is able to localize fishes under water and to grab them with the sharp claws of their huge feet. Many bats have important functions in the ecosystem as pollinators or dispersers of seeds. The Common Vampire, object of gory stories, is a small bat that is able to land and run on the ground to approach large mammals. It makes a small incision with its sharp teeth, unrecognized

**Left**, a Three-banded Armadillo from Brazil's Pantanal region. **Above**, the Vampire Bat is reviled, but fairly harmless in reality.

America, of which about 30 are associated with Amazonian forest. Related, but often different species occur west of the Andes or in southern Brazil's Atlantic forests.

Higher primates evolved fairly late, namely about 35 million years ago. It is believed, that the New World species are related to Old World Monkeys, and consequently must have reached the isolated South American continent by some unknown means, possibly on natural rafts from Africa. This event was much more likely in the geological past, when the Atlantic was still fairly narrow.

Marmosets and tamarins are small, normally of about 300 g weight. Many species are quite handsome: they have manes and moustaches, which just like their limbs can be colored different from the rest of the body. In contrast to other monkeys, they have sharp claws, an adaption to their permanently arboreal life-style. Unusual for most monkeys, they are monogamous, and a founder pair heads a family group. They eat fruits and flowers and forage for small animals. Some species, have a predeliction for plant saps and gums. Only a single marmoset species can inhabit any particular area, but sometimes several tamarin species fit into divergent ecological niches of the same habitat. The center of diversity is upper Amazonia, and no species occurs in northern South America.

subspecies. Some species are mostly black, like the Black-mantle, the Saddle-back, and the Red-handed Tamarin, which can be told apart by having either reddish legs, or reddish arms and legs, or just bright red hands and feet. The Emperor Tamarin has gray fur, a reddish tail and long white moustache.

The marmosets and tamarins of the Amazon are fairing well as long as extensive tracts of their habitat exist. The limited distribution of some species between particular river systems makes them very vulnerable, however. Because of reduced habitat, some relatives west of the Andes or in southern Brazil, like the Cotton-top Tamarin or the Lion Tamarin, may face extinction.

Three species of marmosets are found in the region. The brownish Pigmy Marmoset distributed in the western Amazon is with a weight of 190 g the smallest New World monkey. The Tassel-ear Marmoset occurs between the Madeira and the Tapajós rivers, while the Bare-ear or Silvery Marmoset, a nearly white species with a pink face, lives only south of the Amazon. Related to the marmosets is the blackish Goeldi's Monkey with a western distribution.

The eight tamarin species of the region have evolved numerous subspecies with quite divergent colors. Rivers form frequently impassable barriers to particular species and

The capuchin monkeys are much bigger animals than the marmosets and tamarins, most species weighing one to several kilograms. They are also much more diverse in their appearance and their ecological adaptions. Twenty-four of the 30 capuchins are associated with the Amazonian rain forest, but they are more widespread than the marmosets and tamarins and occur in many parts of Colombia and Venezuela. Some have extended their range from the lowland into the upper montane forest above 2,000 m.

The capuchins have evolved the world's only night-active higher primate, the Night Monkey, a fairly small monkey with a grey or brown back and a light or orange underside, but no prehensile tail. The eyes are surrounded by white patches, which are separated and surrounded by black stripes. To make contact with mates, the male calls in an owl-like manner in moonlit nights. The Night Monkey may have evolved its nocturnal habits as a protection against diurnal predators, and to use food sources where it would have competition by day.

Among the many sub-species, the Squirrel Monkey is probably the most common primate of the region and outcompetes other monkeys in search of food by sheer group size. With its black snout and black upper head and white fields around the eyes, it has

gave rise to these monkeys' vernacular name.

The six species of sakis and two species of titis are particularly handsome animals. Male and female of the Pale-faced Saki have different coloration, the male has black fur with a white, somewhat owl-like face and a black snout, the female being uniformly brown. The Monk Saki with a long grey and white-mottled fur, is encountered south of the Rio Negro and Amazon. In contrast to the long-haired sakis and titis, the woolly monkeys have a very dense and short-haired fur, as well as a prehensile tail. Widespread is the Common or Humboldt's Woolly Monkey, while the Yellow-tailed Woolly Monkey is restricted to Peru.

a somewhat Mickey-Mouse-like face, although the big ears are of light color rather than black.

Widespread is also the Brown Capuchin, while Weeping Capuchin only occur to the east of the Rio Negro and north of the Amazon. The White-fronted Capuchin is to be found to the west and south. Most capuchins combine a white face with a brown or black upper head, a pattern that reminded early naturalists of cap and habit of monks, and

**Left**, a wide-eyed Night Monkey. **Above**, it's not difficult to see why the Emperor Tamarin got its name.

Rare and restricted to the flooded forest are the Red and the Black Uakari, the only New World monkeys without a long tail. The former one has white fur, a red and wrinkled face, and a bald head which give an uakari in his best years a somewhat senile appearance. It occurs on islands of the upper Amazon, and many riverboat tours make a special effort to find this species, since it appeals to many observers.

Amazonia's three howler monkeys are unrelated to Southeast Asia's gibbons, but form with their inspiring though somewhat frightening vocalizations similar part of the nature experience. Like sloths, they are successful through the ability to feed on the amply available leaves of the forest. The Red Howler Monkey is distributed north of the Amazon, the Red-handed Howler Monkey to the south, and the Black Howler Monkey further south. In a different way reminiscent of gibbons are the Black and the Long-haired Spider Monkeys in their ease to move through the canopy setting hand over hand suspended below branches.

**Rodents:** Hares, distantly related with the rodents, occur only with a single species, the Brazilean Rabbit. But the rodents proper are with more than 100 species, the biggest mammalian order in Amazonia. Many rodents are of mouse- and rat-like stature, and hence

In contrast to the mouse like rodents, squirrels, which occur with seven species, are the favorites of human observers, since their appealing bushy tail, and their open life in habitats remote from Man overcomes the handicaps of mice and rats. The Fire-vented Tree Sqirrel with its bright red head, underside and limbs and a body-tail-length up to 60 cm may be the most attractive representative. In general, squirrels are rarer than in the Old World or in North America.

Mice, rats and squirrels were found in South America only during the last few million years, and don't differ much from North American forms. But South America became colonized by rodents nearly 35 million

don't attract too much human interest. Their handicap is to be secretive and small, and considered to be pests, a prejudice which is only appropriate concerning some few species like the House Mouse or the Black Rat, which have been introduced from Europe, inadvertantly with the first sailors. They stay close to human settlements and do not penetrate into the forest. As a matter of fact, since many cities of the region, like Manaus, are quite dirty, it is difficult to avoid this kind of wildlife observation. In contrast, numerous native species of mouse-like rodents and the big spiny rats and tree rats are indigenous denizens of the forest.

years ago. Some of these ancient rodents like the New World porcupines and the capybara, have become quite different from a *bona fide* rodent. In the absence of ungulates, some evolved into some of the ungulate's ecological niches, the capybara, for example, falls into a niche filled elsewhere by pigs and hippopotamuses.

The capybara is the world's largest rodent, weighing up to 66 kg. Some extinct relatives could reach the size of small horses. Sometimes capybaras are referred to as water pigs, since they live much of their life close to water. They are, of course unrelated to the pigs, but with compact, barrel-shaped bodies of similar habit. Capybaras live in groups of up to 40 animals. The male has a large scent gland, called *morrillo*, on top of the snout, and both sexes have two scent glands on each side of the anus. These glands produce in each individual, a different mixture of chemicals. The odor provides an important means of individual and sexual recognition. They are considered attractive game just like agoutis and acouchis, and the paca. The latter has a shape similar to capybaras, but a reddish brown fur with white lateral stripes and lines of dots. The orange or brownish agoutis and acouchis have the appearance of a short-eared hare. An example is the Red-rumped Agouti of the eastern Amazon.

bears. The latter group doesn't enter the lowland forest, and its only South American member, the Spectacled Bear, is restricted to Andean forests.

The jaguar, with a weight that can exceed 100 kg, is the only big cat of the New World. Its fur is golden brown and spotted with black dots and rosettes. It resembles the leopard of the Old World, but, when seen together with this species in a zoo, is decidedly the bigger animal, looking more impressive also through its broader forehead. Just like the leopard, the jaguar has a completely black variety. These are individuals which only differ in some gene, but not being a separate subspecies. The jaguar lives in

New World porcupines, in contrast to their Old World relatives, spend much of their time on trees. Several species have evolved like some monkeys, opossums and anteaters, the tail as a fifth limb to facilitate their arboreal life. This convergent evolution shows the usefulness of this property in this habitat.

**Carnivores:** About 30 species of carnivores occur in the region, and they belong to five families, cats, dogs, raccoons, weasels and

**Left**, a young Red Uakari: this species is close to extinction. **Above**, the Agouti is a rodent that lives only in unflooded forest.

dense forest, often close to water, and preys on all large mammals. It rarely attacks Man, but preys on domestic stock. This adversary relationship with Man, and the value of its fur, has led to dramatic reduction of its numbers.

Seven species of smaller cats include the puma, a highly adaptable species, that occurs throughout the Americas in habitats as diverse as desert, steppe, northern or tropical forests. The puma is uniformly brown, while the jaguarundi and the ocelot, are spotted like the much bigger jaguar. Like all attractive cats of the world, they are endangered by fur and pet trade. Like the jaguar, they are very secretive and can rarely be observed, and the best way to monitor their presence is to identify their tracks in sand and mud of river banks.

Six species of dogs include four with a fox-like appearance such as the Small-eared Dog. Particularly attractive is the Maned Wolf. It is reddish brown with black, unusually long legs, which give him a somewhat antelope-like gait. It occurs at the southern rim of Amazonia, and its long legs are an adaption to move over the top of the tall grass of savannas.

One species of raccoon, the Crab-eating Raccoon, is related to three strangely looking denizens of the forest, the South Ameri-

this species. It has protection in Peru's Manu National Park, where it is one of the sanctuary's prime tourist attractions. The grison is of grey color with a black face, breast and limbs, the tayra looks like a black marten with a pale head.

**Ungulates:** Hoofed mammals are poorly represented in the region, particularly, since many groups, like wild cattle, antelopes or goats, never made it from North into South America. Four camels such as the llama occur on the alpine grasslands of the Andes, while the rain forest holds only the tapir, peccary and deer family.

There are 11 species of deer in South America, and six of these occur in Amazonia.

can coati, the kinkajou and the olingo. Although being carnivores by relationship, these three mammals mostly live on fruits and insects. The coati is a lovely animal with a long, upturned snout, while kinkajou and olingo look like the sad outcome of a breeding experiment, a cross between a monkey and a weasel.

To the weasels belongs the Amazonian Skunk and two otters, the South American River Otter and the Giant Otter. The head-body-tail length of the Giant Otter can reach two meters, and exceeds the size of the Sea Otter of the North American Pacific Coast. The fur trade has dramatically decimated

All species are quite similar to deer elsewhere. The White-tailed Deer has extended its range from North America into the north of the region, and occurs in most parts of Colombia and Venezuela. South of it occurs the Marsh Deer, which can be observed in the Pantanal. The Red and the Grey Brocket Deer have only half the shoulder height of the previous two species, and very small antlers.

The Collared and the White-lipped Peccary are the closest relatives of the Old World's wild pigs and of similar stature. Living in herds of up to 100 animals, they are one of the jaguar's principle prey. Peccaries are famous for an unusual altruistic behavior:

if a group with young cannot evade a predator, a single adult individual confronts it, of course often with fatal outcome for the heroic peccary.

Tapirs are the only odd-toed ungulates of South America. Widely spread is the Brazilian Tapir, while the Mountain Tapir is restricted to Andean Forests and the Baird's Tapir to the Northwest. South America's tapirs are brownish animals in contrast to the black-and-white Malayan Tapir. The stout body is considered to be an adaption to the need to push through the forest underbush. Their snout is a short trunk, which can be used to grasp food such as leaves, like the way elephants can use their much longer trunks.

They are strictly herbivoric, and feed during high water seasons in the rich plant life of floating meadows, and they fast after retreating into the river during the low water season. The manatee was once abundant in the Amazon river system, but has not reached the Orinoco basin. With the arrival of the Europeans, it became heavily hunted for food, and is today rare, endangered, and still inefficiently protected. A species related to the Amazonian Manatee occurs in coastal waters.

The Amazonian River Dolphin or Boto occurs throughout the Amazon and Orinoco river system. It is a member of the long-snouted river dolphins, and its only relatives

Three mammalian groups have adapted to live more or less permanently in the water: While one of the groups, the seals, is absent from the Amazon, the river is home to two species of whales and two species of manatees. Amazonian Manatees are with a length up to three meters and a weight up to 500 kg, South America's largest mammals. Unrelated to the seals, they have a similar body shape, but no hind limbs, and the tail is developed in a whale-like manner to a horizontal fluke.

**Left**, a pair of Bush Dogs on the foraging trail. **Above**, the Puma, or Mountain Lion, can be found from Canada to Argentina.

today occur in the Indus, Ganges and a lake in China. A second whale species, the Tucuxi or Grey Dolphin, is related to marine dolphins, the coastal forms sometimes being considered a separate species. The Boto is with a length up to 2.6 m by far the larger animal, the Tucuxi being only half this size. The dolphins have lost in the turbid river water some of their eye-sight, and for navigation and for hunting fishes, they depend on echolocation. This sense is so precise that Botos can even tell differently shaped fishes apart. As the objects of local mystics and fairy-tales, they are not hunted and hence common, even in very small tributaries.

The world's richest bird life is to be found in the vast expanse of the lowland rain forests of Amazonia. Estimates of the number of species present depends on the geographical criteria employed but lowland Amazonia itself certainly boasts well in excess of 1,000 species. Hundreds more species can be added if we consider all areas within the Amazon catchment, including the alpine *paramos* of the Andes and the subtropical and temperate forests of the foothills. It is, perhaps, more meaningful to express this enormous diversity in terms of specific areas. Within western Amazonia (the richest area of all) studies at more than one site have shown that within just a few square kilometres of lowland forest it is possible to record more than 500 species. These are figures unequaled anywhere else in the world.

It is a matter of considerable biological debate why the lowland Amazon rain forest has such a rich avifauna. Partly it may be explained by the inherent stability of the tropical forest which has, in evolutionary terms, brought about a high degree of specialization with a very intricate level of animal-animal and animal-plant interactions. That way, it is possible to pack a large number of species into the same area, because each will have very narrow and specialized requirements. There are also opportunities for certain traits to evolve which would be difficult in temperate or strongly seasonal environments: for example, those animals that have diets almost wholly consisting of fruit. It is, however, a misconception to think of the Amazon rain forest as being a vast, uniform, unchanging environment. On the contrary, the forest is a very dynamic place. The vegetation is dictated precisely by soil or moisture levels, so that within a small area one can encounter distinct types: areas dominated by particular palms, patches of bamboo, sandy areas, seasonally flooded areas. Over this framework, changes are constantly happening, varying

in scale and time from the movement of the flow and course of rivers to natural clearings caused by treefalls. All of these will bring corresponding changes to the vegetation and resulting changes to the bird life. Thus there are species that may not only just be restricted to zones of forest which flood seasonally (*várzea* forest), but to particular types of vegetation within those zones. A visitor who recognizes that the forest is indeed a mosaic and allocates time to explore the various parts of the mosaic, will see more species of birds than one who restricts his attentions to a narrow range of habitats.

As in tropical forests everywhere, birdwatching in Amazonia can be a frustrating business. Within the forest itself the visitor confronts a bird community which can be subdivided into various components. Bird of the forest floor are often quite skulking, more often heard than seen. Even when birds are glimpsed, the poor light conditions can make the recording of plumage details difficult. However, should a birdwatcher encounter a swarm of army ants, the attendant can become almost oblivious to one, so intent are they to collect the fleeing cockroaches, crickets and spiders as prey! Most birds are in the under-storey, canopy or at mid-levels in the forest profile. Observing these birds will cause an aching neck and a new visitor cannot expect to identify every small bird briefly seen between gaps in the foliage over 40 m above. Many species of insectivorous birds travel around in mixed species flocks. This has the effect of creating clumps of birds, so that a birdwatcher walking through the forest may spend most of his time not seeing anything. Once a flock is encountered, however, there can be great confusion as perhaps 20 or 30 species pass by all too rapidly. Each species is normally represented by pairs or small family parties, so that a flock may contain almost a hundred individuals. These flocks occupy distinct territories and if a visitor is fortunate enough to spend more than just a few days in an area, it is usually possible to identify these territories and relocate the same flock on subsequent days. There are also opportunities to try "sit and wait" techniques. Fruit-eating

**Preceding pages:** the Black-necked Red Cotinga is one of the forest's feathered beauties. **Left,** the Straight-billed Hermit, a hummingbird, attaches its mud-nest to the tip of a palm leaf.

birds of many different species will often visit the same type of tree or shrub. If a fruiting tree is found to attract birds it is well-worth sitting nearby in a concealed place to wait for a range of fruit-eating birds to arrive. Probably the most difficult of all birds are those which habitually fly above the canopy: these include birds of prey and swifts. To see these requires facilities to get above or close to the canopy using towers and tree ladders, or exploit areas where a good view of the sky is possible, such as on the banks of lakes or rivers or beside clearings.

Throughout the neotropics, there are 85 families of birds of which a staggering 30 are endemic, or nearly so, to the region. This is

being of haunting quality and usually comprising tremulous whistles. These may be single notes or in a rising or falling series, depending on the species. They are almost often heard at, or just before, dawn and during the late afternoon. Some sites in Western Amazonia may support about five or six species.

The changing pattern of the watercourses in Amazonia creates oxbow lakes, which in their turn produce weed-choked pools and swamps. This variety of wetlands supports a diversity of herons, egrets and ibises. Some of these are forest-dwelling, feeding along small creeks, and rarely seen out in the open. These include the tiny and rarely seen Zigzag

partly because South America as a continent remained isolated until just three million years ago. The majority of the bird families found in South America are represented in the Amazon basin and each of the important ones will be described below.

Tinamous are ground-dwelling birds which feed mainly on fallen fruits and seeds, although some insect prey may also be taken. They are rather furtive birds, often "freezing" when approached, although sometimes taking flight in a loud and clumsy way when flushed. Their plumage is dull grey or brown, some species having barred or spotted upper parts or flanks. The calls are most distinctive,

Heron, so called for its intricate pattern of barred plumage. The elegant Agami Heron, with a very long, dagger-like bill also feeds close to cover whereas birds like the Cocoi Heron are readily seen along river banks. The Green Ibis feeds quietly in damp areas on the forest floor, but at dawn and dusk is extremely vocal, flying across rivers on stiffly-held wings, giving a loud rollicking call. Over 14 species of herons and egrets may be found within a single locality.

Of all tropical areas, the Amazon basin has a surprisingly small number of species of waterfowl. They include the Orinoco Goose and the Muscovy Duck, which was the ancestor for the familiar farmyard variety. One distinctive species is the Horned Screamer. This is a huge, heavily-built bird which feeds on waterside vegetation and is most often seen perched atop bushes or trees. Its far-carrying call is an extraordinary loud donkey-like braying sound. Despite its bulk, once airborne it flies powerfully and can soar and glide like vultures.

Four species of New World vultures can be seen over the rain forest. They feed on carrion such as dead sloths and other mammals and it is likely that at least one of the species, the Turkey Vulture, detects the food by an acute sense of smell. The large King Vulture tends to soar at great heights and

world. It is remarkably agile, twisting and turning through the canopy to seize large monkeys, its principal prey. It is however, difficult to see since it very rarely soars above the canopy. Other sub-canopy hunters include the Forest-Falcons which mainly feed on small birds. At dusk, appear Bat Falcons, small elegant raptors which as their name suggests, feed mainly on bats, although they also take large flying insects. A very common bird along the water's edge is the Yellow-headed Caracara, which is mainly a scavenger.

The cracids are made up of the chacalacas, guans and curassows. They are all fruit- and seed-eaters, spending most of their time in

may watch for the movements of the Turkey Vultures descending to a carcase.

Diurnal birds of prey form a large part of the avifauna and over 30 species can be recorded from a single locality. The globally distributed Osprey is a familiar species throughout the Amazon. It does not breed in the region, but non-breeding individuals are present throughout the year. The Harpy Eagle is one of the most powerful raptors in the

**Far left**, the Boat-billed Heron can be seen near forested rivers. **Left**, a Rufous-necked Puffbird. **Above**, a Greater Yellow-headed Vulture about to take flight.

trees. Chacalacas are noisy birds, often in small parties, and appear fairly tolerant of areas in which Man is active. In contrast, the curassows are extremely vulnerable to hunting pressure and require forest with little or no disturbance. This being the case, they are excellent indicators of the conservation status of a forest area. Curassows are large, fowl-like birds which have characteristic whistling or humming songs.

Another fowl-like bird are the curious trumpeters. They are terrestrial, long-legged birds, dark-plumaged with lax, pale plumes on the back. They occur in small parties, feeding on reptiles and amphibians and are much prized by Indian tribes as pets because, like domestic geese, they produce loud calls on the approach of intruders.

In swamps and lakesides occur Purple and Azure Gallinules, and Wattled Jacanas. The latter species is polyandrous, the female laying more than two clutches, with different males tending each one and the female playing no role in parental care. A small number of crakes and rails occur in grassy fringes of the forest. The Grey-necked Wood-Rail is particularly well-known, producing a loud dueting chorus at sunset.

Along shady creeks, close to overhanging vegetation, may be seen the shy Sungrebe. This is a member of the Finfoot family, a

levels of the rivers, caused by rainfall patterns far away in the Andean catchments. This translates to river level fluctuation of over 10 m in many areas. When the water level is low, great sand and mud banks are revealed. These provide important nesting areas for Large-billed and Yellow-billed Terns, Black Skimmers and some species of waders such as Collared Plovers. Migrant waders from North America may also congregate to feed on such sites.

Pigeons and doves are represented by fewer than ten species within any site and include canopy species like Ruddy Pigeons, ground-dwelling quail-doves and birds of forest edges and open sites such as Ruddy Ground-Doves.

pantropical group represented by single species in each of the three great tropical zones. A family unique though to New World and comprising a single species is the Sunbittern. This elegant, dainty bird is widespread but difficult to see. It slowly picks its way along forest streams, sometimes climbing into the branches to call: giving a slightly wavering whistle. Its plumage is beautifully patterned with intricate barring of golds, browns and blacks, but only when its wings are outstretched revealing brilliant orange patches can its full splendor be appreciated.

The strongest seasonal effect throughout most of Amazonia is the changing water

Parrots are well-represented and over 20 species can be found in single locality. These range from the spectacular large macaws, such as the Blue-and-Yellow, the Scarlet and the Red-and-Green to Sparrow-sized Parrotlets. The parrots are some of the most obvious of the bird groups, often seen in the late afternoon flying in flocks to roosting sites on river islands. Parrots are attracted to exposed earth banks in certain areas which they visit to consume various minerals, the most spectacular of such sites is found in the Manu National Park in Peru where large numbers of several parrot species congregate.

Sites in western Amazonia may support

nearly a dozen species of cuckoo, including the migrant Yellow-billed Cuckoo from North America. Typical birds of the forest are the Squirrel Cuckoo and Black-bellied Cuckoo, long-tailed birds which creep through tangles of vegetation and hop along branches in a fashion resembling squirrels. They often join mixed species flocks of insectivorous birds, preying on large caterpillars. The Greater Ani frequently occurs in flocks of over a hundred strong, often beside watercourses and in apparent association with troops of monkeys. The Red-billed Ground Cuckoo is a timid terrestrial species which preys on lizards, frogs and large arthropods. It sometimes preys on the animals

sows, owls, nightjars, nighthawks and potoos. The latter are similar in appearance to the frogmouths of Asia and Australia. Cryptically colored, they spend the day perched at the end of branch stumps. At night they undertake sallying flights to catch moths, usually returning to the same perch after each flight. The Common Potoo has an eerie series of mournful whistles, an unforgettable sound, mainly heard on moonlit nights, which local people traditionally believe is given not by the potoo but by a sloth.

Over 15 species of hummingbird may occur within the same area, although each tends to have fairly specific habitat requirements. The hermits are large hummingbirds which

flushed out by raiding swarms of army ants. Related to rail and gallinaceous birds is the extraordinary prehistoric looking hoatzin. Living in groups beside lakes and swamps, its crop resembles the digestive system of a ruminating animal. Pairs are aided by helpers at the nest and the young, if alarmed by a potential predator, will leap from the nest into the water. There they will swim back to the bushes, clambering back through the branches using claws on their wings.

Night birds include the nocturnal curas-

**Left**, a Sunbittern. **Above**, a Wattled Jacana foraging at the edge of an oxbow lake.

are forest-dwellers, traveling sometimes a kilometer or more close to the forest floor, feeding on the hanging flowers of *Passifloras* or on *Heliconias*. Other species like the Glittering-throated Emerald are more pugnacious, defending flowering shrubs on the forest edge. Jewelfronts and Black-eared Fairies are also forest birds but feed mainly in the canopy or subcanopy, darting around flowering trees and also collecting small insects from the tips of leaves.

A pantropical group which shows its greatest diversity in the neotropics are the trogons. The Amazon forest supports the largest number of all. Seven species may coexist, including the Pavonine Quetzal. Trogons have hooting calls, a characteristic sound of the Amazon forest. Normally perched at mid-storey in the forest, they peer from side to side before taking flight to pluck a large caterpillar or cricket from nearby green foliage. All are brightly colored, yet their unobstrusive behavior can make them difficult to spot. Five species of kingfishers are widespread. Three of them, the Ringed, Amazon and Green, are mainly to be found along rivers, whereas the Green-and-Rufous and the Pygmy frequent smaller watercourses, usually within the forest itself.

The jacamars are an exclusively neotropical family which in many respects fill the habit shared by some of the other puffbird species, including the Swallow-wing which nests in sand banks and is common in open country. Forest puffbirds include the Spotted Puffbird which nests in termite colonies.

Barbets are found throughout the tropics and reach their greatest abundance in Asia. In most areas of Amazonia, usually no more than three species will occur together. Although mainly fruit-eaters, barbets also take arthropods. These are often captured acrobatically with the bird hanging upside down to tear open furied dead leaves in which animals like cockroaches seek refuge.

Toucans are also mainly fruit-eaters and are characteristic birds of the Amazon forest.

same niche as Bee-eaters in the Old World. They normally occur in pairs, perched beside open situations such as small clearings, darting from their perches to catch flying insects in their long bills. Up to seven species may be recorded within the same area.

Also restricted to the neotropics are the puffbirds. These include the Black-fronted Nunbirds which are readily observed in semi-open areas around villages or small settlements. Often three or more of these smart black birds with their long, slightly decurved red bills, will gather and produce excited choruses. They are unusual among Amazon birds in nesting in burrows on the ground, a

No fewer than seven members of the toucan family may coexist, including the smaller aracaris and toucanets. Large toucans such as the Yellow-ridged Toucan have croaking calls and are usually seen in pairs or small parties, flying from tree to tree, making a great "whoosh" sound with their wings.

Woodpeckers range from the large crimson-crested and red-necked to the tiny piculets. They share the tree trunks with woodcreepers, a group only found in the neotropics. Unlike the woodpeckers, which are generally boldly patterned, the woodcreepers are extremely similar in appearance. With 15 or more species present in the same area, they can cause considerable identification problems. Whereas wood-peckers have powerful bills which are used like chisels, the woodcreepers have decurved bills, used to prise into crevices to pick out small animals. The most extreme of all are the Scythebills which have extraordinarily long, decurved bills.

A very diverse group are the ovenbirds or furnariids, of which over 25 may be found within one site. These include the Pale-leg-ged Hornero that does indeed produce a nest made of mud which resembles an old brick oven. The group also includes spinetails, which are generally long-tailed dull brown-ish birds which skulk in long grass, although

name derives from the habit of some to follow swarms of army ants, collecting large arthropods as prey items, flushed by the raiding swarm. However, the group as a whole has diverged to fill a wide range of niches. They usually occur in pairs or family groups, although antpittas are normally soli-tary. Some like many of the antwrens and antshrikes habitually enter mixed species flocks of insectivores, indeed *Thamnomanes* antshrikes are key flock species present in most mid-storey flocks. Antbirds occur from the forest floor (antpittas are essentially ter-restrial) to the canopy (some of the antwrens). Most species are habitat specific, restricted to certain types of forest in preference to

there are also forest spinetails and some which creep along branches. Ovenbirds of-ten carry names which describe their behavior, thus Foliage-Gleaners collect prey by searching through leaves. Leaftossers feed on the ground by overturning leaf litter and Palmcreepers creep in palms.

The antbirds, like the furnariids, are exclu-sively found in the neotropics. Fifty species of this group can be encountered in some of the prime western Amazonian sites. Their

**Left**, Hoatzins can lay claim to being one of the strangest birds found in the Amazon. **Above**, a Yellow-ridged Toucan.

others. Some occur only in certain vegeta-tion such as bamboo clumps, others exist only on river islands and rarely cross to the river banks. Of all the forest species, the antbirds are perhaps the most vulnerable to forest fragmentation. The ant-following spe-cies in particular seen to be the first to disap-pear from isolated forest patches. This maybe, in part, due to the spatial requirements of the army ant colonies.

Some of the most attractive of the small forest birds are the manakins. In almost all species the males are gaudily plumaged and assemble in courtship leks where they per-form elaborate dances. These may take place, in the example of White-bearded Manakins, in a special "court" where the birds have taken care to clear the area of small leaves and twigs. Others, such as the Golden-headed Manakin perform their displays in the canopy. Probably the most curious display of all is that of the Wire-tailed Manakin. Both sexes have long filaments in their tails, those of the male being particularly long. During the dance, the male strikes the female across the sides of her head with these projections. One of the most complex of the dances belongs to the Blue-backed Manakin in which two or even three males perform a tightly synchronised bouncing dance.

Related to the manakins, and also essen-

zonian Umbrellabird which has an umbrella-like crest and a long hanging pendant wattle.

Probably the biggest group of all is the Tyrant Flycatchers. Over 60 species have been recorded from single localities. They vary in size from wren-sized, pygmy-tyrants and spadebills to the thrush-sized Boat-billed Flycatcher. Strictly speaking, the longest species is the Fork-tailed Flycatcher, an elegant species which can sometimes be seen in large roosting flocks beside lakes. The kiskadees are a readily recognizable group which perch out in open situations. The most common is the Great Kiskadee which incessantly calls its name and is known locally as "Victor Diaz". The smaller tyrannulets are

tially fruit-eaters are the cotingas. It is another family endemic to the neotropics and show their greatest diversity in the lowland rain forests of the Amazon. Most are polygynous with the males brightly colored. The Screaming Piha, however, is a very drab greyish bird and males form widely dispersed leks in which display is manifest by loud, explosive calls. These are one of the most characteristic sounds of the Amazon forest and are quite unforgettable. Bare-necked Fruit-Crows are one of the largest cotingas and are often to be seen flying across river in small flocks. One of the strangest members of the family is the Ama-

the biggest headaches for any visitor anxious to identify all sees. Many are canopy-dwellers and can only be identified with experience and knowledge of their calls.

Very few North American passerine birds winter in the lowland forest of the Amazon basin. Exceptions are the Barn Swallows and Bank Martins which arrive in August. The former in particular can be a common sight along the rivers. Migrants also arrive from the southern part of South America such as Southern Martins which can form huge roosts in towns like Leticia and Iquitos.

The loud duets of the Thrush-like Wren are a distinctive sound throughout the region. This is a large wren, spending most of its time close to clumps of bromeliads and other epiphytic plants in the canopy. Probably related is the Black-capped Donacobius, a very smart dark brown and cream bird of the tall grasses growing beside rivers and lakes. Pairs will frequently perch high on the grass stems, producing loud whooping calls. In the forest, the small crake-like Nightingale Wren is a denizen of the forest floor and produces a captivating series of notes, delivered singly with ever-increasing pauses between them. The Musician Wren also keeps mainly to the ground and produces a song remarkably like that of a human whistling.

Thrushes are poorly represented, but may

close to settlements often being chosen, presumably to reduce the risk of raids by monkeys and even toucans. The species is polygynous, the male performing a song richly mimetic as it shakes and bobs its plumage. This display is taken to a more extreme level with some of the larger oropendolas. The Russet-backed Oropendola produces a song akin to the sound of water leaving a bottle.

The honeycreepers and tanagers are the group which may comprise over 40 species in a single locality. Many of these species are gems of birds which occur in twittering mixed species canopy flocks, these may include the aptly-named Paradise Tanager, the Opal-

include one or more migrant species from North America. One of the resident species, Lawrence's Thrush is an accomplished mimic, imitating the calls of a range of other species, including manakins and antbirds.

Another important group of birds are the icterids. These are quite diverse, including troupials and orioles, caciques and oropendolas. The Yellow-rumped Cacique is a familiar bird beside water. It is a colonial nester, isolated trees surrounded by water or

**Inspiring names and inspired appearances: <u>left</u>, the Wired-tailed Manakin and <u>above</u>, the Blue-naped Chlorophonia.**

rumped Tanager or the Gold-and-Green Tanager. Others are common species of open or secondary habitats such as the Silver-beaked Tanager or Blue-grey Tanager. Others, such as the Ant-Tanagers are birds of the forest shrub layer and understorey.

Finally, the finches are mainly birds of the river banks, forest edges and cultivated areas. The Red-capped Cardinal is an attractive riverside bird and grassy patches will support parties of small seed-eaters such as Chestnut-bellied Seed-eaters. One species will be encountered in the forest, the Slaty Grosbeak, which has a fluty, thrush-like song, delivered from high, canopy perches.

# REPTILES

Some reptiles like anacondas, caimans, or large river turtles are among the most impressive creatures that inhabit the Amazon and play a prominent role in this ecosystem. Equally interesting are the many smaller species of lizards and snakes that have developed an incredible diversity of life-styles and have colonized all available habitats ranging from subterranean, aquatic, to the forest canopy. In fact, 550 of the 1,200 South American species of reptiles occur in rain forests, and about 270 in the lowlands of the Amazon basin. In the lowland forests, most species are widespread, whereas the eastern slopes of the Andes contain a high number of local endemics. These Andean foothills also support the most species rich reptile faunas in the world, with more than 90 species of reptiles found within a few square kilometers in some areas. Like many mammals and birds, some reptile species suffer tremendously from direct exploitation by humans. They are hunted for their skins, their tasty flesh, or simply because they are considered harmful.

In Amazonia, many vertebrates are difficult to encounter, because they occur in low densities and live a secretive solitary life. However, some reptiles, such as the large aquatic turtles and caimans, make an exception. Caimans are readily observed from a boat at night; when spotted with by the eyes' reflection of a strong torch light. Additionally, the Spectacled Caiman (*Caiman crocodilus*) and the Black Caiman (*Melanosuchus niger*) take extensive sunbaths on beaches. Both species construct nests of rotting debris, which provide constant temperature and humidity for their eggs. Upon hatching, the females and occasionally the males help their offspring out of the nest and guard them for some additional weeks. Caimans feed on a variety of fish, mammals, and birds, which are generally captured and eaten in the water. Attacks on humans are exceedingly rare, but caution should be taken with children and when swimming, especially

**Preceding pages: eat or be eaten – Piranha meets its match in the form of a Black Caiman. Left, an Emerald Tree Boa (*Corallus caninus*) shows off its curves.**

at night. Spectacled Caimans grow up to three meters, whereas Black Caimans may reach more than five meters. Two small (1–1.5 m) species of caimans, the Musky Caiman (*Paleosuchus palpebrosus*) and the Smooth-fronted Caiman (*Paleosuchus trigonatus*), inhabit small forest streams and ponds. Unusual for crocodilians, the Smooth-fronted Caiman is semi-terrestrial and feeds largely on terrestrial vertebrates. Neotropical caimans together with alligators form the family *Alligatoridae*; true crocodiles (*Crocodylidae*) are absent from Amazonia, though two species occur in northern South America.

Nesting aggregations of the Arrau Sideneck Turtle (*Podocnemis expansa*) are one of the most fascinating wildlife experiences in South America. While females of this species attain a carapace length of 89 cm and a weight of more than 90 kg, it's fossil relative, *Stupendemys geographicus*, from the Miocene of Venezuela, was the largest turtle that ever lived; the Museo Nacional de Ciencias Naturales in Caracas owns a carapace of 230 cm length. The Arrau inhabits the main rivers of the Amazon and Orinoco river systems during the dry season, and enters floodplains during the wet season, where it feeds on fruits that fall into the water and other plant material. Nesting occurs at night during the height of the dry season when the sandy river banks and islands are fully exposed. Females generally bury 80 to 90 eggs. After an incubation of about 45 days, 5 cm-long hatchlings emerge and race for the water. Since the European colonization of Amazonia, a drastic decline of the Arrau has taken place. In 1814, Alexander von Humboldt estimated that 33 million eggs were harvested annually on the Orinoco, and in 1863, Henry W. Bates reported the annual export of turtle oil from the upper Amazon amounting to 48 million eggs destroyed. This heavy exploitation continued until the turtles and their eggs were finally protected by laws in this century. Today it remains doubtful, whether the populations of the Arrau are stabilizing. Presently, the only remaining nesting concentrations are on the Rio Trombetas, the Rio Tapajós, and a few places along the Rio Orinoco in Edo, Apure, Venezuela. A smaller relative of the Arrau, the Terecay (*Podocnemis unifilis*), still is abundant in most tributaries of the Amazon. With a length of up to 46 cm and 10 kg in weight, it is now the major source of turtle meat in the area. The Terecay basks regularly on beaches or logs in the river during the low water period. Single females nest on sand or clay beaches all along the inhabited rivers. Probably the most curious turtle, the Matamata (*Chelus fimbriatus*) is easily recognized by its flat carapace with rough conically raised scutes and its grotesque broad head with fleshy appendices. Typically resting in quiet shallow water, a snorkel-like snout allows this species to breath unnoticed by stretching

larly offered on Amazonian markets and their flesh is highly appreciated. For this reason, they have become exceedingly rare around all populated areas.

Some of the most distinct lizards and snakes are also found in or near water. Once heavily persecuted for its skin, the magnificent anaconda (*Eunectes murinus*) now is frequently displayed at tourist lodges. Despite their gigantic dimensions of more than nine meters and 200 kg, the largest snakes in the world, will not attack humans. They kill their prey, large vertebrates, through constriction. Like all other boas, anacondas give birth to live young, with a single litter containing 30 to 80 baby snakes of about 70 cm. The only

its long neck towards the surface. Small fish which pass close to the head are sucked in by the sudden opening of the mouth and expansion of the throat.

A single land tortoise inhabits the Amazonian forests, the Yellow-footed Tortoise (*Geochelone denticulata*). The length of this tortoise varies from 30 to 40 cm. During the early morning and late afternoon, these tortoises search for food, which consists of fruit and other plant matter, mushroom and carrion. Local people collect tortoises around certain fruit trees or attract them with carrion, and keep them in enclosures as live meat preserves. In fact, tortoises are regu-

venomous aquatic snake, the Surinam Coral Snake (*Micrurus surinamensis*), inhabits all of Amazonia and feeds on eel-like fish. Caiman Lizards (*Dracaena guianensis*) are well-adapted to life in water and somewhat resembles their namesake. Another distinct lizard living along the rivers is the Common Iguana (*Iguana iguana*), a large (2.2 m) green "dragon" with a prominent dorsal crest, spiny scales on the head and a huge dewlap. Iguanas usually perch on limbs of trees and will jump into the water when threatened. In some places, local people tame free living iguanas by regularly feeding them fruit, in other cases they are hunted for their meat.

Common Iguanas are a member of the mainly neotropical lizard family *Iguanidae*, which is represented with 40 species in Amazonia. Most rain forest iguanids are arboreal and each species is specialized for inhabiting certain parts of trees. Various slender anoles (*Anolis*) are abundant on leaf litter, trees and even huts near forest. Male anoles display their huge colorful dewlaps and rapidly move their heads up and down during courtship and agonistic encounters. Climbing on thin twigs of trees and bushes, the iguanid, *Polychrus marmoratus* in many respects resembles an old world chameleon. Another iguanid, *Plica plica*, dwells only on the lower parts of big trunks and feeds on tree ants. Tucano Indians of Colombia

extremely fast and seek shelter in ground holes. Their smaller, partially green relative, the Ameiva (*Ameiva ameiva*), is abundant around human dwellings and plantations.

Tropical Geckos (*Hemidactylus mabouia*) often are the first reptiles encountered by tourists, because they live in houses even within the large cities. Following the colonization of Amazonia, the Tropical Gecko has spread along all the major rivers and roads. The only indigenous nocturnal gecko, the Smooth Gecko (*Thecadactylus rapicauda*), occurs on trunks inside primary forest, but also may be found in palm huts close to the forest. Geckos are harmless and help inside houses by feeding upon roaches, yet they are

worship this lizard as the "*Vaimahse*" (Lord of animals), for its distinct large penis with outward curled tips, which it probably displays during courtship.

The second large neotropical lizard family are the teiids (*Teiidae*). Most teiids in Amazonia are small and secretive inhabitants of the leaf litter. A large teiid, the Northern Tegu (*Tupinambis nigropunctatus*) prefers open sunflooded patches in the forest and clearings. The omnivorous tegus run

**Reptilian splendor on display. Left, a Vine Snake (*Oxybelis argenteus*). Above, a Parrot Snake (*Leptophis ahaetula*).**

feared as venomous by many local inhabitants. Geckos are adapted to nocturnal life with their cat-like vertical pupils, and have broad toe-pads enabling them to climb even smooth surfaces like glass.

The majority of lizards are habitat specialists and diet generalists, whereas most snakes are food specialists able to survive months without nutrition, but utilize a rather broad range of habitats. Snakes are rarely encountered in rain forests, although their highest diversity occurs there and even specialists need months to find a substantial fraction of the sometimes more than 60 species at a single site. Promising times to look for snakes are after heavy rains or storms. Behind the anaconda, the boa constrictor ranks as the second largest snake in Amazonia, with more than five meters maximum length. Boa Constrictors can be aggressive and will emit loud hisses when harrassed. Probably the most common boa in Amazonia, the beautiful Rainbow Boa (*Epicrates cenchria*) is hunted by local farmers because it frequently preys on chicken. Rainbow Boas are rivaled in beauty by Emerald Tree Boas (*Corallus caninus*), which have a bright green dorsum with white vertebral marks and a yellow venter, providing a perfect camouflage in the foliage. Their brownish relative, the Garden Boa (*Corallus enydris*), frequently lives in

when molested. Two bulky colubrids, the Mussurana (*Clelia clelia*) and the Cribo (*Drymarchon corais*), may reach 2.5 m but are considered beneficial to humans, because they feed largely on rats and even venomous snakes.

Remarkably thin Blunt-headed Snakes (*Imantodes cenchoa* and *Imantodes lentiferus*) with extremely large eyes slowly glide through trees and bushes at night searching for small lizards and frogs. Snail-eating Snakes (*Dipsas*) are similar in appearance, but specialize on snails as prey.

Among colubrids, several species exhibit a remarkable mimicry to venomous coral snakes. Imitation of the coloration of coral

palm-thatched roofs of Amazonian houses, where it hunts mainly bats.

Typical Snakes (*Colubridae*) are the most species-rich family of snakes; more than 40 different species may occur at a single site. Most colubrids are nonvenomous, but some bear enlarged grooved teeth in the back of the jaw and may cause slight envenomization, especially if allowed to chew on their victim. Among the slightly venomous forms is the Green Whip Snake (*Oxybelis fulgidus*), a slender arboreal snake feeding on frogs and lizards. Also green and arboreal is the non-venomous Parrot Snake (*Leptophis ahaetulla*) but will open its mouth fiercefully

snakes supposedly protects the nonvenomous or slightly venomous species from predators. The False Coral Snake (*Erythrolamprus aesculapii*) is one of the most stunning examples, because of the species range of variable patterns, each resembling a specific species of coral snake that it overlaps with in distribution. Real coral snakes (*Micrurus*) resemble colubrids in body shape, but have very short tails and tiny eyes. The most common coral snakes typically have a pattern of distinct rings with alternating colors, usually black or with two bright colors and may reach over one meter in length. Their comparatively short fangs inject a potent neural venom, that causes paralysis of the skeletal muscles. The earliest symptom of envenomization is double vision and death may occur from respiratory arrest if no measures are taken. Fortunately, bites from Amazonian coral snakes, are exceedingly rare.

The most feared snakes throughout the New Word are pit vipers (*Crotalinae*), which account for the majority of deaths caused by snakebite. Pit vipers have large sensory holes (pits) between the eyes and nostrils that serve as heat receptors. The membrane in these holes registers infrared radiation and is sensitive to temperature changes of only 0.003° C. Thus, the pit-organs help localize prey as well as potential enemies with body tempera-

ure, or intra-cranial hemorrhage. Responsible for the majority of envenomizations in Amazonia are Common Lanceheads (*Bothrops atrox*), which occupy cultivated areas as well as rain forest. This predominantly terrestrial snake is about 1.5 m long and feeds on mammals, frogs, and birds. Exclusively arboreal, the slender Two-Striped Forest-Pitvipers (*Bothriopsis bilineata*) and Speckled Forest-Pitvipers (*Bothriopsis taeniata*) are usually found coiled around twigs of bushes and trees. Neotropical Rattlesnake (*Crotalus durissus*), the only rep-resentatives of their kind in South America, avoid forests, but occur in some savanna enclaves within Amazonia. All pit

tures different from the environment. Pit vipers are also characterized by large triangular-shaped heads and very short slender tails. All vipers possess highly developed teeth (fangs) for venom injection, which are erected from their horizontal resting position during a strike. Pit viper venom affects primarily the blood, and first symptoms include local pain, vomiting, sweating, headache and swelling. Without treatment, mortality is about seven percent, caused by hypotension, renal fail-

**Color tells all in the animal world. In this case, its the difference between a male Iguan (*Iguana iguana*), <u>left</u> and <u>above</u>, its female counterpart.**

vipers mentioned here, give birth to live young, whereas the giant Bushmaster (*Lachesis muta*) lays eggs. Meeting the largest viper in the world (up to 3.7 m), in its natural habitat, is an unfulfilled dream for many naturalists but a nightmare for most local people and tourists. Bushmasters are reported anecdotally to be extremely aggressive, but this is rarely the case and in fact, few bites occur. Bushmasters have short activity periods at night in which they hunt for mammalian prey. Mostly, they hide under roots and logs, and after having fed, will remain there for the time of digestion which may take two to four weeks.

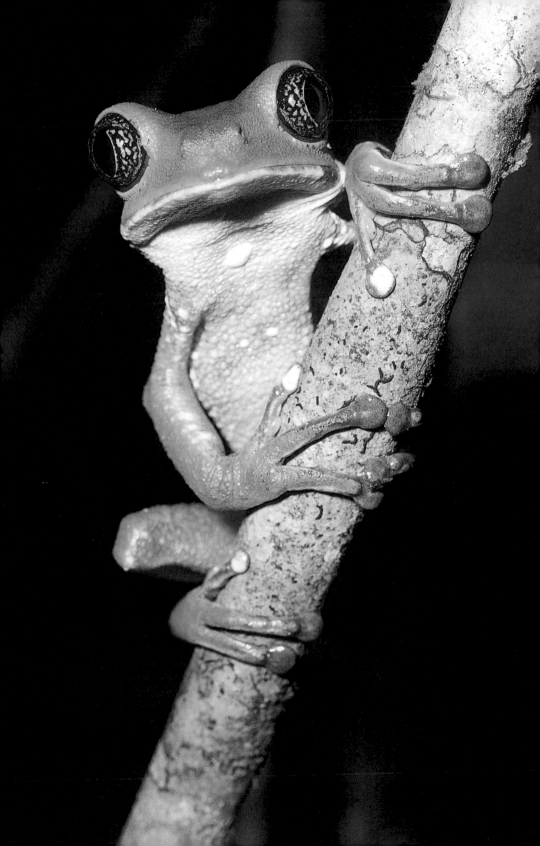

The Amazon basin with its high humidity and warm temperatures throughout the year provides excellent conditions for amphibians, whose body temperature is dependent upon that of the immediate surroundings and whose skin is not well-protected against water loss. Two groups of amphibians, but salamanders (*Urodela*) and caecilians (*Gymnophiona*), are rarely encountered. The limbless caecilians are either terrestrial burrowers or aquatic. Due to their slender, worm-like appearance they are occasionally mistaken for snakes or large earthworms. Salamanders are surprisingly scarce (in spcies as well as in individuals) and represent an exception to the general tendency for taxonomic units to show a high diversity in the tropics.

Frogs (*Anura*) are by far the most abundant amphibians. Over 300 species of these amphibians with jumping ability occur in the Amazonia and new species are being described every year. With a density of more than 80 species within a few square kilometers, the western Amazonian lowlands host the richest frog fauna in the world.

Over 75% of the Amazonian frog species are nocturnal. Nightly choruses heard along river banks, oxbow lakes, flooded forests or forest ponds are dominated by frogs. Advertisement calls emitted only by males, guide the mute females to their mating partners. These species-specific vocalizations sound like low rumbles [Cane Toad (*Bufo marinus*)], low metallic beats [Gladiator Frog (*Hyla boans*)], whistles [Whistling Frogs (*Leptodactylus* sp., *Adenomera* sp.) some Poison-dart Frogs (*Dendrobatidae*)], mooing cattle [Boatman Frog (*Phrynohyas resinifictrix*)] and rasping barks (several treefrogs of the genus *Hyla*) or high-pitched "clicks" (small *Hyla* sp.). These calls are often mistaken for those of birds, insects or even mammals. Calling males reveal where frogs live (or at least breed): they are found virtually everywhere as they have become very effective predators on the ubiquitous arthropods. These amphibians occur in burrows dug by rodents, in cavities beneath the superficial root system, on ground level or within leaf litter, in the canopy. Even though most species prefer *terra firma* primary forest, frogs are also abundant in large water bodies, floating meadows, flooded forests, in open fields and human settlements.

Reproduction in Amazonian frogs does not necessarily follow the generalized scheme involving aquatic eggs and larvae as in most temperate species. In fact, less than 45% of the Amazonian lowland species are known to lay their eggs in water. About one third lay eggs outside of water bodies but still undergo an aquatic larval phase. Hatching larvae reach water either through flooding, dropping from arboreal egg-laying sites to the water below, or are even carried to the water by the male parent. Nearly 20% of the Amazonian frog species are independent of water bodies throughout their developmental phases: eggs are laid either in unflooded foam nests or undergo direct development within the egg capsule. Rather than depositing their eggs, a few species carry them in dorsal brood chambers and pouches until metamorphosis or at least until the larvae hatch.

Frogs associated with human settlements are most frequently encountered. The Cane Toad (*Bufo marinus*) is regularly observed near huts and houses, where it leaves its daytime retreats soon after dawn and starts to feed on insects attracted to artificial light. This toad has a thick, glandular skin and grows to a body size of 20 cm. It is a frequent guest at dog- and cat-food bowls, where it consumes any kind of food (rice, noodles, meat balls etc.) offered. Thus, the Cane Toad is the only Amazonian frog known to be an occasional vegetarian and to occasionally feed on non-moving objects. Secretions from large ear glands may cause irritations or brief, excruciating pain when coming into contact with the eye or open wounds. Even though considered to be highly venomous by some locals, this large Amazonian toad is actually innocuous to humans; it may cause serious problems for dogs and cats who pick them up with their mouths. The most common toad in forested areas is the. Leaf Litter

**Preceding pages**: the Reticulated Poison-dart Frog (*Dendrobates reticulatus*) carries its tadpoles to water. **Left**, this tree toad (*Phyllomedusa tarsius*) lives up to its name.

Dweller (*Bufo typhonius*), which dorsally resembles the shape and color of a dead leaf. Due to its cryptic coloration, individuals are usually detected only while moving. This species is a so-called explosive breeder: breeding assemblages and chorusing males can be found only on a few days of the year. Members of the diurnal Harlequin Frogs (*Atelopus* sp.) are small colorful toads which establish territories near forest streams. When ready to mate, the female is dorsally embraced by a male; both partners then move to the stream where the eggs are laid.

Fast flowing streams – common at the flanks of the Amazonian basin but scarce in its lowlands – are also the breeding habitat

is not surprising. Out of the 130 analyzed lowland frogs, more than 80 show climbing ability. The most species-rich group of Amazonian frogs are the Tree Toads (*Hylidae*). Like in all frogs with climbing capabilities Tree Toads possess enlarged finger and toe discs, which aid in attaching to stems and leaves. The large eyes of these nocturnal frogs allow good vision and give them a myopic appearance.

Tree Toads show a great variety in life history. The medium- to large-sized genera *Osteocephalus*, *Phrynohyas* and *Phyllomedusa* live primarily high up in the canopy and may descend to the ground only to breed and some even breed in water-filled leaf

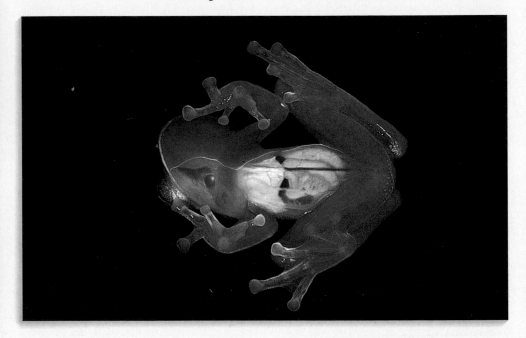

of the nocturnal Glass Frogs (*Centrolenidae*). Their transparent skin, which reveals the intestinal system, the beating heart, and the green bones, has given these frog their name. Eggs are attached to the underside of leaves overhanging running water. The clutch is attended by a male until the larvae hatch and drop into the stream below. The tadpoles are adapted to bury themselves in the detritus or gravel of the stream bottom; this minimizes predation by fish and aquatic insects and a prevention from drifting away.

In an (originally) almost completely forested ccosystem such as the Amazon basin, the dominance of tree-dwelling species

axils or tree holes at varying heights. A canopy species frequently heard in the Amazonian *terra firme* primary forest but hardly ever seen is the Boatman Frog (*Phrynohyas resinifictrix*). This frog breeds exclusively in tree cavities at great heights. Male frogs are spaced at large distances (>100 metres !!) and call from water-filled treeholes. These function as acoustic resonators and allow long distance communication. Its loud "queng-queng" call is repeated three or four times. This explains its common name Boatman Frog: the croaks imitate the tapping of oars against the side of the canoes, which are used by the Indians to maintain the rhythm of the stroke when rowing. Morphologically similar to the Boatman Frog, yet behaviorally and ecologically different, is its relative, the Poison Tree Toad (*Phrynohyas venulosa*), which breeds in water bodies at ground level. This medium-sized frog extrudes a sticky substance from dorsal glands when caught.

Maki Frogs (*Phyllomedusa* sp.) are easily identified by their bright green color, vertical cat-like pupils and their somewhat sluggish behavior. These frogs possess a wide array of biologically active peptides in their skin; these effectively thwart attacks by predators. Some of these peptides are related to mammalian hormones or neurotransmitters and play an important role in pharmacologi-

of water. Here the male and female mate, about 3,000 eggs are laid and early larval development occurs. These pools measure less than a meter wide and are created at the border of a water body during the wet season. Subsequent flooding usually allows the tadpoles to leave their confined home. A close relative, the Wavrin Frog (*H. wavrini*), is one of the few frogs heard in the *igapós* in central Amazonia (Manaus and the Anavilhanas) during high water level. Widely-spaced males perch on leafless branches up to four meters above water level and alternately produce low, continual croaks of one second duration.

The floodplains of whitewater rivers are the breeding ground of many colorful Tree

cal studies. Amazonian Maki Frogs deposit non-pigmented eggs on leaves or in folded leaves, which at least partly conceal the eggs.

Some Tree Toads, such as the members of the brightly colored and small-sized *Hyla leucophyllata* group deposit their eggs on leaves or mosses above water without forming specific nests. Yet most members of the genus *Hyla*, lay their eggs directly into water. The large Gladiator Frog (*Hyla boans*) heaps up a barricade of mud to enclose a pool

**Left**, the transparent Glass Frog. **Above**, not an obese specimen, merely the Narrow-mouthed Toad (*Synapturanus mirandaribeiroi*).

Toads, which compose the main part of the spectacular nightly chorus at the floating meadows. *Hyla leucophyllata*, a yellow frog with orange webbings and chocolate brown dorsal markings, frequently calls in the vegetation dominated by water-hyacinths. Greenish frogs of the semiaquatic Tree Toad genus, *Sphaenorhynchus* are also found here or on floating carpets of aquatic ferns.

Water ferns are the habitat of *Lysapsus limellus*, a silvery lined little green semiaquatic frog frequently used by fishermen as bait and belonging to the family, *Pseudidae*. The enormous larvae of the medium-sized (70 mm) frog, *Pseudis paradoxa* measure up to 250 mm (!), the largest tadpoles known.

A common guest in bathrooms and water closets – even in large cities – is the brown Tree Toad (*Ololygon rubra*); it runs along vertical structures similarily to its frequent coinhabitant, the Tropical Gecko.

Open formation habitats such as roadsides, grasslands and human settlements in the Amazonian lowlands are frequently inhabited by the foam-nesting members of the nocturnal Whistling Frogs (*Leptodactylidae*). In *Physalaemus* sp., foam nest construction takes place on the surface of small puddles. Floating in pairs close to the shoreline, the male beats oviductal secretions of the female into a frothy mass. During the beating mo-

rows, males of *Leptodactylus fuscus* protect offspring from parasitic flies and predators. At the entrance of its burrow, a male attracts a female to the nesting site by continuous whistle-like calls. At the onset of the raining season these calls are frequently heard during evening hours along street and roadsides. Once a female approaches, the male guides her into the chamber and embraces her from behind. The mating pair remains within the burrow until the foam nest is finished.

The Broad-mouthed Horned Frog (*Cerato-phrys cornuta*), with its terrifying appearance, does not resemble a typical frog. This large, green or brown colored leaf litter frog is a typical sit and wait predator and capable

tions, unpigmented eggs are fertilized and distributed within the foam. Within the foam, eggs and early larval stages are well-protected from aquatic predators. After a few days the foam nest dissolves and the rapidly developing larvae are released into the water body. The large South American Bull Frog (*Leptodactylus knudseni*) produces a three weeks lasting foam nest in a depression on land next to water. Although protected from aquatic predation, eggs and the long-tailed tadpoles are frequently destroyed by parasitic maggots or taken by wasps (*Angiopolybia* sp.) which pluck the eggs or even tadpoles out of the viscous foam. By excavating small bur-

of devouring large amphibians as well as small mammals.

Members of the genus *Eleutherodactylus* possess enlarged toe and fingertips and are found in forests from the ground floor up to the canopy. This species-rich genus (more than 400 species and the richest genus among the vertebrates!) is characterized by direct development (small replicas of the adult frog leave the egg capsule!).

Probably the best known frogs of the Amazonian amphibians are the Poison-dart Frogs (*Dendrobatidae*). These colorful small frogs are active only during the day, feeding on termites, ants and other small insects.

Most dendrobatids live in leaf litter. Males establish territories and announce their readiness to mate by continuous calling from elevated sites (e.g. fallen tree trunks). The Penemkwitsi (*Allobates femoralis*), a brown frog with golden lateral stripes may occupy breeding territories of up to 200 square meters continuously for up to four months. The Secoya Indians of Peru have named this frog after its four-note call "Pe-nem-kwi-tsi". Series of these frequency-modulated (whistle-like) calls prevent males from entering their territories; and the mute females are attracted to the calling males and follow them to a nesting site. Eggs are laid on land in small shelters between leaves. The male

these enter the blood stream, the potent poison affects nerves and muscles, producing paralysis and respiratory failure. The largest Amazonian dendrobatid, the Striped Poison-dart Frog (*Phobobates trivittatus*), uses call sites up to 1.5 m above ground; its monotonous, minute long "tinc-tinc;tinc…" call series can be heard throughout the day, most frequently during early morning hours. *Dendrobates quinquevittatus*, characterized by bright golden stripes on a dark background releases tadpoles in bromeliads or small water-filled tree holes. In contrast to all other Amazonian Poison-dart frogs, it lives high up in trees. Liana paths are frequently used to reach the breeding sites.

frog remains with the fertilized eggs for two or more weeks and carries the tadpoles on his back to the water after they hatch. Small brown dendrobatid frogs without conspicuous markings belong to the non-poisonous genus *Colostethus*. Individuals with vivid warning colors belong to genera (*Phobobates*, *Dendrobates*) containing more or less dangerous skin toxins. The Colombian Choco Indians utilize the alkaloids from frog's skins to produce the poison for hunting darts. When

**Left**, amorous tree toads (*Hyla brevifrons*) leave their eggs on a leaf tip. **Above**, not a pretty face: the Horned Frog: (*Hemiphractus* sp.).

Not uncommon, but hardly seen are the Narrow-mouthed Toads (*Microhylidae*). Most of these squat, small-headed termite- and anteating species are explosive breeders, and hundreds of males simultaneously calling over hours are found at forest ponds few days after heavy rains during the wet season.

During complex courtship eggs are embedded in the dorsal skin of the female in the Tongueless Toads (*Pipidae*). Eggs develop directly into froglets in the dorsal brood chambers. The large, Surinam Toad (*Pipa pipa*) is frequently found in standing waters of open habitats and was even reported from sewage systems of Amazonian cities.

The only systems on earth with a fish diversity comparable to that of the Amazon Basin are the oceans. The rate at which new species continue to be found suggests that there may be as many as 3,000 in the Amazon rivers and lakes. Among them are physical forms and ways of life unique to the fauna of Amazonia, many of which evolved in response to the poverty of their habitats and the flooding cycles in the Basin. In the white water and black water rivers of the Amazon's catchment, the mineral content of the water is so low that, during the dry season, many fish species have to rely chiefly upon their fat reserves for survival since aquatic plants are almost non-existent.

It is when these rivers flood and the waters spread many kilometers into the surrounding forests, that the fish find most of their food. The fish, moving out of the main channels, shoal among the branches of the trees, taking the place of the birds, the mammals and the invertebrates that the floods displaced. Among these fish, the fruit-eaters have a niche peculiar to the Amazon.

The most striking of these fish are the fruit-eating characins. Like many species in the Amazon, their strange dentition reveals much about their diet. The largest species, the tambaqui (*Colossoma macropomum*), resembles a thick-set, bull-nosed European bronze bream, until it opens its mouth. It reveals a set of teeth similar to the molars of a sheep, and they, driven by the impressive muscles of the head, can crack the seeds of the rubber tree and the jauari palm.

The tambaqui and other characins, such as the pacus (*Mylossoma and Myleus*) wait beneath certain trees in the flooded forest for the fruit to fall. They destroy much of the crop produced by the larger-seeded trees; but for those with small seeds they are important agents of dispersal, as the seeds pass through their digestive tracts unharmed, while the fruit is consumed. While the survival of much of the fish fauna depends upon the forests, the fish in turn are an indispensable part of the forest's ecology.

The flooded forest's contribution to the

**Catfish is the catch of the day.**

aquatic ecosystems is such that if it is removed the freshwater foodwebs are likely to collapse. Regrettably the Brazilian government has been promoting the clearance of the floodplains for agriculture, believing that food production is likely to be increased in this way. As many of the riverside soils, are infertile, clearance is likely to produce just the opposite effect, as the critical protein supplies provided by the region's fisheries are replaced by unproductive cattle pastures.

Many of the fish species feeding in the forest migrate to take advantage of the seasonal fluctuations. Some of the fish in the Amazon system, by contrast to those elsewhere in the world, migrate downstream to spawn, laying their eggs in or around the white water channels before returning to the forests surrounding the black water rivers. Many of the species making use of the flooded forest appear to be highly specialized feeders, and concentrate upon the food they are best equipped to eat.

One of the strangest phenomena involving the fish of the Amazon system is the *friagem* which takes place some years during June. Cold winds from the south of the continent chill the surface of the rivers, and the cold water sinks. This displaces the water close to the riverbed which, in the white water systems, has been depleted of its oxygen by bacteria. This water rises through the river, and the fish coming into contact with it float gasping to the surface, a bonanza for the fishermen. The phenomenon does not occur on the black water rivers, as there is insufficient organic material on the riverbeds for intense bacterial activity to take place.

The characin family has diversified in Amazonia to a remarkable extent. Besides the fruit-eaters, its members include the Hatchet Fishes (*Gasteropelecidae*), are surface-living characins which have deep keels and greatly extended pectoral fins that enable them to glide above the water when pursued, much in the way of the marine flying fish. The tetras, among which are iridescent neon and cardinal tetras, are heavily exploited in the Amazon for the tropical fish trade. They are also an important part of the subsistence of many river dwellers. But undoubtedly, the most famous and perhaps the most bizarre of the known characins are the piranhas.

The ferocity of the piranha has, like so many of the hazards of the natural world, been greatly exaggerated. There is no doubt that some species, under certain conditions, have killed people and other large mammals; but most of them at most of the time, are harmless. The behavior of the predatory species seems to depend on the state of their habitat. In the main river channels and the large lakes of the Amazon, especially during the flooding seasons, they appear to leave swimmers unmolested. But when the shoals are confined to limited bodies of water with scarce food supplies, such as the pools left behind by a falling river, they can become

The catfish of the Amazon are just as diverse as the characins, having evolved to fill some extraordinary niches. The largest are predatory, and grow to more than two meters and 150 kg. Some very small species, by contrast, are parasites feeding upon larger fish, either consuming the mucous on their skins, or sucking their blood, penetrating the skin by means of long fine teeth. The bloodsucking species known as Candirus (*Trichomycteridae*) have been known to parasitise bathers, entering the nose, ears, anus or urethra, from which they are notoriously hard to extract. None of this, by the way, should discourage swimmers: swimming in the rivers of the Amazon is normally pleasant

dangerous. The Black Piranha (*Serrasalmus rhombeus*), which reaches a length of 40 cm, has a particularly evil reputation, and has been blamed for several recent deaths.

But several of the piranha species are harmless to people under any conditions. Some, like the Opossum Piranha (*Serrasalmus elongatus*) have adapted to eat the fins and scales of other fish, which is why so many of those on sale in the markets look worn. Others eat fruit and seeds and sometimes even leaves, while several species are partly insectivoruous. Any visit to a fish market will show that there are many more piranhas eaten by Man than men being eaten by piranhas.

and safe. Other catfish are armored with the most baroque adornments. The Bacu Pedra, (*Lithodorus dorsalis*), its entire skin surface covered with heavy plates rather like those of the prehistoric ankylosaurs, its fins protected by thick spines, resembles nothing so much in shape and texture as a weathered rock. Quite why these catfish should arm themselves so heavily has yet to be determined.

The species said by some authors to be the largest freshwater fish in the world is the pirarucu (*Arapaima gigas*), a member of the primitive air-breathing osteoglottid family. While the giants reported by some of the earlier explorers in the Amazon are now rare, specimens of over two meters in length, weighing 125 kg or more, are still commonly caught. The other well-known osteoglottid in the Amazon is the Arawana (*Osteoglossum bicirrhosum vandelli*), known to fishermen as the "water-monkey". It can leap up to two meters from the water – twice its maximum body length – and snatch invertebrates, birds, bats and reptiles from the overhanging vegetation. Besides the piranha, the Amazonian fish best known to foreigners is probably the electric eel (*Electrophorus electricus*), only one of several species which can generate and detect electrical impulses, but the only one which can kill or stun large prey with its discharge.

aquarists. The Tucanare or Rainbow Perch (*Cichla ocellaris*), which can reach 10 kg, is one of the most important food fish of the Amazon.

Some of the most striking recent work by fish biologists in the Amazon has involved the leaf litter banks found in the meanders of black water streams. Many of the surveys conducted have revealed as many as 46 species within 100 m of leaf litter. Most of them are strange looking and dwarfed: the average adult length of one species, the second smallest in the world, is one centimeter. Among them are some of the most outlandish of all the Amazon's species. The Leaf Fishes imitate dead leaves with remarkable fidelity, being flat-

The Amazon is also remarkable for the number of fish species belonging to families normally considered to be marine. Bullsharks have been found 3,000 km inland, and sawfish frequent some of the turbid tributaries. Stingrays are common in the Amazon, and their spines are sold for magical purposes in some of the riverside medicine markets.

As in Africa, there are lungfish in the Amazon – cocooning themselves in mud during the dry season – and Cichlids, well-known to

tened, blotched and tattered, with an extended lower jaw mimicking a stem. The camouflage is so effective than even when disturbed they are hard to distinguish from the substrate, sinking with a motion similar to that of the other settling leaves.

The Wormfish, a catfish which has yet to be classified, lives at the head of the litter bank. Such fish is blind, scaleless and bright scarlet in color, and it breathes through its skin, sharing its habitat with earthworms. Such a discovery hints at what there may still be left to find, as scientists begin to investigate the smaller species of the world's most diverse freshwater fauna.

**Left**, even out of the water, this Black Piranha looks dangerous. **Above**, the bizarre head of an Armored Catfish.

**Small Creatures, Big Numbers:** During the evolution of life on earth, the clear winner of the competition for species diversity was the humble insect. There are so many species that even the quest to estimate their correct number is a topic for scientific controversy. To date, several hundred thousand species have been identified, about 10 times more than all mammals, birds, reptiles, amphibians and fishes put together.

But while nearly all vertebrates are known to science, we may know as little as one percent of all insects. Because of the huge numbers involved, science relies not on the traditional method whereby each individual species is recorded, but rather on more indirect calculations. Some are based on the catching of insects in particular habitats: the number of species that is caught (without necessarily becoming individually identified) in a particular habitat within a set time frame, multiplied by the estimated number of different insect habitats, leads to approximations of several million insect species.

Some experts even believe that there may be several tens of millions of insect species on earth. Proponents of this theory assume that a major proportion of all insects is specific to one particular plant species. The number of insect species collected after the fumigation of a tropical tree with insecticide, multiplied by the number of tree species, leads to these high estimates. Whatever the final outcome of this controversy may be, it is accepted that the Amazonian rain forests are the richest habitat on earth in terms of insect diversity.

Taxonomists help us to simplify the diversity by placing each of the millions of species into one of about 30 orders. Only about a dozen of these orders have conspicuous and interesting species – the attribute "interesting" is used here in reference to the function of insects as tourist attractions: few people, after all, travel to the tropics to view fleas and earwigs. And other groups, like mosquitoes and flies, attract attention only when people are trying to avoid them.

**Preceding pages:** what else but a Waxy-tailed Lantern Bug. **Left**, transparent-winged Cicadas.

In contrast to mammals and birds, insects had already evolved to a large degree of diversity 100 million years ago, before Gondwanaland broke up to create the isolated continent of South America. This is one of the reasons that no higher insect taxa – like the edentates among the mammals – are restricted to this continent. But some lower taxa, like the Morpho butterflies and the leafcutting ants, evolved in and are typical for the neotropics.

**Termites and Ants:** Termites and ants are the insect groups with the largest number of individuals and they represent the largest insect biomass in the rain forest. It is even believed that their combined biomass in a tropical rain forest exceeds that of all vertebrates. Termites, although often referred to as white ants, are a fairly primitive order of insects in contrast to the advanced ants, and they are actually completely unrelated to ants. Termites are nearly colorless due to their life underground or in rotting wood. They live mostly on the cellulose component of plant material. They cannot digest cellulose with their own enzymes, but do so with the help of bacteria that live in their intestines, a similar dependence as that of the ruminants. Termites are social insects and some species live in societies of millions of individuals, all being descendants of a single royal couple.

Ants live similarly in large societies and show a baffling variety of social structures and cooperative and antagonistic interactions between different species, which even includes the complete dependence of some species on slaves captured from another ant species. Possibly the most conspicuous ants of the neotropics, and endemic to them, are the leafcutting ants. Their large columns are easily visible, as each individual carries a piece of leaf several times its size to its underground dwelling. There, the leaves are chewed into a paste, which forms the substrate to grow a fungus, on which the ants live. The dependence of fungus and ants is mutual: the fungus will only grow well in the ants' dwelling, and the queen, when founding a new colony, has to assure the transfer of the right fungus by carrying a little piece of the fungul mycelium with her.

Most ants have a permanent dwelling, but army ants move about carrying their eggs and pupae with them. They advance in military-like formations, sometimes several meters in breadth and a hundred meters in length, and devour every living thing that happens to be in their way. Birds often follow these army ants and prey upon insects that try to escape. Ants are part of the reason for the overwhelming success of Cecropia trees, the most abundant vegetation in neotropical secondary growths. Cavities in the branches of these trees are colonized by ants that attack every animal – including humans – that touches the Cecropia.

Ants belong to the order of the *hymenop-*

disturbing to most people, with the possible exception of retired chain-saw-operators. Only the males call in both insect orders, but with quite different musical "instruments". Grasshoppers follow the principle of playing a violin, using a scraper that is rubbed across a file; cicadas, meanwhile, have exploited the principle of drumming. The noise-making instruments of cicadas are thought to be the most complex ones among insects, and involve a resonating apparatus of air-sacs, membranes, and covers that can be used to modulate the sound output. Grasshoppers mostly chew plant food, while cicadas suck plant sap.

Planthoppers belong to the same order as

*tera,* which they share with wasps and bees, to whom they are related. Taxonomists normally list the *hymenoptera* as the last order of insects, meaning that they were the most recent insect species to appear in the evolution process.

**Grasshoppers and Cicadas:** Without doubt, two unrelated insect orders, grasshoppers and crickets and cicadas, have the biggest impact on the tropical nature experience, as they provide most of the acoustic backdrop of rain forest landscapes. The combination of trilling, thumping, groaning, or outright roaring sounds that rolls through the forest like a bizarre outdoor concert can be a little

cicadas and can boast some of the most bizarre looking specimens of the insect world. Perhaps the most well-known is the Lantern Fly, which, in spite of its name, is neither a fly nor a luminescent insect. The jutting forehead of this insect looks like the head of a crocodile: dark and light marks seem to simulate eyes, nostrils, and teeth, and these colors may be part of its natural "disguise" to scare away hungry birds. Another peculiarity of some planthoppers is their ability to modify their appearance with wax excretions: a typical representative, the Waxy-tailed Lantern Bug, is easily found in the flooded forest. The ribbons of wax that resemble a trailing tail protrude from the ducts of the insect's wax glands, and may protect the animal against enemies.

Cicadas and planthoppers are closely related to bugs, and often unified with them in the same order. Many members of this large insect group are very colorful and they are some of the most attractive insects to be found. Most feed on plant juices or aphids although some, such as the bed bug, are a bane to humans. Bugs of the group *Triatominae,* meanwhile, are responsible for Chagas' disease, a tropical malady caused by the protozoan *Trypanosoma cruzi,* and which is widespread in South America. This bug hides during the day in cracks in the

visible to other insects, is predatory, and the forelegs transform into a deadly trap once potential prey wander within reach.

In contrast to the mantids, walkingsticks and leaf insects are vegetarians with a nocturnal life-style. During daylight, they do what they do best, namely mould into their surroundings. Some imitate twigs, brownish or greyish drawn-out creatures, with sizes that may exceed 30 cm, while others are green and flattened to resemble leaves, similar to those of the plants on which they reside. They move very slowly and if detected by a predator, have to rely on a foul-smelling liquid deterrent that they squirt out. There are many species of grasshoppers that com-

walls of human habitations, and attempts to get its blood meal at night.

**Mantids and Leaf Insects:** Two orders of insects, the mantids and the walkingsticks or leaf insects are restricted to warm countries, though not strictly to the tropics. The Praying Mantis gets its name from sitting motionless, or slowly swaying with the wind, on plants, its forelegs held up as if folded in prayer. Quite in contrast to this seemingly harmless posture, the mantis, virtually in-

**Left**, nature produces some weird and wonderful things: the grasshopper *Maskia nystrix*. **Above**, a Lantern Bug (*Fulgora lanternaria*).

pete with leaf insects in their astonishing abilities to disguise themselves, and the wings of these insects may not only imitate the color and shape of particular plant leaves, but even contain specks and fringes reminiscent of fungal growth on a decaying leaf. Other members of these orders have spiny-like protrusions and other bizarre-looking features that make it difficult for a predator to swallow.

**Butterflies and Moths:** Without doubt, butterflies and moths are the single most important insect attraction for visitors. In particular, amateur entomologists aspire to collect members of the Morpho group, large butterflies with shining blue wings, that are restricted to the neotropics. Among them, *Morpho hecuba* has a wingspan of 18 cm, making it the biggest butterfly of the Amazon. Only the male Morphos have this vivid coloration, the females being rather dull brown. Morphos are frequently sold as souvenirs, and fortunately, most of them are now bred commercially. Three more butterfly families – among the more than 100 that belong to this order – are restricted to the neotropics: the *Ithomiidae,* the *Heliconiidae,* and the *Brassolidae.* The latter are called owl butterflies, since they have a large eyespot on each wing, probably a means of protection against birds. Among the moths, the Great

gans, such as ovipositors, that only occur in wasps.

Butterflies and their caterpillars also have developed brightly colored patterns to signal inedibility, which can, for example, result from the caterpillar's diet of poisonous plants. Based on identifying these colors, the insect's enemies eventually learn to leave particular creatures alone. This phenomenon has triggered chains of convergent evolutionary processes: "edible" insects imitate colors and shapes of their inedible counterparts, and even fly together with their look-alikes. And since a predatory bird's ability to identify an unsavory item is limited, insects seem to agree on certain limits, such that

Owlet, with a wingspan of 30 cm, is probably the largest moth on earth.

Nature presented the insect world with the gift of disguise to help a species outlive the dangers of the rain forest. And while leaf insects and grasshoppers are the masters of disguise, butterflies are certainly the champions of mimicry. One apparent strategy is to look like members of other arthropode groups that are avoided by birds: some moths take on the appearance and movement of bristly spiders, and some butterflies have body shapes where details of the wings, waist, and mouth resemble certain wasp species, even to the point of seemingly growing false or-

even unrelated species adopt similar warning colors. For example, more than a dozen species of unrelated but similar-looking butterflies often join together to form something called a mimicry ring.

**Beetles and Mosquitoes:** With about 300,000 known species, beetles are the biggest order among all living beings. Beetles cannot compete with butterflies in terms of attractiveness, or with ants in complexity of social structure and their contribution to the biomass, or with cicadas for noisemaking, but they are probably the insect group with the greatest diversity of life forms exploiting various ecological niches.

What's more, some are extraordinarily beautiful, and because of this beetles of the *Buprestidae* are sometimes used by Amazonian Indians for ornamental purposes. The biggest beetle of the neotropics, the Hercules beetle, has a length of 15 cm. Its home, however, is Central America and the West Indies, rather than the Amazon. The South American Palm Weevil, with an average length of four centimeters, is unusually large. This insect species did not do much damage as long as palms like the local Mauritius palm stood naturally in small stands or as isolated individuals in the forest, but it can be very harmful in today's extensive stands of the coconut palm, which is not native to the

cans use the term *zancudos* instead for these insects.

Mosquitoes are a problem along the white water rivers, but they are virtually absent from the black water rivers such as the Rio Negro, whose waters are too acid for larval development. And nowhere in the moist tropics are their sheer numbers as much of a problem as in the northern latitudes. Only female mosquitoes bite, since they need a blood meal as a nutrient source to complete their egg production. Male mosquitoes only indulge in drinking the nectar of flowers. Mosquitoes are justifiably feared as transmitters of various dangerous diseases. For instance, Aedes mosquitoes transmit the vi-

region. Humans have discovered more than one way of fighting back, however: the Palm Weevil's large white larvae are sometimes fried and eaten by the local population.

Together with the flies, mosquitoes form the order *Diptera,* the "two-winged" insects whose name originated from the change of their hind wings into halteres, an organ for balancing. Although the word "mosquito" comes from the Spanish language, and means "little fly", Spanish-speaking South Ameri-

**Left**, this Bush Cricket (*Cyclopetra speculata*) is a master of disguise. **Above**, leafcutting ant after a successful outing.

ral diseases yellow fever and dengue fever, Anopheles mosquitoes the Plasmodium protozoa that cause malaria. This Amazonian scenario is not quite as dangerous as one might perceive: there is complete protection against yellow fever through vaccination, and some, but not total protection against malaria by preventive drugs. The best defence, however, is protection against the bite through insect repellents during the day and mosquito nets and cooling fans at night. Only a very small fraction of all mosquitoes carries any pathogen, and their numbers decrease the further away one gets from human habitats.

The Amerindian population of Amazonia today consists of the remnants of a large group that numbered well over three millions in 1500, before the advent of the European influx. Since then, the Indian population of the Amazon has greatly decreased – especially during the 20th century. In 1900, in Brazilian Amazonia, there were 230 tribal groups. By 1957 only 143 remained. Many of these remaining tribes have only a few hundred individuals left, and today there are only an estimated 50,000 Indians left in Brazilian Amazonia.

**The Decimation of the Indians:** What are the reasons for this tremendous decrease in population? The Indians of Amazonia have been abused since Europeans conquered the region. Large tribes were annihilated in warfare and others were driven into slavery even as recently as the turn of the present century in the rubber-gathering boom. The Indians also have no resistance to many western diseases, and thus colds, influenza, chicken pox, and measles are often fatal. In addition, western Man has introduced tuberculosis, venereal disease, and smallpox, which have further decimated the population.

In 1910, an Indian protection service was founded by the great humanitarian Marshal Cândido da Silva Rondon, whose motto was "Die if necessary, but never kill." However, the Indian protection service gradually became corrupt and was eventually replaced by FUNAI, the Foundation for Indian Affairs, which has offered protection to Indians in some places and allowed atrocities in others.

The pre-1500 populations of Indians were most dense along the main Amazon River. Large tribes such as the Omagua and the Tapajós Indians lived in the floodplain area and were perfectly adapted to it. During the dry season they cultivated fast growing crops of manioc and corn. They caught much more fish during periods when the water was low. The fish was dried and stored in preserving oil from turtle eggs or manatee. These Indians

also grew wild rice, from which they brewed a nutritious wine, also stored for use in the flood season. With their supplies, they survived the flood season. However, as western Man also preferred to live on the floodplain, these riverine tribes have become extinct.

**Daily Life Today:** The 130 remaining tribes today are mostly small units driven back into the remotest parts of Amazonia. Most of them live in the upland forest rather than along the riverbanks. These Indians are well-adapted to life in the forest and are independent of western commodities. Their houses are mostly built with logs, thatch, and mud from their surroundings. They use poison bows and arrows blowguns for hunting. Fish are caught by poisoning, and they have their own medicines, which in some tribes, even include contraceptives. These Indians are so familiar with the forest that they can walk freely through it without getting lost. Westerners could learn much from the Indians, but the tragedy is that they are disappearing faster than this information can be catalogued.

The main language groups are the Arawak, Je, Carib, Pano, Tupi and Xiriana (Yanomamo) language families. Although there are many tribes – each speaking different languages or dialects – all of their dialects are derived from one of the above groups.

Pano is confined to the western margin of the Amazon forest. The Tupi speaking tribes are in southeastern Amazonia east of the Rio Madeira and in the lower Amazon. The Carib in the northeastern part delimited by the Rio Negro and the lower Amazon, and the Arawak tribes are mostly in the western part south of the Rio Negro and west of the Rio Madeira. The Xiriana are confined to the Venezuelan-Brazilian frontier region.

Today some Indian groups are in large Indian Reservations. The best known such reservations is the Xingu National Park in the south of Pará in Brazil. There, the well-known Brazilian anthropologists, the Vilas Boas brothers, have protected a large numbers of tribes in that region. However, even such parks are not inviolated, since 8,213 square km of the Xingu park was taken away by Presidential decree in 1972 to allow highway building. And although an addition was made in the southern portion of the park, the relocation of Indians from a forest to a savanna area did not work well.

One of the most interesting features of Indian culture is the variety from tribe to tribe. It is difficult to make generalizations about them because each group has significant cultural differences. Some tribes live in large community dwellings or *malocas*, others live in small individuals family houses. Some tribes have isolated houses for males only, whereas others do not. A few truly nomadic tribes such as the Paumari remain, however, most Indian tribes are semi-nomadic, practicing a shifting agriculture. They cut the forest, burn it, and plant crops for a few years before moving on

ship relationships, taboos, and other customs play an important part in their culture. These taboos and customs are as varied as the number of tribes.

Some tribes produce ceramics. This has left a scattered archaeological record of their past. Pottery is made from a local clay, often reinforced with the ashes of the *caripé* tree.

The largest group of Indians remaining near the banks of the Amazon are the Tikuna Indians who live on the Brazil, Colombia, and Peru frontier region. The Tikuna are now in constant touch with western civilization and even have primary schools in some of their village.

The Tikuna have their own language group

to new areas. Due to the small size of their fields, this shifting agriculture has never done permanent damage to the forest.

The staple crop of the Indians is manioc, from which they make a coarse flour. The Yanomamo staple crop is plaintain bananas. Other crops include sweet potatoes, yams, taro, corn, pineapple, papaya and peach palm. Various groups grow different crops and different varieties of the well-known crops. For example, the Jamamadi on the Purus River have 12 varieties of manioc, each yielding at different times after planting, and each best suited for a different purpose.

Indians life is well-organized. Family kin-

and, with a population of over 10,000, they are still one of the larger tribes. Tikuna objects are for sale in Leticia in Colombia, and their shoulder bags and hammocks made from the fiber of the wild pineapple are particularly interesting and practical. The fiber is separated from the rest of the leaf, dried and dyed with natural dyes to make extremely strong and durable bags.

Near Pebas in Peru, downriver from Iquitos, there are large groups of both Huitoto and Bora Indians, many of whom were relocated from Amazonian Colombia in the 1930s. The Huitoto around Pebas are largely acculturated and only the older men remember their folk medicine, hallucinogens, and ancient customs. The Boras, however, are less acculturated and preserve many of their older tribal customs. A recent study of the Boras' use of hallucinogens was made in 1977 when the oceanographic vessel *Alpha Helix* visited the area with a group of ethnobotanists and pharmacologists.

**The Riverside *Cabolos*:** Today there are many peasants or "*caboclos*" living beside the major rivers. In some cases, these settlers live and depend entirely on the river and *várzea*, of flooded area, but many cultivate some of their crops on *terra firme*, or non-flooded areas. The riverine *caboclos* have adopted many of the same techniques as the Indians. Their life is strongly influenced by the annual cycle of the

height of the flood season, the houses on stilts have water almost at floor level. Sometimes in an exceptionally wet year, such as 1953 or 1976, water actually enters the houses.

Floating houses are built on a number of large tree trunks. There used to be a floating community as part of the city of Manaus, but in 1961 it was abolished for safety and health reasons. Along the major rivers there are still a large number of floating houses which rise and fall with the river level. In the *paranás*, one can even find floating stores, churches, gasoline stations, and ports.

A common feature of such houses is a raised garden, usually planted in an old canoe shell. This garden is protected from the floods, as are

river. In the dry season, many will cultivate beans or corn on the exposed river margins. In some places, especially near Manaus, cattle have been introduced and in flood time they must either be removed to *terra firme* or kept on rafts and fed with grass. In spite of the adverse conditions, there are about half a million head of cattle in the floodplains of the Manaus-Careiro area.

**Housing:** The simple houses of the *caboclos*, usually built with a palm-thatched roof, are either on stilts or are floating. During the

**Left** and **above**, tribal custom dictates the wearing of body paint instead of clothing.

the chickens and other domestic animals.

The riverine people depend heavily upon fishing for their livelihood, and use lines and hooks, as well as nets (called "*tarrafas*") to catch the fish. In the dry season they catch a lot of turtle.

Life on a white water river is quite different from life on a black water river. In the black water river floodplain, there is no rich alluvial soil beside the river. The flooded areas are usually on white sand. The fish population is also smaller than in the white water areas. The result is that there are far fewer settlers along the black water rivers than the white water areas. Most of the settlers are along the Amazon itself, and its major white water tributaries. The settlers along the Rio Negro, for example, depend more on the high ground away from the floodplain. One compensation for settlement in a black water area is the absence of noxious insects, as mosquito larvae do not breed in black water areas.

**Rubber and Jute:** Another reason for settlement along the rivers is the presence of the rubber trees. The main source of rubber, *Hevea brasiliensis*, is a tree of the *várzea* area. Many riverside dwellers live where they do because there is a sufficient number of rubber trees worth tapping. Although the peak of the Amazonian rubber boom passed when rubber was introduced into Asia, rubber is still an

riverine population of the interior of Amazonia, for the factory workers of Manaus, and for Brazil as a country. Enough jute is now produced to export, in addition to fulfilling Brazil's domestic demand.

Jute is good crop for Amazonia because it grows in the floodplains (*várzeas*) and does not require artificial fertilization. The fiber is easily extracted without complicated machinery. Jute fiber is beaten out of the stem of the plant at about the time when the river level rises, facilitating transport to the factories. Jute takes about four months to grow from planting of the seed to harvesting: this makes it an ideal crop for *várzea* land when the river level is low.

important crop in Amazonia. In 1971 the production of crude rubber from Amazonia gathered from wild trees in the *várzea* forest was 32,973 tons.

An important plant to the riverine population of Amazonia is jute. Jute, a native of Asia, was introduced to Amazonia relatively recently (1939) by a Japanese farmer, Kotaro Tuji. Jute is grown near the rivers. The various ponds which remain after the water level recedes are used to soak and beat out the fibers. The fibers are taken to Manaus where there are modern well-equipped factories that process the fiber into burlap. Jute is now one of the most important sources of income for the

Many others fibers are known and used in Amazonia. The Indians have used fibers for many years. One which has entered into western commerce is the *piaçaba* fiber from the trunk of the *piaçaba* palm tree (*Leopoldina piassaba*), a native to the banks of the upper Rio Negro region. The *piaçaba* fiber is used mainly for brushes. It is common to see the conical-shaped packs of *piaçaba* on launches descending the Rio Negro, or in the trading houses of Manaus. The fiber is mainly exported to Europe.

**Transport:** Most of the settlers use small dugout canoes as their main means of transport. The more prosperous *caboclos* power their canoes with outboard motors, but the paddle is still the most common means of power. As the Europeans gradually settled along the riverside, a transport network soon developed.

At the beginning of the present century, mechanized river transport was by steam-powered paddle-steamers with rear paddles. Today a large fleet of launches, barges and ships of all sizes move up and down the river system, buying rubber or Brazil nuts from the *caboclos* in return for supplies. Often the price of products is exorbitantly high and the people of the interior are in debt to the boat owners. This system, a relic from the days of the rubber boom, still exists, although today

**Life-style:** The life of the *caboclos* is simple. In general, they are content when they have caught enough fish for the day's meal and do not intend to catch more to sell. Their food consists mainly of cassava flour mixed with fish, and they grow some fruit and keep a few chickens. The *caboclos* have many superstitions and legends originating from Indian culture, including a rich history of folk medicine. Where there are groups of people sometimes primary schools may be found. However, the vast majority of riverside people live too far from such schools to receive any education.

**Legends of the Amazon**: Many of the mythological stories of the Indian tribes and

there are many more honest traders on the river.

In the more frequently traveled sections of the rivers, especially around Manaus and Santarém, there are numerous small launches, called "*recreios*", making regular trips up and down the river. They stop at many small ports and houses and form a regular bus system which transports people, produce, animals, and almost anything else that will fit on a launch.

**Left**, boats come in handy during the rainy season. **Above**, waiting for a chance to spear the Arapaima fish.

legends about supernatural beings, which many *caboclos* say appear in animal form such as wild pig (*caititu*) or dolphin (*boto*), survive today in the folk tales of Amazonia.

**The *Uirapuru*:** The "*uirapuru*" is an olive-green bird with a red tail whose song is so melodious that the other birds in the forest are said to stop singing and listen.

It also provides an enchantment of another kind. Those who wish to win the love of a person who is indifferent to them, burn the body of a *uirapuru* and prepare a lotion from its ashes. This magic spell, taught by primitive tribes, is still performed in some parts of the interior.

The Peruvian jungle village of Llullapichis (also spelled "Yuyapichis") is situated on the Rio Pachitea, a river about 300 m across. It was founded some 30 years ago to house the workers of a sugarcane plantation and distillery. In the early days, the only way to reach the village from the closest town, Pucallpa, was by traveling for several days by boat. In the 70s, light aircraft started making regular landings on a grass landing strip in the middle of the village. A road was then put through in the course of one of the many projects which are

which is not attacked by termites. The roof is made from three dried fronds of the *yarina* palm (*Phytelephas* sp.), each of which is about four meters long; these are folded lengthways and lashed with lianas to the rafters like tiles. These roofs can remain watertight for ten years and more, and they make the inside of the hut pleasantly cool. Newer huts with roofs of corrugated iron are often unbearably hot. The floor is raised because of the floods of the rainy season. Very few of the huts are completely enclosed with wooden walls and di-

designed to further the economic development of this jungle region but whose effects can be devastating for large areas. Although this has shortened the journey to Pucallpa to one day, it becomes impassable for several months each year during the rainy season.

The village today comprises 25 to 30 huts, housing a population of around 80 adults and 250 children. They are all of mixed Indian and European blood, and the term they use for themselves is "*cholos*". Most of the huts are set along the bank of the river, linked by muddy paths which turn into a slippery morass after every rainfall. The huts are open on several sides, and most are constructed of palm wood,

vided up into rooms. The windows have fly-screens and can be closed with shutters, and there is not one pane of glass in any of them. The huts are small (between 20 and 35 square m), but thoroughly practical and comfortable in design: there is a veranda hung with hammocks, a large table and stools or benches, and a corner where beds or mattresses are screened off if the hammocks are not used for sleeping. Meals are prepared over a wood fire or on a petroleum cooker in a free corner of the hut. Very few of the huts have a place which can be locked up, despite the fact that petty thievery is quite common. The huts are usually surrounded by banana plants, lemon trees, coconut palms and mango trees, which afford shade during the midday heat.

A government school was built in the village in the 70s. Standing at the edge of the jungle, it is the only building with a cement floor, and comprises two rooms in which altogether 11 classes are taught. Depending on the wishes and the income of their parents, children start school at between five and ten years, and most leave before they have finished primary level. The state teachers are poorly paid and have minimal training, and so

Births, marriages and deaths are registered in a municipal office which was opened in 1988. Although almost everyone in the village is Catholic, there is no church or chapel. A priest comes from the town once a year to baptize children, often with extremely unusual names such as "Venus" or "New York". Probably no more than a tenth of the couples are legally married (although the mayor can perform the civil ceremony), and there has never been a church wedding in the village. There is a small cemetery about half an hour's

the very authoritarian teaching never goes much beyond reading, writing and basic addition, subtraction, multiplication and division. Very few parents can afford to send their children to a better school in town or pay for them to learn a trade. A lot of young people set off for the town in the hope of earning a living, but very few manage to find any kind of apprenticeship or a job which pays a reasonable wage. The rest of them either return to the village or eke out a living from underpaid occasional work.

**Left and <u>above</u>, the harsh realities of village life in the Amazon.**

walk away where the dead are buried without any particular ceremony. The graves are marked by a simple wooden cross, with no other form of ornamentation. On All Saints Day, sweets and lemonade are given to all the children who visit the cemetery.

There are no police, and no courts. When small items of food, clothing or crockery are stolen, the victim complains about it in the village, but no attempt is made to seek out the culprit. With more serious crimes such as the theft of cattle, the victim gets a group of villagers together and confronts the suspected thief. If the stolen item is found in his possession, the owner simply takes it back again – but not before he has given the thief a good talking too, of course! If disputes cannot be resolved among the parties themselves, for example where payment of debts or compensation is involved, then the case is brought before a person in a position of authority. This "judge" – who is usually one of the older villagers – reaches a decision which is accepted by all the parties. Once the situation has been cleared up, then relationships between the villagers go on as before.

A small shop sells food and beer, and there is another with household goods and clothing – at least, as long as there is no breakdown in the supplies from the town. Two huts serve as hotels, providing accommodation mainly for

wooden boats to much larger boats of over 15 meters with decks and cabins. The boats are poled or rowed, or driven by powerful outboard motors.

The river is teeming with fish, which are caught with casting nets or hooks on a nylon string attached to a wooden rod. Using these casting nets, which weigh up to 15 kg, involves standing upright in a hot very stable dugout, and it requires a lot of skill to make a catch. Fish are the main source of protein for the villagers, as there is now virtually nothing left to hunt in the neighboring jungle. Expeditions are seldom undertaken to more distant spots to hunt armadillos, large rodents, collared peccaries, monkeys, wild fowl or deer. The only

prospectors who have come to buy fresh supplies with their gold and spend the rest of it at village festivals. The focal point of the village is the *bodega* – part tavern, part general store; it is here that people go to celebrate birthdays, the end of school or other special occasions, and where political meetings are held. The owner, the stoutest and jolliest individual in the village, is the man to ask if you want to know anything about the health or whereabouts of anyone in the village. From the *bodega*, a steep, slippery path leads over the embankment to the "harbor", a muddy landing place.

There is a great variety of craft to be seen in the village, ranging from simple dugouts

animals which are raised are chickens, ducks and pigs, and these roam around the village and sleep under the huts. Guinea-pigs are bred in cages for food. There are also domestic animals such as cats and dogs, but apart from being fed they receive no special care or attention. Meat is eaten only on feast days, and cassava, cooking bananas and rice are staples. Vegetables, which with a little care will grow well at this latitude, are not often eaten. Despite the abundance of fresh fish, tinned tuna and sardines are considered a great delicacy. Fruits such as papayas and sweet bananas are eaten, and coconuts, citrus fruits and mangoes are available in season. There is no fresh milk at all, nor any milk products, since the villagers themselves do not raise cattle; but dried milk is used. The only form of coffee is Nescafé. The most common drink is water, but beer is preferred.

For cooking, they use the murky brown river water, which the women carry to the huts in plastic buckets on their heads. Washing is done in the river itself, using hard soap and scrubbing brushes. Every evening, families go down to the river individually or in groups to bathe. The children play around naked or in their underwear, but the adults wear at least a vest and a pair of shorts or a skirt when they bathe.

Sadly, virtually all the sewage from the

wooden splints, and the usual procedure for snakebites is simply to let nature take its course. There is very patchy knowledge of traditional healing methods and plants with medicinal properties, and these are used with reluctance. When a baby is born – often to a mother who herself may be no more than 14, the older more experienced women act as midwives.

The birthrate in the village is high, and families with four to six children are the rule. Babies form the center of attention in the family, and older brothers and sisters, other relatives and neighbors take great pleasure in looking after them and playing with them. As soon as they reach their first birthday, which is

village ends up in the river as well. Since water is usually not boiled before it is drunk, villagers regularly fall prey to hepatitis and typhus. Diseases such as amoebic dysentery and intestinal parasites like mawworm are frequent, as are leishmania, maggots and itch-mites affecting the skin. However, accidents are relatively rare. Since there is nobody with any medical training, let alone a doctor, the patient either gets no treatment at all, or he simply takes whatever medicine happens to be around at the time. Broken limbs are bound up with

**Left**, hammocks are often used instead of beds.
**Above**, an Amazonas-style barbecue.

an occasion for a great celebration, this situation changes drastically: the child is learning to walk and to eat and must now learn to fend for himself.

The head of the family is the husband, even if his wife is older than he is and most of the children in the household are not his. He is the breadwinner. Just a fifth of the men provide for their families by tending a little field they have cleared by cutting and burning jungle vegetation, in which they grow bananas, cassava, maize (for chickenfeed) and sometimes sweet potatoes. Because the soil is poor in nutrients, such fields are really productive only in the first year, and after two or at most three years, the crop is so poor that a new patch of jungle has to be cleared and a new field planted. Most of the village's food comes from neighboring farmers. None of the villagers can raise cattle because they do not have the grazing land this would require. A few families make a living from trade, buying goods in town and selling them at two or three times the original price. Despite this huge mark-up, however, these small-time traders often lose money as a result of the constant inflation. This is also the reason why larger purchases such as motors, cattle or radios are paid for with gold, and why all savings are in gold. Since not all the villagers own a casting net or have the patience to use a rod and line, two or three families make a

or need it quickly, most of the men will spend two or three weeks panning for gold. They choose a spot on the sandy banks of the Pachitea itself, or go up one of the smaller tributaries, where the pickings are richer. After a two day journey they finally reach the chosen spot, and now, constantly bent, and bitten by swarms of midges, they stand in the water and wash sand from the river bank. The work is certainly worthwhile when compared to what they earn on the *haciendas*, for in a good spot it is possible to find between two and four grams of gold in a day, but it is much harder. So far, not one of these gold seekers has struck it rich.

The most important social occasions in the village are the *fiestas*. Virtually every Satur-

living from fishing. Anyone with the right skill can catch 10 kg of fish and more within two hours. The village's two carpenters make a reasonable living, for they are skilled enough to get work making furniture and boats for neighboring cattle farmers. Yet the vast majority of the men have no regular work. When they need to, they can hire themselves out for the odd day or week on one of the huge *haciendas* which are to be found all along the Rio Pachitea. Here they load cattle, clear the pastures of weeds, help with burning and clearing new land or with building work, or fell trees. They are given full board, and get about 50 dollars a month. If they need more money,

day the villagers will find something that calls for a celebration. What characterizes a good celebration is plenty to drink, plenty to eat, and plenty of dancing.

During a *fiesta*, the villagers drink almost exclusively beer, following a Peruvian drinking custom: a glass filled to the brim is passed round, and each drinker empties it in one draught, shakes out the froth onto the dirty floor, and then passes on the empty glass and the bottle to his neighbor. The amount of beer drunk during a *fiesta* is enormous, and crates filled with thousands of empty bottles are piled up behind the tavern.

One feature of these *fiestas* is the *pacha*

manca, a special meal which takes almost a whole day to prepare. Early in the morning, a pig is slaughtered and cut up into individual portions. The meat is then laid in a spicy marinade whose main ingredients are ginger and *paprika*, and left until the evening. In the afternoon, a hole of about one meter in diameter is dug in the ground. Thin branches are laid across it, and stones about the size of a man's head are brought from the river and piled up on top. The stones are sprinkled with saltwater so that they do not crack excessively when they are heated. Finally, a huge pile of dry wood is built and a great fire is lit. As soon as the fire has died down and the stones are red hot, the actual cooking of the *pacha manca*

sacking is laid, and finally the earth from the original hole is used to cover the *pacha manca* and seal in the heat. An hour or more later – depending on how hot the stones were – the moment which the whole village has been waiting for arrives: the *pacha manca* is opened up, and the special meal is ready. Everything is now taken out, laid on fresh banana leaves and eaten with great relish.

For the last few years the village has had its own band, which plays at every *fiesta*. Although the repertoire is not exactly large, the villagers will dance enthusiastically and without tiring throughout the night. The drink usually takes its toll as the evening progresses, and the number of partners available decreases, but

begins. This is usually done by a villager who has got it down to a fine art, since considerable skill is required. A layer of banana leaves is placed on an even expanse of heated stones; marinated meat is laid on top, and then covered with another layer of banana leaves. More layers are added until all the meat has been used up. In the uppermost layer there is also cassava, cooking bananas and sometimes cobs of corn. Then the whole pile is covered with a thick layer of banana leaves, on which jute

**Left, preparing manioc pancake, known locally as *beiju*. Above, mixing *farinha*, the flour that goes into making manioc.**

this does not deter the women, who have no inhibitions about dancing with each other. As soon as children can walk, they are encouraged to join in, thus they master the steps at an early age. Tired children are put to bed in hammocks or in quiet corners of the tavern. Arguments and petty jealousies frequently lead to fights, but these are never major incidents and peace is quickly restored. At daybreak, the exhausted villagers make their way home. They spend the following day recovering, and reliving the high points of the *fiesta*. On Monday or Tuesday, there will be plenty of time to think about how to earn a living over the next few days – and to plan for the next *fiesta*.

People of Indian origin, that had come to America from Asia approximately 40,000 years ago, have populated the Amazon for several thousand, possibly up to 20,000 years. During this long period they interacted with nature in a sustainable manner, and neither altered the landscape on a large scale nor brought about the extinction of plants or wildlife. This changed when people of European origin entered the Amazon basin: from the very beginning, they had exploitation on their mind. It began with a Spanish expedition in

threatened. There are parts in which a journey of two hours by small plane reveals no more destruction than the clearance of riverside gardens by traditional farmers. While 11% or 12% of the forests have been cut, more than three million square km remain undisturbed.

But none of this is reassuring when the speed of current deforestation is taken into account. In Brazil, Bolivia and Colombia, the forests are disappearing at rates of over two percent each year, while four percent of the Ecuadorean Amazon is deforested annually.

1541 that started from Quito and worked its way eastward into the lowlands and down the Amazon in search of gold. The quest for quick profits became a continuing theme, and it has to be feared that today's mining, logging, damming of rivers, a shortsighted development of inappropriate forms of agriculture, and indiscrimate settlement projects may destroy one of the last great wildernesses on earth. Unsustainable use may leave the region not only devoid of its biodiversity but impoverished to a point that even no sustainable use for people is possible.

Flying down the Amazon, it is hard to imagine that the entire ecosystem could be

There are suggestions that the roads have already been built which make the destruction of 50% a certainty, and the army's plans to open the northern frontier to development threaten to trap the remaining forests between two advancing fronts. Though the rate of deforestation has been constant for the last years, the reasons for the destruction have changed rapidly. Subsidized clearance of huge, unproductive cattle ranches was until recently responsible for nearly all the deforestation in the Brazilian Amazon. The government subsidies are no longer instrumental, and much of the ranching is now funded by other destructive activities, such as timber cutting and mining.

The governments of some of the Amazon nations have promised to restrict the devastation caused by the ranchers. The Brazilian Environment Institute has been identifying pasturelands being cleared without permission and prosecuting the owners, though there remains some doubt as to whether the fines it has imposed will be collected. But it is now the small farmers whose activities are becoming increasingly important, and they will be much harder to control.

While state colonization schemes continue

flow of colonists. New arrivals in Amazonia are finding their own lands, buying territory from earlier settlers, squatting on the property of big landowners, or subscribing to the private colonization schemes which are now becoming a significant cause of deforestation in states such as Mato Grosso.

Most of the colonists are not, drawn to the Amazon by a mistaken belief that they will find fertile lands and a great improvement to their standards of living. They tend instead to be driven to Amazonia, fleeing insuperable

to disburse land to peasants in Ecuador, Peru and Bolivia, the Brazilian government has ceased to fund Amazonian settlement. There are still a few active schemes in the states of Roraima, Acre and Rondônia, but gigantic colonization programs of the sort that accompanied the construction of the disastrous *Transamazon Highway* across the south of the Basin, or the catastrophic BR364 road in Rondônia in the 1980s, seem to have stopped. This, however, has done little to reduce the

**Preceding pages, left and above: 18th-century impressions of the Amazon rain forest imply that nature is greater than Man.**

problems in their home states. Many in Brazil are expelled from their own lands by big landlords working with hired assassins and policemen, attempting to prevent the peasants from gaining political power. Others are dislodged by government policies favoring the expansion of agroindustry at the expense of traditional subsistence farming, or are ruined by the region's economic problems. Dispossessed of their land and their livelihoods, they have no choice but to move either to the shantytowns or to the Amazon.

For most of the settlers entering Amazonia, life remains hard, as they grapple with soils which cannot support the crops, with diseases, a lack of infrastructure and support, and ranchers trying to take over their new lands. All these factors, as well as the profits that some of them make from selling their land to new arrivals, keep the colonists moving to new frontiers. In Bolivia, Peru and Colombia, many of the settlers moved into the forests by the government, have no economic option but to grow cocaine: over 2,000 square km of the Peruvian Amazon are converted to coca fields every year.

Many of those leaving their homes in southern Brazil are now moving to the Amazon not to farm but to mine. Freelance goldminers – or *garimpeiros* – have been working in the rain forests for several centuries

arrived, and by 1990, 15% of the Yanomami had died of malaria, tuberculosis, influenza, mumps and common colds. Now 35% of the survivors are infected with malaria. In this territory, the *garimpeiros* find their gold not in deep mines, but in the beds of rivers. Their excavations of the river valleys cause the water to become so turbid that many of the fish the Indians depend on can no longer survive. Though most of the miners have now left the lands of the Yanomami, moving into tribal territories elsewhere, the effects of their work will persist. Mercury used in the extraction of gold is likely to become a serious health risk to the Yanomami as it accumulates up the food chain.

It is also the big mining companies which

but the great goldrush truly began in 1980, when *garimpeiros* took over an extraordinarily rich deposit in Serra Pelada, near the Rio Tapajós. There are now around 600,000 people in the Brazilian Amazon directly dependent on freelance mining. As they are highly mobile and the government's efforts to restrain them are they feeble, represent a significant threat to the survival of the region's ecosystems and its indigenous communities.

In 1987, goldminers started flooding into the territory of the Yanomami Indians in northern Brazil, where the 10,000 indigenous people had suffered little contact with the outside world. By 1989, 40,000 miners had

threaten the forest in Amazonia. The Grande Carajás Program in the south-east, which could result in the transformation of 18% of the Brazilian Amazon, began with the discovery of the world's largest deposit of iron ore. The mine which was opened there, with the financial help of the World Bank, the European Community and the Japanese government, has been carefully managed; but the developments associated with it, such as ore smelting, cattle ranching, timber cutting and colonization, are proving to be destructive. Several new industries in the Carajás region are fueled with charcoal made from the forest trees, which is likely to result in massive deforestation.

In Ecuador, European and North American corporations are drilling for oil in some of the most diverse rain forests of the Amazon. Inadequate safeguards have led to 30 major spills from one pipeline alone, and some of the companies are discharging the caustic chemicals they use directly into the rivers. The World Bank is now considering lending the Ecuadorian state oil company $100 million to help establish the infrastructure allowing oil drilling to take place in an International Biosphere Reserve.

The industry which is becoming the greatest threat to the forests is timber. Until recently timber cutting was comparatively unimportant, as the great diversity of the trees made

counts for the most destructive timber cutting, even though it handles only a small proportion of the wood being felled. Because of the high value of trees such as mahogany, and their wide dispersal through the forest, the export timber cutters build great lengths of road to reach them, destroying far more than they harvest.

Timber is not the only product whose sale abroad is helping to destroy Amazonia. In the *cerrado* scrublands to the south of the Basin, soyabeans are being grown to sell to European cattle farmers. Oil palm planting in countries such as Ecuador is destroying the territories of several indigenous communities, while an attempt to establish a paper pulp plantation at

their exploitation difficult. Now, however, foreign importers accept a wider range of timber, and the sawmills cutting for the domestic market have moved into the Amazon from other parts of Latin America. The money that colonists and ranches make by selling the timber on their lands enables people to flood to new frontiers. The roads the timber cutters build are bringing development to parts of the Amazon which would otherwise have remained untouched. The export industry ac-

**Left**, confirmation of the rain forest's dominant role. **Above**, road building is a difficult task in the Amazon.

Jari in the Amazon's estuary has produced no concrete results but deforestation. The international trade in pets and skins is now a serious threat to the survival of many spectacular species. One product not exported from the Amazon, however, is beef. Because of the inefficiency of the ranching, the region is a net importer of beef, and the threat of foot and mouth disease means that Brazil is prohibited from selling raw meat to industrialized nations. While jungleburgers from Costa Rica are eaten in the United States, the international Amazonburger is a myth.

The Brazilian national power company had planned to build hydroelectric dams in most of the major river valleys. While some of those constructed so far are producing reasonable quantities of electricity, others have been environmental and economic disasters. The Balbina dam, to the north of the city of Manaus, has flooded at least 2,400 square km, while producing electricity which has to be heavily subsidized by the Brazilian taxpayer. The decision to build was political rather than economic, as the lucrative construction contracts and the employment involved could be used to consolidate political support. In the near-stagnant reservoir, the acids released by the decomposition of the drowned forest are strong enough to eat through the turbines every few years. The upstream migration of turtles and

Atlantic coast to Colombia, to join another, following the western border. It hopes to attract mining companies, timber cutters, ranchers and settlers into the region.

Calha Norte has been hindered both by bad planning and by the well-organized resistance of Indian groups such as the Tukano. But mining companies have started work in some indigenous lands, a few colonization schemes have been funded, and the project facilitated the *garimpeiros'* invasion of Yanomami territories. The army claims that the program is necessary for national security. Critics, who see no significant threats to the northern border, suggest that it is instead designed to consolidate the army's power in the region. There

several fish species has been stopped, depriving the last of the Waimiri Indians of some of their subsistence. The national power company has now postponed and possibly canceled many of the dams it had planned, but is still to build at least 25.

Perhaps the most dangerous of all the government initiatives in the Amazon is the Calha Norte project, administered by the Brazilian army. This intends to open a new development frontier, along the entire northern border of Brazil. This is the region in which many of the remaining Brazilian Indians live, inhabiting some of the least disturbed forests in Amazonia. The army intends to build a road from the

might be several million indigenous people living there before the Europeans arrived. Now, of the sixteen million people in the Brazilian Amazon, only 200,000 are Indians. Though populations of most groups have stabilized, several tribes are still in danger of extinction. The Uru Eu Wau Wau Indians of Rondônia are believed to decline by half since they were first contacted by outsiders in 1981. The population of Nahua Indians in the south of Peru has also fallen since 1984. They have been stricken by the diseases of the colonists. The numbers of the Waimiri-Atroari of the central Brazilian Amazon, are believed to have fallen from 3,500 to 374 between 1974 and 1986.

Despite their considerable problems, however, it is the Indians who are responsible for the most positive events in the region's recent history. In the western Amazon, the organizations representing the indigenous communities have taken responsibility for their own development planning. Mapping and demarcating their reserves; training their own lawyers, teachers, soil scientists and ecologists; lobbying their governments; reinforcing an economy based upon the sustainable use of natural resources, Indian organizations have built a new framework for Amazon development. This, if they continue to succeed, is likely to do more than international initiatives to save the forests.

some of the damage done by the work of other government departments. But no Latin American administration has yet confronted the problems which continue to propel the deforestation, such as the inequality of land and power, which allows the privileged few to act with impunity and deprives the poor of any options but destructive farming and mining.

In several respects the industrialized nations are to blame for many of the problems in Amazonia. The military government responsible for promoting Brazil's disastrous development policies was helped into power by the United States and supported from abroad even during its most repressive and destructive years. Though much of the loans are the result of

The governments of some of the Amazon nations have taken imaginative steps towards reducing the destruction there. Colombia has recently upheld the rights of its Amazon Indians to 10,000 square km of their traditional territories. When President Collor came to power in Brazil in 1990, he appointed one of the government's most outspoken environmental critics as his Environment Secretary, and Jose Lutzemberger has been trying to halt

**Left, the Tucuruí dam on the Rio Tocantins harnesses hydroelectric power for the region. Above, an all-too-common sight along the Amazon.**

irresponsible lending as of irresponsible borrowing, the industralized nations still insist that the Latin American countries repay their debts. The size of the such debts makes a significant contribution to the poverty and dispossession promoting destruction in the Amazon.

Though several northern governments are now making efforts to reform the ways in which they make aid available for projects in Amazonia, lending by certain nations and institutions continues to fund misguided development. If the governments of the industrialized nations are to help save the greatest ecosystem on earth, they must first attend to their own destructive policies.

The picture of vast stretches of natural landscapes unaltered and uninhabited by Man is certainly close to the heart of everyone who likes nature. But in face of a growing human population and 20 million inhabitants of the Amazon presently, conservation strategies have to compromise.

Neither the global environment nor the inhabitants of the Amazon can be helped by attempts to preserve the forests as if there were no one living there. Most conservationists now recognize that the people in the Amazon can become the defenders of the forest, if conservation attempts to attend to the problems of the Amazon by tackling the problems of its people.

Such strategies as buying portions of Amazonia or exchanging areas for debt relief lead to selective conservation successes, but do not solve the basic problem that much of the rural population continues to be driven by poverty and dispossession into the virgin forest. Nature reserves are often targets for settlement, rather than regions to be avoided, as they contain none of the gunmen that ranchers hire to keep people off their lands. Conservation efforts which depend upon foreigners purchasing or taking control of other countries' sovereign territory tend to convince the inhabitants that conservation is solely a concern for outsiders, and affects themselves only negatively. The celebrated debt-for-nature swap, in which the American foundation Conservation International attempted to save the Chimanes forest in Bolivia, facilitated the invasion of timber cutters, as it took control of the reserve out of the hands of the Indians who lived there.

If the flow of colonists into Amazonia and the impunity with which timber cutters, mining companies and ranchers can operate are not addressed, then no conservation area in the Amazon, however much money is spent on it, could be considered safe. While stopping the big corporations is a question of maintaining pressure on governments both in Latin America and in the North, the arrival of colonists can only be retarded by working on the problems driving them. Without land

**The Amazon has paid the price for progress.**

reforms, education, a better distribution of wealth and attention to the needs of small farmers, peasants in other parts of the continent will continue to migrate to Amazonia. Without guarantees that they will not be expelled from their lands and better means of farming in the forest, they will continue to move once they get there. It may help that people in Europe and North America continue to lobby their governments to stop helping Latin American nations to sustain the injustices driving people into the forests, but conservation initiatives within Amazonia must come not from abroad but from the people living there.

All over the Amazon Basin there are communities of Indians and peasants taking the development initiative into their own hands. As they frequently make clear, they recognize that the conservation of the forest is essential to their own livelihoods: it is only because they have had no choice that many communities have been destroying the forests they live in. With little money and no training, farming communities have begun to devise the means of living in the forest sustainably.

These techniques involve the planting of perennial crops – trees and shrubs – rather than annual species, such as rice and beans. These help to conserve the minerals in the soil and provide a steady source of income, while in many cases reducing the amount of labor required. Some settlers are experimenting with beekeeping or the intensive cultivation of vegetables, exploiting gaps in the Amazonian markets. Long established residents, such as the rubber tappers, are trying to identify new forest products, with which they can supplement their low incomes.

The extent to which native forest products, such as nuts, fruits, gums, resins and fibers, can generate income to the people and consequently help protects the forest of the Amazon, is at present difficult to predict. Some estimates of the economic potential of forest products are likely to be too optimistic, but preservation of the forest for sustainable uses is likely to be economically superior to unsustainable timber harvest. Investigation of potential forest products is important to make available as wide a range of development options as possible.

The most exciting of all the Amazon's initiatives come from organizations of indigenous people. In countries such as Peru, Ecuador and Bolivia, Indians have combined programs for improving their economic situation with campaigns to strengthen their political position. In Peru, indigenous organizations such as AIDESEP have overseen projects designed to improve the marketing of their crops: building up their own transport network and cutting out the white middlemen. Another program has begun which enables the people themselves to take charge of the health of their communi-

suits the Indians, rather than the outsiders trying to develop them. In the north-west of the Amazon, the Tukano Indians are trying to fund their campaign against the enormous *Calha Norte* program by selling the pottery and basketwork they produce. In several places, Indian organizations have started regeneration projects, to restore the damage colonists have inflicted on their lands.

Forests in the Amazon regenerate, if the land is cleared only once. The trees which return are not as diverse as those of the virgin rain forest, and the relationships between the plants and the returning animals are much less intricate. However, like those of the primary forest, the new trees store carbon, which would

ties, combining traditional medicinal knowledge with conventional health training. When the government failed to demarcate their lands, the Indians took charge of the process themselves, invited consultants to train their own people in surveying and mapping techniques, and set to work with an enthusiasm never manifested by outsiders.

In Brazil, the Union of Indigenous Nations has set up a university course for the training of Indians in such subjects as law and ecology. On graduating, the students return to their own communities, helping them to defend themselves against colonist and development schemes, and to build up an economy which

otherwise contribute to the build up of greenhouse gases in the atmosphere. If the forest is cut repeatedly, and the seeds and shoots from which it would normally regenerate are killed, it is replaced not – as has often been supposed – by a desert, but by scrubby grassland. This can be replanted at a surprisingly low cost with trees of the closed canopy forest. Soil nutrients that may be leached out of surface layers are still present and regenerating at greater depth. Fortunately, most of the deep rooted Amazon tree species grow with astonishing speed in old pasturelands, so the Indian regeneration projects should flourish to a full forest cover within 25 years.

One of the best conservation investments of outsiders would be to assist the beneficial projects started by the Indians and peasants of Amazonia. Rather than disbursing millions for schemes devised by governments and involving local people only as employees, funding agencies should be supplying the much smaller amounts of money required to help schemes which are clearly for the benefit of the people being developed, rather than that of the developers. In these cases the communities controlling the development projects have every interest in their success, and ensure that the resources are preserved for their future use.

By contrast, such funding institutions as the World Bank are still financing schemes which

enous support groups, such as Survival International and Cultural Survival, is extremely important. All this means that the forest can best be helped by conservation, rather than preservation. This is to say that, precious as the Amazon's natural diversity is, environmentalists must reconcile themselves to the loss of some of it, if the system as a whole is to be saved.

If the fundamental problems of the Amazon can be solved, then there is a hope of preserving within the Basin, some areas which will never be altered. If any of the ecosystems are to be saved, it is essential that tree cover of one sort or another is maintained over most of the Amazon. This is because

are likely to destroy more than they conserve. Whereas some of these programs cost hundreds of millions, many of the Indian initiatives could be funded for less than ten thousand dollars. But while there is no shortage of money for big projects, the small ones are being allowed to wither and die, because the big funding agencies overlook them. So the role of environmental organizations – such as Friends of the Earth, the Gaia Foundation and the Environmental Defense Fund – and indig-

**Left**, Guaraná seeds are used to make local handicrafts. **Above**, harvesting the fruits of the forest.

the trees of the forest generate much of their own rainfall. The water which they put back into the atmosphere by means of transpiration from their leaves condenses above the forest to form rain. As the prevailing winds in the Amazon blow from east to west, the forests in the east are responsible for much of the rain that the forests in the west receive. Without it they are likely to be killed either by drought or by fire.

Therefore, by enabling communities in the Amazon to use their forests without destroying them, conservationists can ensure that the places which are to be totally preserved receive the rainfall they require.

Over the last century, roughly one hundred vertebrate animal species become extinct and today, an even larger number is endangered unless their plight is heard. There are three groups of reasons that may spell doomsday for any particular species. Firstly, a species may require a specialized habitat that initially shrinks and then disappears through human activities, leaving the animal without basis for feeding, breeding, resting. Secondly, many animals are unable to cope with diseases, or with other exotic species introduced by Man, conversion, the Atlantic forests of southern Brazil and the forests of the Western Andean slope in Colombia and Ecuador have disappeared in the last decades with a rate equaled only by few areas on earth. It is here that habitat loss has brought many mammal and bird species to the brink of extinction. Examples in Brazil are the Lion Tamarins and the Muriqui, South America's largest monkey, and in Colombia, the handsome Cotton-top Tamarin.

Researchers address the problem of shrinking habitat in a systematic way: important

a scenario particularly frequent on islands, like Hawaii or larger ones like New Zealand. Thirdly, animals have been indiscriminately hunted for food, as trophies, or for the caged animal trade. Fortunately, in the Amazon proper, no vertebrate animal species has yet been brought to extinction by these reasons. But there are warning signs that should be watched to stop processes as long as they are reversible.

The first reason, habitat destruction, is not yet a source of extinction. Most of the Amazon is still unbroken canopy. Even pessimists accept that about 80% of the forest is still essentially undisturbed. But most South American nations have a record of aggressive land long term experiments aim to determine the minimal size of the habitat that is sufficient to guarantee survival of a small population of any particular species. A few hectares may suffice for a small forest bird of the undergrowth, and a square kilometer of forest for a group of howler monkeys, while a Harpy Eagle may need a hundred square kilometers to find enough prey. Some day, nature reserves may be gazetted or managed according to the research into these minimal critical habitats.

Fortunately for the Amazon, the sheer size of uninterrupted habitats has kept diseases and exotic competitors at bay, at least as far as animals are concerned. Of course, disease and competition with European immigrants are the reason for the decline of the indiginous humans, the Amazonian Indians. Excessive exploitation of animals by settlers had a negative impact on population sizes, but has not led to the extinction of any species – at least not in the Amazon proper: on the southern periphery, the pet trade has driven several species of attractive blue macaws into or close to extinction.

Impact was heavy particularly on those animals that live in or close to the waterways, Like the Amazonian Manatee and the Giant

The skin trade eliminated the Giant Otter from many areas, which survives in healthy populations only in some pockets of the Amazon. One of these areas is the Manu National Park in Peru, where otters are one of the principal tourist attractions, an example how an animal can be a continuing source of income for the economy of a region in contrast to the skin trade. Also for their skin, caimans have been heavily hunted to the point that the observation of a large individual has become a very unusual sight. Some animals which are

Otter. The manatee was hunted by the Indians since the arrival of the Portuguese, but without decline of the populations. The Europeans liked the manatee's flesh, which was termed by the Portuguese "fish-cow", and started to export thousands of these animals to Europe in the 17th century. Those surviving came under great hunting pressure of the skin trade at the beginning of this century. Today, some manatee populations are closely surveyed in some areas, but still killed for food in others.

**Left**, stall in a Belém market sells illegal animal skins. **Above**, this Black Caiman is a victim of the skin trade.

considered food at the Amazon, wouldn't be eaten elsewhere, for example armadillos and monkeys. Thus their populations are depressed, but not to the point to threaten their survival.

All countries of the region make efforts to legally protect wildlife, but most often, neither the political will nor the funds are available to enforce these laws. Hundreds of thousands of people in the Amazon are incredibly poor, and the fur of a poached cat or a trapped parrot may make a week's income. Maybe it is not unrealistic to hope that nature tourism may deliver arguments to use the rare animals of the Amazon has a source of steady income rather than to eliminate them in a one time harvest.

**Parrots**: Within the neotropics, there are 141 species of parrot of which no fewer than 43 are considered by the International Council for Bird Preservation to be globally threatened. Of these, only Golden and Pearly Conures of Brazil are endemic to the Amazon region, but for several other species, most notably the large macaws, Amazonia provides one of the last great refuges. The Scarlet Macaw, for example, is now critically threatened in Central America, but remains quite common in parts of the Amazon basin.

populations of birds, making them extremely vulnerable to other pressures such as trapping. Habitat loss is particularly significant for those species which have limited ranges (such as the endemic parrots of the Caribbean). Important areas for parrots where habitat loss has been extremely severe include the northern Andes of Colombia and Ecuador and the Atlantic Forests of South-East Brazil where, between 5% to 10% of the original forest cover remains.

With wild populations facing such pressures, trapping for the bird trade imposes an

There are two threats to the New World parrots: habitat loss and trade. Some indigenous hunting has taken place over many centuries yet it seems that its impact has been limited. But if an external demand is imposed, for example to supply feathers for tourist trinkets, or when populations are already threatened by other factors, hunting can be more significant. The decline of Hyacinth Macaw in the southern fringe of Amazonia might have been due to hunting for plumes and the Golden Conure is still hunted severely by colonists.

Habitat loss is ultimately the most important threat for most endangered species. A reduction in the area of suitable habitat will reduce

unacceptable burden. Parrots have always been kept as pets by the indigenous peoples but the numbers taken from the wild have been sustainable. The whole perspective changed, once a demand for parrots was established by the developed world, over 700,000 parrots of 96 neotropical species were imported into the United States during the first half of the 1980s. In fact, the number of wild birds trapped exceeded this figure which is just the tip of the iceberg. The total does not include illegal trade, nor does it include all those birds that died during capture and transport.

To capture birds, trappers use a variety of techniques including lined sticks, decoys and robbing nests. The latter often involves the felling of the nest tree. This practice not only risks killing the nestlings but destroys the nest site. There is increasing evidence that even in pristine forest habitat, parrot populations may be limited by nest sites. Thus the action of the trapper inflicts long-term damage on the remaining population. Once captured, birds spend a long time in confinement, densely packed with others, vulnerable to the spread of disease and to physical injury. Inadequate food, water or ventilation imposes fatalities.

certain countries with tight controls and export them from others with legislation less strict. For example, Brazil has banned the export of wildlife, but still loses birds, via routes through Guyana and Paraguay. Species with tiny populations are under great pressure. The Hyacinth Macaw persists in just three regions and illegal trapping is the most serious threat it faces. Spix's Macaw, long a prize specimen for trappers, is now extinct.

How can the bird trade be tackled? Much effort is needed to encourage people, if they

Probably 60% of the parrots die before leaving their country of origin. About 20% of the survivors are likely to die in transport and quarantine. The bird trade involves great cruelty and is incredibly wasteful. This wastage drives the demand on wild populations still further.

Large sums of money change hands in the illegal trade of rare species, although little is earned by locally-based trappers. The traffic continues despite international treaties seeking controls. Smugglers divert birds out of

**Left, Blue-and-Yellow Macaws in captivity. Above, Golded Parakeets are endangered by the pet trade.**

must keep parrots, to purchase captive-bred birds which are not only healthier but make better pets as well. At the moment, international trade is forbidden only if it can be shown that trapping is threatening populations. Many conservationists now believe that a system of "reverse listing" should be adopted – those species whose populations proven to withstand trapping being commercially exploited. Finally, the development of well-planned wildlife tourism can create a source of revenue across local communities, creating conditions where tourism becomes a viable and attractive economic activity. Such schemes are being tried in Peru and Brazil.

The novice to travel and nature observation in the tropics is likely to plan his trip with false hopes and unjustified fears. The most important preparation for a trip to the Amazon is to deal with wrong concepts that range from the expectation of unproblematic access to the wilderness and easy observability of wildlife on the positive side to life threatening diseases as well as massive nature destruction on the negative side.

**Getting There:** The larger cities in the Amazon are served by only very few intercontinental flights, and there are not many more international connections. An exception is Manaus, which has regular service from many South American capitals. To reach the region, it is normally necessary to fly to the capital of the respective country, Caracas, Bogotá, Quito, Lima, or La Paz, and to continue from there by domestic flights.

The Amazon of Brazil is best accessible by flights from São Paulo, Rio de Janeiro, or Brasilia, what means, that the visitor from Europe or North America will normally fly 2,000 km to the south of the equator, before proceeding in the opposite direction. It is important to plan early, since most of these domestic connections tend to be overbooked.

A glance on a road map of the Amazon will give the impression of a network of highways. Many of these roads exist only in the phantasy of development oriented politicians, which are either impassable for much of the year, or have even never been built at all – for the preservation of the forest. Limited transport exists, however, and the daring and budget-conscious traveler may consider to embark on the two to three day long journey from Rio de Janeiro to Belém by bus.

**The Cities:** The biggest mistake one can possibly make is to assume that once having arrived in cities like Iquitos, Manaus, or Belém, all located in Amazonian heartland, one is only a leisurely stroll away from undisturbed forests and from large numbers of wildlife. These cities are surrounded by a wide corridor of developed land, and without guidance by a local naturalist, one has little chance to find access to good nature.

**Public transport on the Amazon.**

The cities themselves should not be approached with much hope to find them being sensitively fit into the environment. Past generations had development for Man on their mind, without considerations for his natural surroundings. This has led to the result that a city like Manaus does not even have a larger park that may reflect the nature of the region, and that may invite to get away from the noise of the bustling city. Mentioning Manaus: even some residents suspect their home town to be the dirtiest city of Brazil.

The nature tourist is left with the insight that he has to get away from people to get a chance for nature observation. The distance is not large, and undisturbed nature is within view of all Amazonian cities, just normally not in walking distance, but rather on the other side of the river, and a combination of high prices to rent a boat and language barriers may limit this access route.

**Language:** English is rapidly becoming a language understood in most parts of the world, but not so in the Amazon. Most people in Brazil speak only one language, Portugese, and those of the surrounding countries Spanish, and it will be useful to establish a small vocabulary in either language. The languages are related, and, while pronunciation is quite different, many words can be understood in either language.

**Budget:** Most nations of the Amazon have gone through two decades of economic decline. A large fraction of the population is very poor, and even people who are better off have much difficulties coping with high inflation rates and irregular supplies. Nevertheless, it is a big mistake to conclude that a tourist with a wallet full of greenbacks or deutschmarks will feel like a rich man here. Hotels – in particular in Brazil – have often higher prices than in Europe or North America, and prices for private transportation, by boat or plane can be exorbitant.

There are two ways to escape the crunch between technical difficulties and limitations of the budget: if one wants to visit multiple areas, one should either arrive by prearranged tour, or hope that one can join such a tour that just happens to have a vacancy. This way, the large expenses for transport are shared among a dozen or so people. If one arrives alone or in a small party, the it will be best to accept the travel arrangements of a local agency for a stay in one of the few hotels or *posadas*, that cater to the nature tourist, or to reach one of the few national parks with accommodation and food supply.

**National Parks:** The largest part of the Amazon falls into the boundaries of Brazil. Unfortunately, this country has only some few national parks and biological reserves, and none of them has lodges, restaurants and campgrounds like national parks in other parts of the world. Access is expedition style, by rented boat, with acommodation on the boat or in tents. Food, drinks, and fuel have to be

Manu in Peru, still requires some time and effort, and Canaima in Venezuela is certainly unique by providing access by jet plane, and accommodation in proximity to the runway.

**Public Transport by Boat:** It is certainly the cheapest way to explore the Amazon by public transport. Boats run regularly between all major and minor population centers, and it is possible to tour most of the Amazon and its tributaries in this manner. One should count on two weeks to get from Belém to Manaus, and another two weeks to continue to Iquitos. This type of trip will certainly be adventuresome, but primitive conditions on the boat like sleeping on a crowded deck in a hammock, noise, and sanitary conditions, will certainly

brought along. Over the last decade, a large fleet of Manaus-based excursion boats that cater to this type of nature tourism has come into being, and many boats offer considerable comfort, fan-cooled cabins with beds, electricity, good kitchen, and cold drinks. Considering the ease of nature observation from such a boat, expeditions into the Amazon are even a possibility for the physically handicapped. In Belém and Santarém, these developments are still in their infancy.

In the countries that surround Brazil, the last decade has seen significant activities in the creation and in the development of national parks. However, access to most of them like

appeal only to few travelers. Occasionally, the authorities even asked foreign travelers to sign forms to confirm, that they are aware of the hardships ahead, before being allowed on the boat. Also, one keeps ending up in population centers rather than close to nature, and during the boat ride, the Amazonian forest is often nothing more than a dark line on the horizon. For the enterprising traveler, these boats a good means, however, to get to small settlements, and to find accommodation with local people in walking distance from little disturbed nature.

**Climate:** Temperatures in the lowlands of the moist tropics are extraordinarily predictable, with averages around 28° C. Under the canopy of the forest, there is little deviation from the average, and forest walks are comfortable at any time of the day. Outside the forest, the visitor from temperate climates tends to feel uncomfortable during the heat of the day, and may wish to restrict his activities from 10 a.m. to 4 p.m. Conveniently, birds and many other animals feel the same way, and activity is maximal in the early morning and around sunset. Inner cities tend to heat up to uncomfortable 35° C, but this should not be mistaken for the Amazonian climate. On a boat, nights are pleasantly cool, and even heat sensitive people get by without airconditioning.

**Dangers:** It is important to put into perspec-

mains for most visitors an unfulfilled dream rather than a dreadful nightmare. Similarly, the indiginous humans, in particular their poison-dart-blowing representatives, are rather in danger of becoming extinct than becoming a nuisance, and they will shy away from the daring tourist rather than being hostile. Vampires mostly attack domestic animals, and mosquitoe nets suffice for protection. Piranhas are said to be dangerous in some localities, but in most places, the locals take a swim in the same area where they fish for piranhas. Rule of thumb: do it the way the Amazonians do it! And leeches, a major nuisance of Southeast Asian rain forests, are actually quite rare in the Amazon.

tive the unfair view of the tropical rain forest, as being a place full of blood thirsty jaguars and vampire bats, poisonous snakes, clouds of mosquitoes, ready to suck you dry and inject a dozen or so deadly diseases, piranhas that chew you up once you stick the tip of a toe into the water, and poison-dart-blowing Indians. It will be the most dangerous experience of most tourists to pass the road that separates Manaus' shopping malls and the jetties on the banks of the Rio Negro.

Jaguars are secretive and to spot one re-

**The Amazon is well worth exploring, whether on foot (left) or by boat.**

Some snakes are certainly poisonous. It is impossible to avoid an occasional encounter, and even less possible, to carry a selection of antisera on a longer trip. The best comfort is the notion that snakes did not evolve to attack humans, and they try as hard as they can to stay away from us. Most snake attacks occur by carelessly running through plant covered territory, digging with bare hands under branches or fallen tree trunks, or by approaching a snake that makes every effort to stay by itself. The best protection is to walk slowly, to stay on trails or on open ground, and to keep the hands away from places where snakes may hide.

It may sound funny, but to get lost is a

serious danger in tropical forests. The tree trunks are very uniform, and, standing dense, quickly close off the view of the horizon, of the river, or of a clearing in the forest. To determine the direction, one may try to look for the sun, which however is barely visible through the dense canopy, most of the day in nearly vertical position anyway, or covered by clouds. The slope of a hill: there are not many in one of the world's largest plains. A call, the motor of some boats? Forget it, the incessant rattling of the cicadees takes care of it. The protection is to enter the forest along trails, leave trails or the rim of the forest only for shortest distances, and generally to take time in the exploration of an area.

born viral infection, but this disease still strikes in some localities. Fortunately, vaccination is cheap, legally required and provides full protection for about 10 years.

Malaria is more problematic than yellow fever. This is a mosquitoe-infected disease caused by Plasmodium, a single cell animal. So far, no satisfactory vaccination has been developed yet. Today, for prevention, most doctors prescrible a combination of Chloroquine and Paludrine, after discovering the drug Fansidar has too many side effects. Unfortunately, Plasmodium becomes increasingly resistant against these chemicals, and the best prevention is to avoid being bitten by mosquitoes. A fan or a mosquitoe-

Globetrotter will agree that places where they were overwhelmed by mosquitoes were in boreal or temperate latitudes rather than in the tropics. As a matter of fact, the regions of the Amazon around black water rivers like the Rio Negro are nearly free from mosquitoes, the acidic water doesn't permit their larval development. Things can be worse on the white water rivers, though.

**Diseases:** While it is rare that exposure to mosquitoes itself becomes really a problem, their potential to transmit contagious diseases should not be underestimated. The times have gone when thousands of people died in tropical America of yellow fever, a mosquitoe-

net at night, long pants and long-sleeved shirts at day, and mosquitoe repellent will provide more protection than a sophisticated drug taking protocol.

**What to Expect in the Forest**: The Amazon is the world's richest ecosystem. It would be a major mistake to take this information such that a stroll through the forest will lead to easy observability of a large number of mammal and bird species, and flocks of large and colorful butterflies. In contary, the first encounter of a tropical rain forest anywhere on earth is that of a quiet, green hall, with little to marvel at except trees. Even in optimal habitat, a whole hour may pass, where one doesn't see a single mammal, and at best may hear the faint call of an unidentified bird.

One should start nature observation with the obvious objects of admiration, the large trees, with their big buttressed or stilt roots. What may be the number of different trees that occur on a single hectare? There may be more than a hundred species. On the forest floor, leaf-cutting ants carry bits of greenery. A sudden flurry of activity in the undergrowth and lower canopy may point to the approach of a mixed foraging flock of birds: there may not have been a single bird for two hours, now, 20 species are passing by within a few minutes, leaving little time for identification. Scanning

congregate, and may become so tame that they feed out of people's hands.

Various strategies help to find different birds: for the observation of parrots, be at sunrise at a place with a wide view, when they change from roots to feeding grounds. For hummingbirds, wait at flowering bushes, for tanagers, at fruiting trees. And one should go out at night with a flashlight, when white, green, or pink reflecting points in the darkness may turn out to be a caiman, a frog, a Boat-billed Heron, or a Spectacled Owl.

**Photography:** The photography buff will wish to read a whole chapter about nature photography in the tropics. The beginner should take three central problems into consideration. Fre-

the upper canopy for the origin of some bird-like calls will identify a group Squirrel Monkeys, may be the only mammal observation of the day, not considering the bats that come out at dusk. To avoid disappointment, it is important to appreciate these details.

How does one go about to increase chances to see wildlife? There are lodges and park headquarters throughout the Amazon, where wild mammals are regularly fed. In these places, up to a dozen mammal species may

**Left**, a comfortable vessel provides tourist trips from Manaus to the Anavilhanas. **Above**, catching 40 winks.

quently, the danger to camera and film in tropical climate is exaggerrated. There is no question, that over a year or two, films will spoil, and even well-coated camera lenses may be overgrown by fungi, and closed containers with water absorbing chemicals provide important protection. Normally, however, over a period of two or four weeks, no such danger deserves consideration. For best photographic results, a sky-light filter is necessary from 9 a.m to 3 p.m, when UV-light is maximal. And for animal and plant photography, a tripod is essential: less than one percent of the sunlight reaches the forest floor, too little to get unblurred pictures without this tool.

There are about four million square kilometers of rain forest in the Amazon, but getting to them may become a more demanding enterprise than expected, simply – and fortunately – because much of the region is still so little transformed by Man. Many of South America's developed and accessible national parks are outside the Amazon region, at the periphery of Caracas, Bogotá, or Rio de Janeiro, where land had to be legally protected to retain any valuable nature. These cities are frequently "stopovers" on the way to the Amazon, and their parks should be a welcome introduction to tropical nature. For this reason, some of these places are included in this book. Excursions to the Pantanal, for example, or the Iguassu National Park, are offered by numerous tour operators in Brazil's cities.

Only very few parks in the Amazon can currently be approached with touristic expectations formed in North America or East Africa for comfortable access, good accommodation, and fine food. But the situation is changing: Canaima in Venezuela can be reached by jet plane, access to Amacayacu National Park in Colombia and Manu National Park in Peru has improved, and similar developments occur in Bolivia and Ecuador. The Brazilian Amazon has no parks with facilities yet, but a fleet of excursion boats stationed in Manaus has developed in less than a decade, and is still expanding. And private hotels are springing up in natural settings, and *ranchos* start to cater to the nature tourism.

The following pages try to describe some of these places and excursions. But given the pace of new developments, and the unpredictability of political and economic developments in most Latin American countries, it is difficult to give faultless advice. In the years to come, nature tourism in the Amazon will require an open mind and a sense for improvisation.

**Preceding pages:** Venuezuelan vista; the falls at Iguassu; meandering through the lowland forest; canoe journey into flooded forest; tropical beach backed by Atlantic forest in Brazil. Left, canyon at the southern rim of Amazonia.

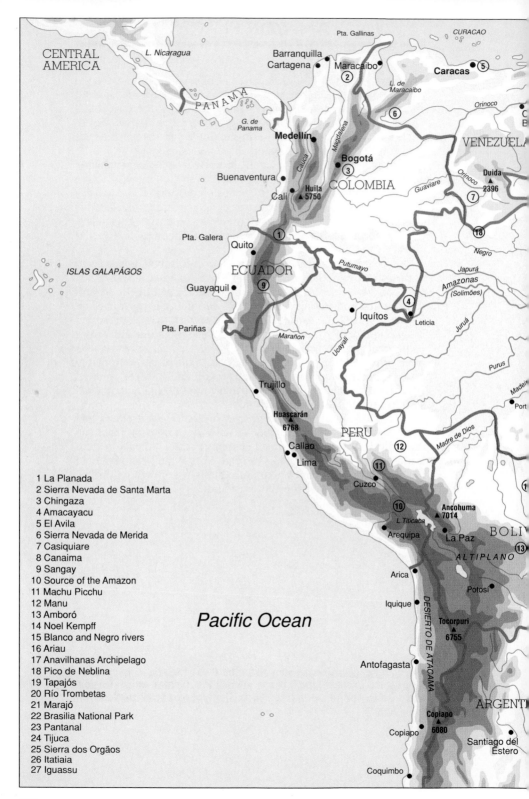

CENTRAL
AMERICA

L. Nicaragua

Pta. Gallinas

CURACAO

Barranquilla
Cartagena
Maracaibo ②
Caracas ⑤

L. de Maracaibo

PANAMA

Orinoco

G. de Panama

Medellín

Cauca

Magdalena

Bogotá ③

COLOMBIA

Buenaventura

Cali
Huila
▲ 5750

Pta. Galera

Quito ①

ISLAS GALAPÁGOS

ECUADOR

Guayaquil ⑨

Pta. Pariñas

Putumayo

Marañon

Iquítos

Ucayali

Trujillo

Huascarán
6768

Callao
Lima

Cuzco ⑪

PERU

Manu ⑫

Madre de Dios

Source of the Amazon ⑩

L. Titicaca

Arequipa

La Paz

Ancohuma
▲ 7014

Arica

Iquique

ALTIPLANO

Potosí

Pacific Ocean

DESIERTO DE ATACAMA

Antofagasta

Tocorpuri
▲ 6755

Copiapó
▲ 6080

Copiapo

Santiago del Estero

Coquimbo

VENEZUELA

Orinoco

Duida
▲ 2396 ⑦

⑥

⑱

Negro

Japurá

Amazonas
(Solimões)

Leticia ④

Juruá

Purus

Madeira

Port

CURACAO

BOLI

ARGENTI

⑬

1 La Planada
2 Sierra Nevada de Santa Marta
3 Chingaza
4 Amacayacu
5 El Avila
6 Sierra Nevada de Merida
7 Casiquiare
8 Canaima
9 Sangay
10 Source of the Amazon
11 Machu Picchu
12 Manu
13 Amboró
14 Noel Kempff
15 Blanco and Negro rivers
16 Ariau
17 Anavilhanas Archipelago
18 Pico de Neblina
19 Tapajós
20 Río Trombetas
21 Marajó
22 Brasilia National Park
23 Pantanal
24 Tijuca
25 Sierra dos Orgãos
26 Itatiaia
27 Iguassu

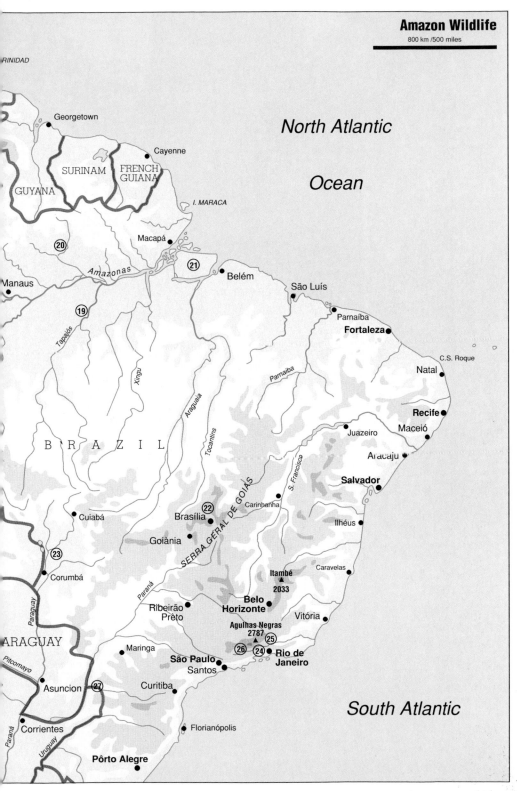

TRINIDAD

Georgetown

Cayenne

SURINAM

FRENCH GUIANA

GUYANA

I. MARACA

North Atlantic

Ocean

Macapá

⑳

⑳

Amazonas

Manaus

Belém

São Luís

⑲

Parnaíba

**Fortaleza**

C.S. Roque

Tapajós

Natal

Xingu

Parnaíba

Araguaia

**Recife**

Maceió

B R A Z I L

Tocantins

Juazeiro

Aracaju

S. Francisco

**Salvador**

Cuiabá

SERRA GERAL DE GOIÁS

Carinhanha

Ilhéus

Brasília

㉒

⑳

Goiânia

Corumbá

Caravelas

Itambé
▲
2033

Paraná

Paraguay

Ribeirão Prèto

**Belo Horizonte**

Vitória

PARAGUAY

Agulhas Negras
2787
▲

㉕

Pitcomayo

Maringa

㉖ ㉔ **Rio de Janeiro**

**São Paulo**
Santos

Asuncion

㉗

Curitiba

South Atlantic

Paraná

Corrientes

Florianópolis

Uruguay

**Pôrto Alegre**

163

# WESTERN COLOMBIA

No one going to **Colombia** who is interested in Wildlife should turn down the opportunity to visit the **Pacific slope of the Western Andes**. Although not geographically part of the Amazon basin, it is included in the scope of this book because its wealth of endemic fauna and flora make the region one of the most attractive in Colombia.

The Pacific slope of Colombia is also readily accessible, thus providing the curious visitor ample possibility to make the fascinating comparison between the vast expanse of the Amazon lowlands with its species rich communities with the wet forests of the Pacific slope with their very high degree of endemism.

This zone is defined biogeographically as the **Chocó** and extends from the Darien Gap to northern Ecuador. The political department of Colombia which also carries this name occupies the northern third of this area. It is a region of very high rainfall which is fairly evenly distributed throughout the year. Indeed, it is the wettest place in the Western Hemisphere, with an average of over 13 m of rain falling per annum in some of the central localities. The heaviest rain falls in the coastal areas, up to about 250 m above sea level.

The natural vegetation is tropical wet forest, boasting a very high diversity of trees, epiphytes and palms. At higher altitudes, long periods of mist and low cloud hang over the forests which drip constantly with condensation: this habitat is aptly named "cloud forest". The trees are festooned with layers of mosses, orchids, bromeliads and other epiphytes. Each tree becomes a garden and presents a visual display of luxuriance.

All the important neotropical bird families are present in the Chocó zone, but the community composition does show some difference compared with that of Amazonia. Most striking is the strong representation of tanagers, indeed the region is popularly described as "the tanager coast". Many of these are of the genus *Tangara*, small and brightly colored birds which travel like gems in mixed species flocks. The Chocó zone is one of the most important centers of high biodiversity in the neotropics and the high level of endemism shown by the birds is reflected by other groups as well.

The best base to use for an exploration of the Pacific slope is the private reserve of **La Planada** that is situated in the Colombian department of **Nariño**. It is managed by FES (Foundation for Higher Education), which is based in **Cali**. It has also received support from a number of international bodies including the WWF.

The journey to La Planada starts in the capital of Nariño, **Pasto**. Pasto is a small city of about 200,000 inhabitants set at the base of the **Galeras volcano**. The journey to La Planada lasts about four hours along the main Pasto to Tumaco road and takes one up to 3,200 m above sea level through a spectacular scenery of gorges, steep-sided mountains and fields of wheat and potatoes, then down into deep valleys with rocky

**Left**, Brazil nut tree of the tropical rain forest. **Right**, a Rufous-tailed Jacamar.

streams and cultivations of sugarcane and bananas.

The reserve lies between 1,300 and 2,100 m above sea level, a distance of seven kilometers above the village of **Chucunes**, near **Ricaute**. Temperatures are moderate, averaging between 10° C and 17° C, with about 4.5 m of rain annually. This mostly falls in the late afternoon and during the night. Most mornings are dry and very clear, with a good view of the **Cumbal volcano**, possible for a short while after dawn.

The infrastructure of the reserve includes an administration building with a restaurant and meeting room, and a scientific center with accommodation in the form of small study bedrooms and a museum. There is also an orchidarium with over 360 species (all collected from La Planada), indeed the orchid flora is one of the richest in the world. La Planada has also been the center for a captive-breeding program for the endangered Spectacled Bear – a two hectare enclosure has been set up and breeding has already taken place quite successfully.

Research on the Spectacled Bear are recorded in the wild state from the reserve and its surroundings.

The reserve contains a range of vegetation types with small pastures, some secondary growth and extensive primary forest which has hardly been modified by Man. At the lower levels, the forest is particularly luxuriant. The flora is a mixture of both low and high elevation elements and shows a high level of endemism.

The avifauna has been well studied by both Colombian and foreign ornithologists and so far over 240 species have been identified. Of these no fewer than 24 are endemic species of restricted range (less than 50,000 square km). The dominant families are the tanagers with 30 species known, of which 11 are of the genus *Tangara,* the tyrant flycatchers with 32 species, hummingbirds with 26 species and furnariids with 15 species. The area around the buildings provides an excellent area to spend the first hour or so of daylight, there is a good range of flowering shrubs **Accommodation at Rio Tatabro.**

that are visited by hummingbirds. Other well-marked trails lead into areas of forest.

Large mixed species flocks of birds are a real feature of La Planada. These can be spectacularly varied and pass rapidly over one in the canopy. Fortunately, because of the relief of the land, the path often follows ridges so that canopy flocks can be watched from above. The parties invariably contain tanagers and usually a suite of fly-catchers, flower-piercers, furnariids and, from late September to April, North American warblers as well such as the Black-and-White Warbler and the Blackburnian. The forest covered gorges provide haunts for the Andean Cock-of-the-Rock. The resplendent males are bedecked in red, black and grey and gather in courtship leks in the subcanopy. Away from these arenas, they lead solitary lives and, being rather quiet and wary, can be difficult to see.

The best birdwatching times are from dawn to late morning. Some activity continues in the afternoon, but the onset of rain puts an end to most productive watching. However, if the rain dies away before dusk, it is important to be out in the field to maximize on any last-minute flurry of activity.

The area around La Planada extends as a forest belt into north-west Ecuador and is the home of the Awa Indians. They mainly live in small settlements, generally below 1,500 m, cultivating crops such as maize, beans, sugar cane and bananas. Their culture has developed throughout the generations in harmony with the forest. Similarly, their great knowledge of the forest enables them to utilize a great range of forest products.

The objective of FES is to develop an international integrated land management project which crosses the frontier into Ecuador, drawing together the interests of both the Awa people and the conservation of this important center of endemism. Overall, such a scheme could protect over 7,500 square km of important forest habitat and the livelihoods of 12,000 Awa Indians. Already the Indi-

The rare Spectacled Bear survives in some Columbian reserves.

ans are closely involved with the management of the area.

Higher up on the slope, above the tree line on the **Chiles volcano**, a second research station has been built. This party has the objective of managing an Andean Condor reintroduction program (there are still a few native condors left) and partly to continue the spread of effort of the scientific research to explore fully and understand the altitudinal variation and movements of the wildlife.

A third cabin, closer to sea level will be completed soon to complete the series. All of these ventures reinforce the philosophy of the La Planada: that by working in close partnership with the local people the effectiveness of conservation can extend well beyond the boundaries of one single reserve. Instead there can be created a zone of influence extending from near to sea level to the tree-line, from Colombia into Ecuador.

Using **Cali** as a base, an exploration of a zone of the Pacific slope somewhat to the north of the La Planada region can

be achieved. Cali is the third largest Colombian city with a population of nearly 1.5 million. It is an attractive, organized city situated in the **Cauca valley** and in the heart of the main sugarcane growing belt. Other important crops include cotton, tobacco, maize and coffee. It is also the Colombian capital of *salsa* music! The climate is very pleasant (with an average temperature of 24° C) and with a good system of buses which can be used to visit many important wildlife sites of southwestern Colombia, it certainly makes a good place to stay.

The start of the Pacific slope can be reached just on the outskirts of the city. 18 km out on the road to **Buenaventura** stands a block of forest on the ridge overlooking the city. This readily accessible site provides an excellent morning's birding and over 50 species are quite possible, including Multicolored Tanager and the hummingbird, the Bronzy Inca. Both are Colombian endemics. There are other patches of forest nearby worthy of investigation,

**Tree fern canopy makes a pretty pattern.**

scattered between small cattle pastures and rather fine houses set in large grounds.

The best known site on this section of the Pacific slope is **Alto Anchicayá**. It lies about 60 km from Cali on the old Buenaventura-Cali road. Access is from the town of **Danubio** where there is a security post at the head of the road which climbs up into the Anchicayá watershed.

The whole area is managed by the *Corporación Autónoma del Cauca* (CVC) to protect the reservoir and its watershed. The reservoir drives a hydroelectric plant. There is a large workforce who are housed in a settlement which includes dwellings, chalets, shops, restaurant and sports facilities. With permission from CVC (obtained in advance in Cali) it is possible to stay in a chalet and buy meals in the restaurant. The staff are extremely friendly and are well-accustomed to visiting naturalists. This is not surprising since the area is extremely attractive, holding a large area of intact forest which remains undisturbed. Scenically it is superb, with forest-clad hills, waterfalls, rocky streams and gorges. The climate is tropical but the high rainfall levels give a feeling of freshness. Most mornings are dry and clear, with the deep valleys filled with clouds and ridge tops exposed. As the sun rises, so do the clouds and by most afternoons there is rain.

From the settlement, most of the bird watching can be done from the roads. These climb the hillsides, affording the walker good views of the canopy in forest below. There are some trails, particularly radiating out from the settlement, which are useful for finding some of the more skulking species of the forest floor and undergrowth. The birds are spectacular and mostly quite easy to see. An experienced observer might well see over 100 species in a day. Like La Planada, a strong component to the avifauna are the mixed species flocks. These are dominated by tanagers, tyrant flycatchers with accompanying woodcreepers and woodpeckers. The tanagers present do not show a very

close overlap with the La Planada avifauna. This is in part because the area around the settlement is at a lower altitude than La Planada, many of the Pacific slope specialities have a very narrow altitudinal ranges and may also show restricted latitudinal distributions along the slope.

Despite being one of the most visited sites of the region, surprises still do turn up. A good example was the recent sighting of the Banded Ground-Cuckoo, a species that had not been seen previously, and was only known from old museum specimens. Alto Anchicayá remains one of the best areas to see the Long-wattled Umbrellabird, similar in appearance to its sibling species, the Amazonian Umbrellabird, but only known from the Chocó zoogeographical province and is considered to globally threatened.

At a lower elevation (200 m above sea level) on the Pacific slope near the village of **Aguaclara**, is a recently built cabin belonging to the Cali-based *Fundación Herencia Verde*. It lies

**A Mantled Howler Monkey at play.**

amidst forest-covered hills, at the confluence of the Rio Tatabro (which gives its name to the site) and the Anchicayá river.

The *Foundación Herencia Verde* is a young conservation body which is seeking to establish schemes of working with the local communities in order to achieve conservation through integrated land use management. Thereby encouraging the sustainable use of forest materials.

Although the forest along the old Buenaventura Road has a superficial appearance of being little disturbed, it has been greatly modified for several decades. Large companies have offered cash for timber for pulp and along the road it is possible to see the neat piles of timber waiting for collection. As the supply of large trees diminishes, the men have to travel further and further inland to find suitable materials. The limit is the distance that can be reached by a loaded mule, beyond that the forest remains essentially undisturbed.

Another source of income has been gold panning. Along streams, there is much evidence of active prospecting, using methods unchanged for centuries. The net effect of all these activities is that the menfolk spend less and less time cultivating their plots and tending animals. The rural economy is changing and much damage is also being done to the environment.

To attempt to stem the tide, *Foundación Herencia Verde* is setting up an initiative to reinforce the family structure and small-scale farming, using forest products in a sustainable way. It is hoped that the economic pressure imposed by external commercial interest can thus be marginalized. So far, they have plot schemes under way with four families.

Although the forest has certainly been modified and the influence of man has led to the local disappearance of large birds like cracids, the area around the **Rio Tatabro** site is excellent for birds, sharing many species with Alto Anchicayá.

Footpaths lead along ridges on the slopes behind the cabin, offering good views of the coastal plain and the Buenaventura bay as well as providing plentiful encounters with bird parties. These are dominated by tanagers such as the Blue-whiskered, the Scarlet-and-White, the Golden-hooded and the Gray-and-Gold (the names themselves epitomize their finery) accompanied by Slate-throated Gnateaters, Lesser Greenlets and Slaty-capped Shrike-Vireos.

Some of the species move around in mainly single-species groups, the common Tawny-crested Tanager and the Dusky-faced Tanagers being good examples. Not all birds will be associating in groups, the Black-breasted Puffbird spends most of its time perched quietly in exposed situations, waiting to collect its large insect prey from nearby foliage or from the ground. A common call is that given by the Fulvous-bellied Antpitta, a very elusive ground-dwelling bird which haunts the thick vegetation around natural tree falls. As well as birds, the region is of great importance for other wildlife. There has been a survey carried out on the butterflies which recorded 520 species along a 30 km transect.

Rio Tatabro is at a somewhat lower elevation than Alto Anchicayá and this is reflected in its higher average rainfall. Practically every day there will be heavy rain. However, as is the usual pattern elsewhere, rainfall occurs mainly during the afternoon or evening, leaving the visitor free to make the most of the mornings.

*Foundación Herencia Verde* is just one of several newly emerged conservation organizations in Colombia, staffed by highly committed and well-trained biologists, anthropologists and economists. They work alongside the government conservation agency, INDERENA, and the universities on studies within the existing national parks and protected areas network. They are also increasingly playing a lead role in exploring the interaction and integration of the community and conservation. The work that these groups are undertaking already receives some international support and recognition, it merits more.

Heliconia, used worldwide as an ornamental plant, originated in tropical America.

# SANTA MARTA

The Amazon basin is frequently described as the lung of the world. If that is the case, then the **Sierra Nevada de Santa Marta** could be described as the heart of the world. At least that is how the Kogi Indians who inhabit this area believe. The Santa Marta massif is the highest coastal range of mountains in the world. The tallest peaks, **Simón Bolívar** and **Cristóbal Colón**, are 5,770 m above sea level and lie just 45 km from the coast.

The Kogi live a life of self-imposed cultural isolation from the outside world, even though they may come into quite regular contact with others in the course of their activities. According to them, they represent the Elder Brother, the rest of mankind being the Younger Brother whose wanton disregard of nature has pushed the planet to the edge of ecological disaster. During the 1970s a lost city (**La Ciudad Perdida**) was discovered in the Sierra Nevada which was built by the Tayrona Indians about 700 years ago. The Kogi Indians are probably the direct descendants of this culture.

The Sierra Nevada de Santa Marta is also a region of great natural beauty and biological interest, holding twelve species of endemic birds including a parakeet and two hummingbirds. To enter the heart of the Sierra Nevada, to travel to the Ciudad Perdida requires a long trek with guides which will last a week to get there and back. It is an extraordinary experience.

Alternatively it is possible to get there by helicopter. Fortunately for the visitor with less time or money to spare, all but one of the endemic bird species can be seen on the northern side of the Sierra Nevada, along the ridge of **San Lorenzo** which is accessible by jeep from the town of **Minca**.

Minca lies in the coffee-growing belt and the road from Minca to San Lorenzo is about 35 km long, very steep and muddy. Therefore, if without a vehicle, **This Tarantula is the size of a human hand.**

it will be a tough tiring haul. This access to San Lorenzo does give on a tantalizing glimpse into the heart of the Sierra Nevada, bringing one into contact with much of the distinctive wildlife, without the risk attached to entering the interior. Unfortunately, parts of the massif can be dangerous to enter on account of the activities of drug smugglers and guerillas.

Advice should be sought from the INDERENA office in Santa Marta, from whom a permit must also be obtained. Accommodation should be available in the INDERENA cabins situated on the San Lorenzo ridge. It is well worth staying overnight. The mornings are chilly at this altitude (about 2,400 m above sea level) but the views are quite incredible. Northwestwards a panorama reveals the Caribbean coast from the city of Santa Marta westwards. Southwards are a series of ridgetops, extending to the snow fields of the peaks. Once the clouds in the valleys have risen, these views are lost for the rest of the day.

Close to the Sierra Nevada is another national park: **Tayrona**. This reserve protects coastal forest, including a gradation of dry to humid forest. This park too contains an ancient town; although not on the scale of Ciudad Perdida, it is easily accessible being just three hours walk from the main park center.

Birdwatchers however should expect to take much longer to cover the distance since the path leads through some excellent forest with birds such as White-bellied Antbirds, manakins and several tyrant flycatchers. The area is a good region for seeing King Vulture. Early morning views are also breathtaking. From the excellent cabins (called "ecohabs") it is possible to see the sea and the snow fields of the Sierra Navado in a single field of view.

Much of the coastal traveling can be undertaken by buses along the Santa Marta to **Riohacha** route, however hiring a car is almost essential if a trip is contemplated to the San Lorenzo ridge and a vehicle does allow greater freedom to explore the arid country to the east towards Riohacha.

It's advisable to avoid the poisonous spines of this caterpillar.

# COLOMBIA: BOGOTÁ TO THE AMAZON

The north-west limit of the Amazon basin is marked by the eastern slopes of the Colombian Andes. The watershed starts high up above the treeline. To take a complete view of the Colombian Amazon, the visitor should first of all pack some warm clothes and head for the Andean moorlands or *paramo*, to find out where it all begins.

The nearest *paramo* to **Bogotá** is in the **Chingaza National Park**, no more than three hours drive from the city center. The Park straddles the eastern *cordillera*, providing on one slope the source of Bogotá's water supply and on the other streams that head eastwards to the great plains of the Llanos. That water, in fact, eventually leads to the Rio Orinoco rather than the Amazon but **Chingaza** provides as good an example as any of the *paramo* landscape and is certainly the most accessible of the sites in the eastern Andes.

*Paramo* is a distinct type of vegetation found in the northern Andes from Venezuela south to Ecuador and parts of Peru. It is open moorland, cold and wet, with tall grasses and characterized by species of composites of the genus *Espeletia,* tall hoary-leaved plants with yellow flowers. The *paramo* starts about 3,400 m above sea-level. Above 4,000 m, the vegetation is dominated more by grasses although some *Espeletia* persists. At lower altitudes and in sheltered areas, are patches of temperate woodland with rather stunted trees, heavily laden with mosses and other epiphytes. Out on the open moorland are boggy areas and small lakes.

It is excellent walking country, although newcomers should take care not to over exert themselves because of the risk of altitude sickness. The climate is variable and can change dramatically in the course of a few hours. If the sun is shining and there is little wind, one can feel comfortable in a tee-shirt. However, with low cloud and a strong wind or rain, it an get very cold and the visitor

is well-advised to carry waterproofs and several layers of clothing. At nighttime, the temperature may drop below freezing.

Within the park is a large reservoir which serves Bogotá. Just outside the park is a cement works supplied by limestone from a large workings within the park itself. However, the park is so large that visitors can enjoy their surroundings quite oblivious of its presence. There are several well-marked tracks that lead off the principal road to the waterworks, these provide excellent routes. Although small paths can be found off the tracks, walking through the moorland vegetation itself can be very tiring.

On the *paramo*, probably the most spectacular birds are the hummingbirds. Chingaza has a good hummingbird community, including the Black-tailed Trainbearer with a tail almost twice as long as its body. The Bearded Helmetcrest is quite common at the higher altitudes and is a regular visitor to *Espeletia* flowers, often feeding from a perched position rather than hovering. This may be a strategy to conserve energy. Unusual for hummingbirds, it can also be seen walking on grass tussocks, making short flights to catch insects. The Great Sapphirewing also visits *Espeletia* or terrestrial bromeliads and is one of the largest hummingbirds, being over 16 cm long and hovers with rather slow, bat-like wing-beats. The most extraordinary hummingbird of all is the Sword-billed Hummingbird, with a long, straight bill (as long as its body). It occurs in some of the woodland patches at lower levels where they border the *paramo*.

The most widespread bird is the Great Thrush, which occurs from the *paramo* well down into the temperate zone: indeed it is a common bird of the Bogotá suburbs. The *paramo* and shrubby areas also supports furnariids such as Many-striped Canasteros and White-chinned Thistletails and seed-eaters such as the Plumbeous Sierra-Finch. Some of the most colorful birds are the Mountain-Tanagers. A visitor should expect to see the Scarlet-bellied Mountain-Tanager:

**An emergent Brazil nut tree rises above the forest canopy.**

a striking black and red bird with a sky-blue rump. The woodland edges provide opportunities of seeing small bird parties including Masked Flower-piercers, Golden-fronted Redstarts, White-throated Tyrannulets, tanagers and Rufous-browed Conebills (a Colombian endemic).

Whilst watching the undergrowth and trees for small birds, it is important not to ignore the sky above one. Black-chested Buzzard-Eagles are impressive broad-winged raptors which take to the wing mid-morning to soar high above the moorland. One should also look out for Andean Condor. Formerly extinct from this part of the Andes, INDERENA (the Colombian National Park Agency) and San Diego Zoo have started a re-introduction programme.

Eastwards the *paramo* gives way to the Andean foothills, thence to the plains of the Llanos to the north and the Amazon forest to the south. Much of the forest of the eastern base of the Andes has been cleared. The southern portion of eastern Colombia is a vast flat tract of forest, the relief interrupted by the massif of the **Serranía de la Macarena**. This isolated range is over 100 km long and, in places, 25 km wide, reaching an elevation of 2,500 m. It is a relict of the mountain ranges which existed in South America before the Andes and is an area of high biological interest, much of it still unexplored. Probably over 450 species of birds occur including several species of cracids as well as boasting a rich primate diversity. It has a spectacular landscape of cliffs, waterfalls and forest.

Access to the National Park is difficult but the biggest obstacle to would-be visitors is the high level of guerilla activity that takes place in and around the park. At present, foreigners are strongly discouraged to travel to La Macarena and no one should consider making a journey without soliciting up-to-date, reliable advice from INDERENA.

The finest introduction to the Colombian Amazon takes place on board the flight from Bogotá to **Leticia**, the

**Andean landscape at Chingaza National Park.**

capital of Amazonas, situated on the northern bank of the River Amazon itself. The flight takes one and three-quarters of an hour. Once the eastern Andes and the Macarena mountains have passed, the plane heads south-east, crossing the equator, and passing over a great sea of forest. An unbroken canopy extends from horizon to horizon. It is then that the true enormity of the Amazon rain forest reveals itself: there is nowhere else in the world that a similar spectacle could be seen.

The plane lands at Leticia in the early afternoon. The plane doors open and a wave of warm humid air floods into the cabin. The temperature outside will be well into the 30° C and the sweat will be running during the short walk from the plane to the arrivals hall. Crowds of people wait just outside to greet the passengers and everywhere, there is the constant rhythm of Latin and Brazilian music.

Leticia is a small administrative town of fewer than 20,000 inhabitants. It sits in the south-east corner of the part of

Colombia known as the "trapezium", a curious piece of territory sandwiched on three sides by Peru and Brazil. It is well worth spending a few nights in Leticia.

The waterfront is the hub of life, a bustling center of commerce where the local communities come to sell vegetables, fruit and fish. From one of the terrace bars, one can relax and watch the comings and goings of the traders, traveling in long, narrow canoes with simple outboard motors. There are floating houses, some of them acting as gasoline stations for the river craft, or as mooring places for larger boats. Across the river lies the frontier of Brazil and Peru, whereas just downstream is the Brazilian border town of **Tabatinga**.

It is only a few minutes walk into Tabatinga and there is great interchange between the two towns. Colombian currency is acceptable in Tabatinga and most people understand and speak Spanish. Many Colombians do their shopping there, taking advantage of favorable prices for many consumer

Tree frog displaying its vocal attributes.

products brought upstream by boat from Manaus, and the Brazilians are certainly happy to accept their currency which is stronger than their own.

From Leticia it is possible to travel by boat to Iquitos (once or twice a week, taking three days) and to Manaus (from the nearby Brazilian port of **Benjamin Constant**, leaving twice a week and taking four days). However, Leticia and the Colombian Amazon can provide an excellent and very inexpensive base for exploration.

It is possible to enter forest from Leticia itself, although a 10 km hinterland, mainly given over to ranches, has to be crossed first. This can be done by taxi. A far better and more satisfying way of seeing the forest is to travel upstream and stay at the **Amacayacu National Park**. This involves a boat journey of 60 km which can be done by making an arrangement beforehand with INDERENA. Failing that, boats can be hired from the waterfront or a ticket bought for one of the public transport boats which ply the Colombian side of the river two or three times a week. This is the cheapest way of making the journey and perhaps the most interesting, although it is also the slowest, taking about anything between six and eight hours.

The Amacayacu National Park is a lozenge-shaped area situated in the center of the trapezium, reaching down to the Amazon between the tributaries of the **Matamata creek** and the Amacayacu river.

At the mouth of the Matamata is the Visitors Center, which has facilities to accommodate up to thirty visitors, although very rarely are there more than a handful of people staying. The center boasts a restaurant, a museum and library, craftshop (run as a cooperative venture between the park and the Tikuna Indian communities living in and around the park) and meeting room. There are also facilities for scientists including living quarters and a laboratory. The site is the main administrative center for the park INDERENA offices and accommodation for park officers. There **A Blue-backed Manakin.**

are three other stations elsewhere in the park which provide bases for officers, all are in radio contact with one another.

The Visitors Center is situated on the riverbank with an excellent view across the river to **Mocagua Island**, one of series of large mid-stream islands. Between the Center and the Park offices is an elevated walkway, between three and four meters off the ground. This is to allow access during the four or five months of the year when the river level is high enough to flood adjacent low-lying ground. This period usually starts in January and is a response to heavy rains far away in the Peruvian catchment. During this period a belt of forest along the riverbank, about a kilometer wide, is flooded. This seasonally flooded forest is called *várzea*. In parts of Peru, the area of *várzea* can cover many hundreds of square kilometers. During the inundation period, access into the forest is only possible by dugout canoe, which can be hired at the center.

The elevated walkway provides an excellent vantage point for birdwatching in the environs of the Center. This region has secondary vegetation, now allowed to regenerate. The more open aspect of this woodland provides a wonderful opportunity to become easily acquainted with some of the birds and butterflies. Indeed, any visitor, however experienced, should devote part of the day here. It offers the best views of parrots and macaws, particularly in the early morning and late afternoon, when birds are making flights between their roosting and feeding areas. Late afternoons, in particular, are very good when Plum-throated Cotingas, Umbrella birds and Bare-necked Fruit-Crows may be seen and, as dusk descends, Short-tailed Nighthawks, Bat Falcons and Common Potoos.

The best time of the year for this area is from late July to early September. This is the flowering season of the *Erythrina* trees. The peach-orange blossoms attract large numbers of hummingbirds, troupials, caciques and parrots such as the Dusky-headed, Tui and Cobalt-winged Parakeets.

**Sitting pretty: a Red-legged Honeycreeper.**

Northwards from the park offices is a path which leads through a kilometer of *várzea* forest, before rising somewhat into *terra firme* forest, which never floods and is the classic diverse lowland Amazon rain forest. The *várzea* has a fairly open ground layer and the trees are bare of epiphytes for the first two or three meters, marking the height of the annual flood. There are also frequent gaps caused by natural tree falls. These fill rapidly by shrubby thickets. Although not as faunistically or floristically rich as the *terra firme,* there are nevertheless a number of species to be looked for here which are rarely if ever found in *terra firme* forest. These include a number of antbirds such as the Bare-spotted Bare-eye, a "professional" ant-swarm follower whose nasal call is a distinctive sound. As the trail rises into the *terra firme*, there is a side path leading to a platform overlooking an open vegetation-choked swamp. This is an excellent place to watch Horned Screamers, Yellow-tufted Woodpeckers, Bat Falcon, and a variety of parrots and macaws. The *terra firme* forest provides the most challenging birdwatching of all and is the area where time spent will pay off in terms of the number of species seen.

Several species of monkeys should be seen in the park, including Squirrel Monkeys, tamarins, sakis and titis. Look out for the puncture marks of sap-tapping on some of the tree trunks, these are made by Pygmy Marmosets. Visitors will almost certainly seen agouti, various squirrels and other small mammals. Tapirs, two species of peccary, deers, otters, Lesser Anteaters, two species of sloth and cats including jaguar, jaguarundi and ocelot are all known in the park, although good fortune is required to see any of these. The butterflies are spectacular, especially the huge Morpho butterflies which patrol sections of the path with slow, heavy wingbeats.

The path runs for about 12 km to the Amacayacu river and there are numerous side trials. Back at the park offices there is a short path leading to a huge

*Ceiba* tree which bears a ladder. Scaling this, it is possible to climb 45 m to the crown of the tree where it emerges from the canopy, affording breathtaking views of the canopy and across the Amazon river to Mocagua Island.

It is possible to arrange with a guide from the neighboring village of Mocagua to visit the island. It is made up of a series of ridges or *várzea* forest with swamps and open water between them. There are also two large lakes. On the island it is possible to find the *Victoria Amazonica* (water-lily), known to be the largest in the world with a leaf diameter of over a meter and a half wide. The island also supports a population of the extraordinary hoatzins. It is also an important roosting areas for parrots and a range of distinctive birds not readily found on the mainland. The swamps also contain remnant populations of the threatened Black Caiman, whilst in the lakes live the world's largest fresh-water fish *Arapaima gigas* or pirarucú, which can sometimes be seen coming to the surface to gulp air.

Fifteen kilometers upstream of Amacayacu is the small town of **Puerto Nariño**. It is accessible only by boat and has about 1,000 inhabitants, a hotel, shop, restaurant and clinic. It marks the entry to the **Loreto Yacu** river along which the extensive system of the **Taropoto** lakes can be explored. Again, arrangements can be made through INDERENA for hiring a boat, or through the hotel at Puerto Nariño. Not only are the lakes scenically very beautiful but they do provide the best opportunities to observe both species of freshwater dolphin that occur in the Amazon basin, as well as a number of waterbirds, raptors and macaws.

A team from the British Ornithologists' Union recently recorded over 490 species of birds during a two-year study. Almost all of these were recorded in the area around the Visitors Center and along the first few kilometers of the forest path.

The best period to visit is between mid-July and September when there is less rainfall, the trails are not flooded and the *Erythrina* trees are in blossom.

**Left** and **below**: the Amazon has butterflies of every shape, color and variety.

# VENEZUELA

**Venezuela** has over 60% of its surface covered with the original vegetation, still untouched by agricultural practices or urbanistic projects. Pristine forests are plentiful as over 10% of its surface are protected as National Parks. Venezuela offers three radically different regions: **the Orinoco-Amazon basin**, **the Caribbean Coast**, and **the Andes**. The Caribbean Coast and the Andes are not considered to be closely related to the Amazonic biogeographic region. But the visitor to this country may wish to explore these areas from **Caracas** before proceeding further south. Caracas is the door to Venezuela and receives most of the international flights arriving to South America from the USA and Europe.

**The Venezuelan Coast and Islands**: Venezuela has 2,200 km of coast in addition to its innumerable islands which show a diversity of sceneries difficult to encounter elsewhere in the Caribbean area. Two hundred kilometers west of Caracas, the mangrove forests and coral reefs of **Morrocoy National Park** offer white sandy beaches, a diverse bird fauna and superb places for diving. Three hundred kilometers to the east of Caracas lays **Mochima National Park**, a wind protected bay containing tiny islands with some mangrove forests and rocky coasts alternated with small sandy beaches. Mochima has a large variety of sea fauna, allowing for good fishing, and excellent beaches for overnight camping.

The most spectacular Venezuelan islands are undoubtedly the **Los Roques Archipelago**, actually protected as National Park. These islands are located some 150 km north of Caracas, and they harbor lobsters, Botutos (*Strombis gigas*), coral reefs, a huge variety of fish, and some sharks in their waters. Their sandy beaches, alternated with coral reefs, form a paradise seldom found anywhere else in the world. The Brown Pelican, Masked Booby, Brown Booby, Magnificent Frigatebird, and others inhabit some of the many islands. The small fishing village at **El Gran Roque**, next to a small air strip, offers some facilities to the visitor like lodging and boats.

**Paria**, **Araya** and **Paraguana** are three peninsulas at the Venezuelan coast, each with their characteristic scenery. Paraguana, at the west of Caracas, is arid with the exception of its single central volcano-like mountain with endemic forests. Goats and migratory birds inhabit the island-like peninsula. Paraguana is separated from the mainland by a strip of desert with enormous nomadic dunes that offer a unique opportunity for a real desert experience. These lands are rich in archaeological remains of pre-Columbian origin, and in some areas, dinosaur fossils and that of gigantic turtles and crocodiles are being excavated.

Three hundred kilometers to the east of Caracas, Araya, also a desert peninsula, harbors what once was the greatest salt-mine in the continent. In contrast, Paria, some 100 km east from Araya, guards dense forests and a national park, where rivers shoot out from the forest to the sea. The forest at Paria maintain a fauna and flora which is similar to that of the Orinoco-Amazon basin on one hand and that of Trinidad and Tobago on the other, although some endemic species do exist here.

**The Coastal Mountains**: The mountain range in the north of Venezuela, facing the Caribbean Sea, are extraordinary rich in tropical cloud forests. **El Avila**, **Guatopo** and **Rancho Grande** (or **Henry Pitier**) are three of the National Parks guarding some of the diversity in flora and fauna of this area. El Avila, limiting the expansion of Caracas to the north, and separating Caracas from the sea, gives the city its peculiar flavor.

The eastern part of the park has been spared by the frequent fires and therefore keeps an exuberant deciduous forest, and at higher altitudes on the mountain, a cloud forests. Three important peaks may be climbed at **El Avila National Park**: **Avila proper**, **La Silla de Caracas** and **Pico Naiguata**. At the summits of La Silla de Caracas, and Pico Niguata, a *paramo*-like vegetation substitutes the cloud forests. Thus in a single one-day march from Caracas, a succession of tropical ecosystems is accessible, including savannas, rivers, forests and *paramos*.

Rancho Grande, 100 km west of Caracas, offers some facilities for the nature lover and a exceptional richness in avifauna. Rancho Grande is the gateway of most migratory birds and insects entering central Venezuela and has one of the most botanically species-rich cloud forests of the world, rich also in snakes, insects, bats and, of course, birds. Thirty kilometers further to the north of Rancho Grande, the ocean has excavated many small bays, where coconuts are still commercially grown. These bays are excellent sites for an experience with the Caribbean Sea. The most beautiful bays are **Turiamo**, **Cata**, **Cuyagua** and **Choroni**, which are easily reached by car from **Maracay**.

Guatopo, 50 km south of Caracas, formerly an area for coffee plantations, has been only recently declared as national park (1958). The forests at Guatopo have suffered very little human intervention and thus have also very few foot paths. The road to **Altagracia de Orituco** traverses the park from north to south, reaching the Llanos at Alagracia.

**The Venezuelan Andes:** The Andes mountain chain originated in the tertiary and is responsible for most of today's geography of America. The mountains rise up to 6,000 m and extend from Alaska to Tierra del Fuego in Southern Chile, although the mountain range is named Andes only from Panama to Chile.

In Colombia, this mountain range bifurcates, so that a secondary mountain range extends to the east entering Venezuela. The mountains in Venezuela reach a maximum height of 5,007 m at **Pico Bolivar** in **Estado Merida**. The Venezuelan Andes, as elsewhere in South America, are densely populated but some national parks conserve the natural vegetation. The largest park: **Sierra Nevada National Park**, encircles the highest peaks in Venezuela in-

**Pierid butterflies gather at a river bank.**

cluding glaciers, spectacular *paramo* vegetation and high altitude cloud forests. In this park, the highest cable car in the world takes the visitor from **Merida city**, at 1,631 m to **Pico Espejo**, at 4,756 m in about 60 minutes, covering a strip of 12.5 km.

The dominant flora are *Frailejones* (*Espeletia* sp.) which expose hairy and succulent leaves, protected from frost, to the sun. These plants are very diverse and show many endemic species, so that many *paramos* or high altitude savannas (from 2,500 to 3,000 m) have their peculiar endemic species of *Espeletia*. Other national parks are **Paramo Piedras Blancas** and **Paramo El Aguila**, near Merida city and **Paramo Guaramacal** near **Bocono**, **Estado Trujillo**.

The Andes mark the limit of the Orinoquia-Amazonia. To the south and the east, their waters flow to the Amazon or the Orinoco river, but to the north or the west, they flow to the Caribbean or the Pacific Ocean. The fauna of the Andes is not very rich but peculiar. The

**Sand dunes at Meadano de Coro National Park.**

Spectacled Bear which is near extinction, still inhabits some of the Andean forests. The flora and the view over the mountains are certainly the most conspicuous features of the Andes, however colorful charming small mountain villages form now part of the beautiful Andean scenery.

**Perija:** At the northern frontier between Venezuela and Colombia, a 300 km long mountain range, extends over the center of the **Perija peninsula** with which it shares its name. The Perija peninsula, west of **Maracaibo city**, shelters mangrove forests on the coast, deserts with xerophytic vegetation, and mountains with very peculiar forests, which have helped in proposing this area as a biogeographical refuge for fauna and flora, where the Spectacled Bear can still be found.

Explorations in the area are dangerous because of continuous fighting beween drug-traffickers and the Colombian and Venezuelan military. The isolation of the area helps the local aboriginal Indians kept many of their tradi-

tions. Thus, Añu Wayu, Yukpa and Bari tribes still cultivate their language and customs, in spite of their contacts with the *criollo* culture at the markets in **Machiques**, **El Rosario**, **El Mojan** and **Maracaibo city**.

**The Orinoco-Amazon Connection:** One of the unique aspects of the Amazon river is its connection to the third largest river in South America: **the Orinoco**. The Orinoco starts at **Sierra Parima**, 1,074 m above sea level, but already after **Mavaca**, about 200 km from its spring, it runs at an elevation of 250 m. At about half of its length, at **Puerto Ayacucho** some 1,000 km downstream, it has only dropped another 150 m.

The **Rio Negro**, at 180 km from **Ocamo**, runs at the same elevation as the Orinoco, but separated from it by swampy flatlands covered with evergreen tropical rain forest. This peculiar location explains the existence of one of the most unusual hydrological phenomena of the world – the **Casiquiare Channel** – which connects the Orinoco with the Amazon basin. The 354 km long "**Caño Casiquiare**" was discovered in 1744 by Jesuit missionaries, and was further admired and described in detail by Alexander von Humboldt in its 1800 Orinoco expedition.

The Casiquiare flows from the Orinoco to the **Rio Guainia**, which becomes the Rio Negro after joining with the Casiquiare at **San Carlos de Rio Negro** and which in turn flows into the Amazon river at Manaus. Although rarely, the Casiquiare river sometimes flows from the Rio Guainia to the Orinoco, and geologists have still to uncover the history of this natural channel. It is very probable that in the past or in the future, the Upper Orinoco flowed or will flow completely to the Amazon, and the lower Orinoco may start with the **Rio Cunucunuma**.

Thanks to the Casiquiare, the Orinoco and Amazon basins are interconnected, sharing important proportions of their water life. The Casiquiare also serves as a mean for the dispersal of plants and animals, making the two river basins a single macro-ecosystem: the Orinoco-

**Merida, gateway to the Venezuelan Andes.**

Amazon basin. It is possible to navigate from the **Orionoco Delta**, to the Amazon river, by reaching the Upper Orinoco, crossing to the Rio Negro thanks to the Casiquiare, and then continue navigation down to the Amazon, with only one obstacle stopping the commercial river traffic between the mouth of the two rivers, the rapids of **Atures** and **Maipures**, dividing the upper from the lower Orinoco at Puerto Ayacucho.

The Orinoco basin with a size of nearly one million square kilometers is divided into four distinct geological and ecological areas: The **Venezuelan Amazonia** and the **Upper Orinoco**, including the Casiquiare and Rio Negro; the **Guayana Highshield** at the south and east of the Orinoco; the lowlands or Llanos to the north and west of the river; and the Orinoco Delta, 2,150 km from the spring.

**The Casiquiare and the Upper Orinoco:**
True tropical lowland rain forest is found in Venezuela only around the Casiquiare and Rio Negro. An irregular commercial flight or a more reliable charter flight connects Puerto Ayacucho to San Carlos de Rio Negro, a small town at the mouth of the Casiquiare and Guainia river, the birth place of the Rio Negro. The flight takes over one hour and cruises over 400 km of largely undisturbed forests, possibly one of the major areas of the world which has resisted human invasion, as most of its surface is still untouched by human. Communication in the area is limited to air or river transport.

Explorations by foot through the jungle are practically impossible, not only for humans but also for any large animal, which may explain the conspicuous absence of large mammals in these forests. Although jaguars, deers and tapirs inhabit the region, they are very rare and much more common in the unforested Llanos. Walking is restricted to some few earth paths of the local Indians who use them to tender their *conucos* or small farms, which they cut in the jungle, tending them for five to ten years, and then letting the forest regain them.

Snow occurs in the high plateau regions.

Indians and *criollos* cultivate mainly banana and manioc, but maize, tobacco and a series of native tubercles and roots are commonly found in *conucos*. With the exception of the Yanomami, all Indian tribes have mastered boat (dugout canoes) building and river navigation. The Yanomami inhabit the southern borders of this lowland forest, up to the mountains south of the Casiquiare and to Sierra Parima, where they maintain foot paths connecting them with other Yanomamies in northern Brazil.

Insects and birds are the owners of this rain forest world. Large swarms of noisy macaws and parrots glide over the forest and clouds of mosquitoes protect this sanctuary from most visitors. Since Humboldt's times it is told that the Siapa and Casiquiare are the world factory of mosquitoes. Some of the animals to be sighted included the birds: toucans, Guianan Cock-of-the-Rock, hummingbirds, parekeets, curassows and mammals like tapir, paca, jaguar, ozelot, and oncilla, and more than a centenary of bats. The **Siapa**, a tributary

of the Casiquiare, has till now avoided its complete exploration, and the first expeditions exploring part of it were only carried out in 1986 and 1989. The upper Siapa, not accessible by boat, is the last refuge for the Yanomami (or Waicas) escaping obligatory civilization by the Afro-European colonizers in Brazil and the Upper Orinoco.

The village from which the area can be accessed is San Carlos de Rio Negro. From here river trips using a *curiara* (dugout canoe) or *voladora* (aluminum speed boat) can be arranged. The *curiaras* are built by Indians who carve a single tree trunk, which is then further opened with fire which softens the wood. The few crevices formed during the dying and processing of the trunk are then sealed with natural tars.

At **San Carlos**, the indigenous inhabitants are mainly Bare, Corripaco and Baniwa, all three with strong Spanish and Brazilian influence. Yanomami are rarely seen at San Carlos and have to be found at the Siapa river, a day tour from San Carlos. River trips from Puerto

**A well-camouflaged stick insect.**

Ayacucho (Puerto Samariapo) may take over two weeks to reach San Carlos, and a further week down to Manaus.

San Carlos is near the frontier of three countries. In front of San Carlos, just across the Rio Negro, lays Colombia. Brazil can be reached in a few hours by *curiara*. The **Piedra del Cocuy**, a day trip from San Carlos, is an impressive granite stone or "Laja" which marks the common border of Brazil, Colombia and Venezuela.

At the other extreme of the Casiquiare, we find **La Esmeralda**, a small village inhabited by catholic missionaries and Yecuana Indians (also called *Maquiritare*), at the border between Yecuana and Yanomami land. River transport upstream on the Orinoco reaches **Ocamo** and **Mavaca**, where catholic missions are trying to christianize Yanomami culture for nearly two centuries.

All places south of Mavaca are accessed exclusively by air, where the protestant New Tribe Mission substitutes the catholic missionaries. After Mavaca,

river transport on the Orinoco is practically impossible due to innumerable strong rapids. In fact, the spring of the Orinoco has been rarely visited, in the first occasion in 1951 by a Venezuelan expedition. Today, access to the area is further hindered by strict regulations, controlling visits by foreigners to protect the Yanomami from devastating illnesses like common flu, malaria and syphilis. Illegal gold mining by *garrimpeiros* or far west type jungle adventurers, penetrating the area from Brazil, is the major cause of environmental destruction. It is difficult to control, and frequently, new illegal *garrimpeiro* camps with small airstrips are discovered and dismantled in and around Sierra Parima.

South of the Casiquiare extends one of the most pristine and isolated parts of the world. The most spectacular feature here is the **Neblina National Park** with a size of 13,600 square km, covering the Venezuelan part of the most southeasterly and highest *tepui* (3,000 m) in the continent. Neblina, a Precambrian sand

**This Palm Viper uses its tail to lure unsuspecting victims.**

stone table mountain of the *tepui* type (see the description of the Guayana Shield below) is accessed by helicopter or by a week long expedition from Brazil. The area shows many other impressive *tepui* like **Aracamuni**, granite monuments like **Aratitiyope**, mixed granite and sandstone formations like **Tapirapeco**, and mountains from the tertiary like the **Sierra Parima**. Sierra Parima, at the birth place of the Orinoco, is a beautiful, non *tepui*-like mountain formation, which reaches a height of 1,500 m, emerging from the surrounding forests. Its climate is excellent and it is the residence of most of the Yanomami. Access for tourists is limited by the military and the government.

West of La Esmeralda, **Duida-Marahuaka National Park** protects three practically unknown *tepuis*: the **Duida**, the **Marahuaka** and the **Huachamakare**. The Duida was first explored in 1928-29 by the Tyler Duida Expedition, whereas Venezuelan expeditions using helicopters, visited the Marahuaka in 1975 and 1983–85. The steep vertical walls around Marahuaka and Huachamakare-tepuy resisted climbing till 1984 when Venezuelan climbers reached both summits for the first time without the help of helicopters. The Duida is easier to reach as it has many gentle slopes, one near to La Esmeralda at the Orinoco river, which makes it possible to reach the plateaux of the *tepui* in a one-day excursion. These three *tepuis* are sacred places in the Yecuana mythology, where Marahuaka-tepuy symbolizes the tree of life, the origin of all species on earth.

Any exploration to this area has to start in Puerto Ayacucho. The city has a good airport and is the melting pot of all northern Amazonian Indians, Africans and Indo-Europeans. The native Indians of the area are the Piaroas, which still inhabit the surroundings. Puerto Ayacucho offers good excursions to the **Rio Cataniapo** (inhabited by Piaroa Indians), to the "**Tobogan de la Selva**" (a natural swimming pool), or further to the south to **San Fernando de Atabapo** where four important rivers meet: the

High rocky outcrops are a feature of the Venezuelan landscape.

**Atabapo** (black water river), the **Inirida** and **Guaviare** (white waters from Colombia) and the **Orinoco**. A river excursion from **San Fernando de Atabapo** to **Santa Barbara del Orinoco** takes the visitor to the mouth of the **Rio Ventuari**, a navigable river crossing beautiful forests and savannas, which is inhabited in its upper part by still non-westernized Yecuana Indians.

Any visitor to the area will be impressed by the color of the black and the white water rivers. Black water is a clear and clean but acid water (pH < 5) which has a color familiar to Coca Cola. The rivers show a red color if not very deep, and are dark black with white foams when over one meter deep. White water is opaque, rather brown, and filled with sediments.

White waters carry large amounts of clay from the Andes and the western Llanos, which made up most of the sediments of the Orinoco river. Black waters come from the Guayana highlands and from the *tepuis*, and own their color to tannines and other plant chemicals, which also produce the high acidity. The lack of nutrients in the sand stone formations, washed for over two billion years by rains, make it difficult for bacteria and other biodegrading organisms to survive in these waters, allowing the accumulations of these chemicals.

**The Guayana Highshield:** One of the greatest spectacles of the continent are the *tepuis*. These are sand stone table mountains with kilometer long vertical walls. Rivers that precipitate from these plateaus form gigantic waterfalls, the highest waterfalls in the world. Their flat summits are covered with endemic flora and fauna, including four carnivorous plant genera, many endemic birds, amphibians and reptiles, and of course exceptional arthropods, including water crickets and gigantic non-poisonous red spiders. The *tepuis* are difficult to visit without a helicopter, which is not easy to get in the area. Visits to *tepuis* are restricted by INPARQUES (the national institute responsible for national parks), but **Roraima-tepui** and

Many waterfalls also dot the region.

**Kukenan-tepui** may be climbed by foot, on a two-day long march from **Peraytepui**, near **San Ignacio de Yuruany**. **Santa Elena de Uairen**, on the south western end of the Venezuelan Guayana, near the Brazilian and Guyanan border, and **Canaima**, at the center of the Guayana Highshield, accessed only by plane, are two of the places to admire *tepuis*.

**Canaima camp** or **La Laguna de Canaima**, is a small Pemon Indian village converted in a tourist resort. It has regular air connection with comfortable jets, hotel facilities (two hotels) and organized tours. A three-day expedition navigating the **Rio Carrao** upstream, or a 15 minutes flight takes the visitor to see the highest waterfall in the World: **Churun-Meru** or **Angel Falls**, where the water drops 972 m from the **Auyan-tepui**, where it formed a second smaller cascade, so that altogether the water falls over 1,000 m in a spectacular setting into the jungle. The fall was discovered by the North American pilot Jimmy Angel, in 1923. The Pemon Indians call the fall more correctly Churun-Meru, or the fall of the Churun river.

Canaima, as most places in the Highshield, is surrounded by savannas and by forests. The mean height of the lowlands of the Guayan Highshield is 800 m above sea level, making the flora and fauna to be different from the rest of the Orinoco-Amazon basin. All rivers here carry crystalline black waters and have spectacular beaches and sceneries. These waters are the source of most tropical aquarium fishes.

Canaima lays in the heart of **Canaima National Park**, a 30,000 square km large protected area including a variety of the more spectacular *tepuis*. Auyan-tepui, the biggest of them, carrying the Angel Fall, raises its steep walls just behind the camp. The national park is reservoir of the water resources which could supply most of Venezuela and neighboring countries for the next centuries. Its waters are used to produce electricity at the **Guri dam**, where the huge Guri lake allows for excellent fishing.

The fauna is not very abundant. Ta-

Living with a mountain in their backyard.

pirs, armadillos, Giant and Tamandua Anteaters, deers, jaguars, turtles, snakes (some of them very poisonous), lizards and frogs, may be sighted with some luck. Birds and insects are plentiful and butterflies show exuberant colors and forms. Plants are the most visible inhabitants and for the naturalist, the varieties of orchids (notably *Cathleya* and *Catasetum*), bromeliads, lianas, flowers, fungi and ferns will be the most delightful friends. Over 10,000 plant species are estimated to inhabit this area, and most of them are known, thanks mainly due to a 40-year effort guided by Julian Steyermark (1909–1988), but many more species remain to be discovered. Rocky areas, savannas with views over the *tepuis*, and diverse kinds of forests cover the area.

Explorations of the Guayana Shield are fairly recent; some of the first extensive explorations were carried out by Robert Schomburg starting from British Guayana, around 1910; and Auyantepui was first climbed and explored by a Venezuelan scientific expedition in 1956. At Canaima, the *Fundacion Terramar* – a charitable private foundation – promotes research on nature conservation and experiments with diverse management techniques in search of a self-sustained conservation plan, which should guarantee the preservation of this unique area for humankind.

**Santa Elena de Uairen**, founded in 1931 by Capuchino monks, is today the southern door to the Guayana Highshield. It is a center of tourists, miners and merchants from Brazil, 15 km away, Guayana and Venezuela, that reach Santa Elena by air or terrestrial transport. Two roads connect Santa Elena to the rest of the world, one coming from **Boa Vista** (Roraima-Brasil) 200 km away and another reaching Santa Elena from **Ciudad Bolivar**, 600 km to the north.

Santa Elena offers beautiful trips to Roraima-tepui, diverse river and *tepui* sceneries such as **Quebrada del Jaspe**, **Kavanayen**, **Rio Aponguao**, **Kamairan**, **El Pauji**, etc. and to diamond and gold mining sites in Kilometre 88, **El**

**Scenes like this have inspired artists through the ages.**

**Dorado**, **Las Claritas** or **Icabaru**.

Roraima-tepui and Kukenan-tepui are undoubtedly the most impressive features in the surroundings. Roraima-tepui was first climbed by Everard F. Im Thurn, curator of the Museum of Georgetown, and H.I. Perkins in 1884. They started from Georgetown and the route they used for climbing Roraima is still not completely clear.

**Roraima**, or "**The Lost World**" in the words of Arthur Conan Doyle, is certainly one of the most beautiful *tepuis*. The sand stone mole has an average height of 2,500 m and forms rocky labyrinths containing lakes, cliffs and spectacular views. Its vegetation is characterized by the carnivorous plants of the genera *Heliamphora* (*Sarraceniaceae*), *Droseras* (*Droseraceae*), *Genlisea* and *Utricularia* (*Lentibulariaceae*); by *Odocarpus* trees, *Speletia*, orchids, etc. The flora of Roraima, and to a lesser extent that of Kukenan, are suffering heavy from depredation by tourists, and may disappear completely, like the quartz crystals which once completely covered the ground of several valleys on the *tepui*.

The savannas near and around Santa Elena extend over sandy soils covered with termite hills whose inhabitants offer an exquisite food: the head of the termite soldiers can be eaten by just biting them off from the thorax and separating them from the mandibles. They taste very refreshing and aromatic. These termites feed on grasses, concentrating nutrients in their hills that allow seedlings to start or expand the existing forests. Another interesting ecological feature is the existence of ant gardens, ant-plants and carnivorous plants. That is, ants build nests on trees, carrying seeds of epiphytes which will grow only on these "garden-nests". Other plants like the *Melastomatacea Tococa* form specific structures to attract ants which will inhabit them, feeding the plant and protecting it from herbivores. Ant-termite associations are also present, besides endemic spiders, a variety of *Coleoptera*, and poisonous butterfly larvae which cause heavy

Camping in the high plateau with nature as your only neighbor.

196

hemorrhage after contact with the skin and which may even lead to the death of the patient.

North of Santa Elena, leaving Canaima National Park, one sees gold miners devastating the landscapes. Some of the mining towns, on the road to **Puerto Ordaz** and **Ciudad Bolivar**, like **El Callao**, **Tumeremo** and **Guasipati** have Anglo-Caribbean influences, due to the fact that labor from Trinidad and Barbados were imported by an English mining company in the last century. From these towns, **Puerto Ordaz** or **Ciudad Guayana**, a city founded in 1961, concentrates the heavy industry which uses cheap electricity from Guri, unlimited water and convenient fluvial transport through the Orinoco, the enormous iron deposits at **Cerro Bolivar** and the aluminum deposits at **Los Pijiguaos**. The surroundings of Ciudad Guayana offer some nature spectacles such as **Parque Cachamay**, **Parque Loefling**, **La Llovisna** and **Los Castillos de Guayana**.

Two navigable rivers, the **Rio Caura** and **La Paragua**, offer excellent excursions into the south of **Estado Bolivar**, with views over the exuberant flora and over some of the fauna. Special river camps for tourists are flourishing as a consequence of a growing ecotourism, which together with gold mining represent the main private economic activities in the area.

**The Orinoco Delta:** The Orinoco Delta is a fan of alluvial deposits bounded on the south by the Rio Grande and on the west by the Manamo River which branch off the main river at **Barrancas**, the apex of the Delta. From Barrancas it is about 180 km along the Rio Grande to the Atlantic Ocean, and an equal distance along the Manamo channel to the Atlantic Ocean. Most of the vast tidal swamp of the delta is irrigated by channels that are called *caños* that form a network of navigable waterways over the entire delta.

The climate is hot (average 26° C) and humid (average 70% relative humidity) and, together with the fertile soils, that are inundated each year by new sediments from the Orinoco, gives

An endemic plant life has adapted to the poor soil and climate.

rise to rich and exuberant forests. Most of these forests are constituted by red mangroves (*Rhizophora mangle*). At the center of the drier islands the mangrove dies and permits the invasion by useful palms such as *Moriche* (*Mauritia flexuosa*) with its exquisite fruits, *Manaca* (*Euterpe* sp.) and *Temiche* (*Manicaria saccifera*). This area is inhabited by a distinct Indian tribe – the Warao – who possibly arrived from Florida, through the **Antilles**, at the delta. The Warao live over wooden *palafites*, built over swamps or rivers, allowing them to escape the annual floods caused by the rising of the water level of the Orinoco during the rainy season, from May to October.

The forests are very dense and river transport is the only means for communication. The forests suffer increasing human pressure due to rice plantations, water buffalo breeding, and to commercial exploitation of natural fibers. *Palmito*, the heart of certain palms, is commercially harvested, slowly decimating the natural palm populations.

Many birds, especially macaws, parrots, and toucans are illegally exported to French Guayana, Aruba, Curazao or Trinidad, from where they enter the European markets or the USA.

Mosquitoes are plentiful in the deltas, in contrast to the Guayana Highshields. But the area has a delightful beauty on its own. **Tucupita**, the main city in the delta, is accessed by car or regular air flights from Caracas. From there, river expeditions give access, through the numerous *caños* to small Warao villages and eventually to the Atlantic Ocean.

Water life is very rich in the delta; various turtle species and crocodiles inhabit the rivers and channels. Fish is plentiful and fishing requires expertise and knowledge of the different *caños*, as each *caño* keeps specific fish species including river catfish (*Meglonema* sp. and *Cetopsis* sp.), pavon (*Cichla* sp.), and palometa (*Pygopristis* sp. and *Mylossoma* sp.). Insect and bird live is very diverse and little explored. Several mammal species may be found in the

**Female Marsupial frog (*Gastrotheca ovifera*) releases froglets from its pouch.**

delta: some of the South American marsupials, monkeys and rodents are common terrestrial representatives, whereas freshwater dolphins, Giant River Otters, and manatees, at the border of extinction, represent mammals in the water. At the mouth of the *caños*, ocean life starts to flourish in brackish, sediment rich waters.

The **Guanoco asphalt lake**, north of the delta, is the largest asphalt lake in the world. A black surface of over two square kilometers, spiked with plant-islands, and surrounded by forests, offers a unique scenario with its exclusive flora and spectacular visual contrasts. Access is difficult and takes a three-day long march by foot or with mules, from a village called **Guanoco**, near **Caripito**.

South of the delta are little explored forests which are slowly being felled for their precious woods. The **Serrania de Imataca** separates the Orinoco basin from that of the Esequibo. This area has a reputation for its numerous jaguars, but gold mining and extensive deforestation are degrading its nature, and illegal hunting further reduces the fauna.

**The Llanos:** The plains north and west of the Orinoco, covering the central parts of Venezuela and Colombia, form a unique ecosystem rich in vertebrate fauna. The Llanos may be classified in upper and lower Llanos. The lower Llanos are located in the south of Colombia and Venezuela but spread mostly north of the Orinoco. They are inundable lands with very few trees.

The upper Llanos, further north and nearer to the Andes and the Venezuelan coastal mountain range, harbor more trees and even small pseudo-circular forests called *matas*, which are thought to be initiated by collapsing leaf-cutting ant nests. The Llanos are best visited during the dry season (December to May), as dirt roads will be transitable and the fauna will be concentrated around rivers and lakes.

The most conspicuous and attractive feature of the Llanos are the bird colonies formed by thousands of Great Egrets, Scarlet Ibises, Wood Ibises, Jabirus, and small numbers of Great

An endemic *Drosera* plant from the high country.

Tinamous, Little Blue Herons, chachalacas, curassows, hoatzins, etc. The rivers of the Llanos offer excellent fishing of pavon (*Cichla* sp.), piranha, cachama (*Colossoma* sp.), or river catfish species (*Meglonema* sp., *Rhamdia* sp., *Cetopsis* sp.), besides easy opportunity to make acquaintance with caimans (*Caiman crocodilus*), iguanas (*Iguana iguana*), anacondas and river turtles (*Podecnemis vogli*). Caimans, locally called *babas*, are reasonably tame, and small ones may be captured by hand. They lay their eggs in nests made of grasses near lakes and rivers. Sadly, they are commercially exploited under the supervision of the Venezuelan Ministry of Natural Resources. The manatee may still be found in the few nature sanctuaries in the Llanos, the national parks: **Aguaro-Guariquito** and **Santos Luzardo**, also called **Cinaruco-Capanaparo**, and the "**Esteros de Camaguan**" near **San Fernando de Apure**. Lesser Anteaters, Red Howler Monkeys, *Cachicamos* or Nine-banded Armadillos, common opossums,

capybaras, White-tailed and Great Brocket Deer, and Brown-throated Three-toed Sloth are only a few of the mammals that are common in these plains.

The Llanos are of sedimentary origin, formed mainly from sand, brought down by rivers from the Guayana Highshield by the Orinoco and from the Andes. Geologically this was a very recent event and happened over the last few million years. The area is characterized by two distinct seasons: the dry and the rainy season. During the dry season, water exists only in a few rivers and lakes, obliging the fauna to concentrate around it. Extensive savanna fires are common, and the Llanos may be crossed with or without dirt tracks. During the rainy season, new rivers emerge and large areas are inundated, isolating many villages during two to five months. Fish reproduces in this season. Animal life disperses making nature observation more difficult.

The Llanos harbor huge grasslands which are used for extensive cattle

**A village in the Orinoco Delta.**

farming. Extensive farming in the Llanos implies that cattle is collected once or twice a year for veterinary inspection and left free to feed and breed without further human intervention. This kind of management has little effect on the natural ecosystems. Modern management techniques, allowing intensive cattle breeding, are becoming popular and change the natural vegetation conspicuously. *Chaguaramas* (royal palms) and colorful trees, most of them flowering during the dry season, give a picturesque look to these extensions.

The most conspicuous mammal in the Llanos is the capybara, the largest rodents of the world. Capybaras, locally called *chiguires*, have developed a complex social behavior allowing them to successfully defend themselves against jaguars, pumas and crocodiles, however, not against humans. The aggressive and famous piranha (also called Caribbe fish), present south of the Orinoco in less aggressive varieties, is the owner of most rivers and lakes in the Llanos, making bathing uncomfortable

**Savanna of the Parucito River valley.**

and even dangerous.

**San Fernando de Apure**, **Maturin**, and **El Sombrero** are the cities to access the southern low Llanos, the eastern middle savannas, and the central upper Llanos respectively. Communication by air is practical to most places, although ranchers often have no alternative but air transport to reach their *haciendas* during the rainy season and even part of the dry season. Some of these *haciendas* have now good facilities for the nature lover. The owners of these farms created small sanctuaries containing minimal facilities with variable degrees of comfort, where the Llano fauna can easily be observed and enjoyed.

**Aguaro-Guariquito National Park** is certainly the most representative place for fauna and flora of the Llanos. Located between the two rivers that gave rise to its name it has been somewhat protected from excessive hunting, thus keeping important populations of monkeys, caimans, deer, anteaters, dolphins, etc. Visits to the Park require camping as no hotels are available.

# THE GUIANAS

The three Guianas – **Guyana** (formerly British Guiana), **Surinam** (formerly Dutch Guiana) and **French Guiana** – vary hardly at all in their geography. A narrow strip of flat, swampy land runs along the coast, and this is where the majority of the population lives. Behind this coastal plain is a region of isolated savannas, marked by expaneses of white sand, grassy plains and an occasional stretch of sparse woodland. Further inland are extensive uplands whose higher points are separated by narrow, swampy valleys. This central Guianese shelf is made of ancient crystalline rock, and is only 250 m above sea level on average. Its most striking geological features are the isolated inselbergs, granite peaks rising abruptly to heights of between 300 m and 800 m which have been formed by the erosion of surrounding formations. The grooves and protuberances which mark the bare rocks near the summits and on the steep walls bear witness to the effects of countless downpours. The tops of the inselbergs, many of which are bedecked with a brilliant carpet of bromelia, afford magnificent views of the virgin forests covering the gently undulating landscape. Rising to the southwest is the **Guianese Plateau**, which reaches a height of 900 m in the **Tumuc-Humac** mountains. The highest point in the Guianas is **Mount Roraima** (2,810 m), a table mountain of red sandstone on the borders of Guyana, Brazil and Venezuela.

All the important rivers of the Guianas rise in the mountains and flow in a more or less northerly direction towards the Atlantic. And it is the many rivers with their frequent rapids to which the region owes its name: "Guyana" means "land of waters". No fewer than ten major rivers flow into the Atlantic on French Guiana's 345 km coast alone. In many cases there are rapids not far from the mouth, and since the rivers were originally the only means of transport, this posed a major barrier to opening up the interior. The Guianese hill and mountain regions are still sparsely populated, and this is largely due to the poor navigability of the rivers, since there are hardly any roads leading into the interior. During the main dry season from August to November, and in February to March, when the rainfall is again low, the rapids become gentle streams flowing through innumerable fissures in the rocks. The network of streams and rivers is a hunting ground for two types of otter. While the smaller, light-colored Southern River Otter fishes at night, the gregarious, dark brown Giant Otter – which with a tail of around 80 cm can grow up to two meters in length – stalks the river banks during the day in search of fish, small mammals and birds.

On the smooth-flowing lower reaches of the rivers where tidal influences already make themselves felt, the vegetation is dominated by mangroves. Here, among the mangroves, lives one of the most peculiar fishes of South America, the Four-eyed Fish (*Anableps anableps*). These fish, which grow to about 20 cm in length, swim so close to the surface that their protruding eyes are half out of the water. A division in the eye at the waterline enables them to see both above and below the surface at the same time. Because these Four-eyed Fish have no tear glands, they must regularly dip their heads below the water to lubricate the upper part of their eye. This upper part presumably serves to keep watch for airborne predators rather than for any form of prey, since they have never been observed trying to catch flying insects. It is easy enough to see them at night, as their eyes reflect the beam of a torch. Like its close relative the viviparous, the Four-eyed Fish bears living young, the eggs being carried in the body of the female until they hatch.

Several beaches in the Guianas are well-known laying grounds for marine turtles. Some of these are officially protected (including Wia-Wia and the Galibi Nature Reserve in Surinam, and Les Hattes and Aouara in French Guiana). Apart from four species of the true marine turtles *Cheloniidae*, one might also be lucky enough to find the giant leatherback (*Dermochelys coriacea*) as

**A Malachite Butterfly (*Siproeta stelenes*) at rest.**

it lays its eggs at night; the best time to see this is between June and July. Growing up to two meters in length and with a weight of 600 kg, the leatherbacks are the largest turtles in the world; they spend most of their lives in temparate and polar seas, and feed almost exclusively on jellyfish. Marine turtles only leave the sea to lay their eggs. Certain sections of the beach above the high-tide mark have been used for generations, and here the females dig holes approximately 50 cm deep in which they lay around 100 eggs. They then cover up the hole and smooth the surface before returning to the sea.

**The Raleighvallen/Voltzberg Nature Reserve (Surinam):** The hinterland of the Guianas provides ample opportunities for anybody who wants to get to know the largely tropical rain forest and its fauna. One reserve which is well worth a visit is the 560 square km **Raleighvallen/Voltzberg Nature Reserve**. There is a direct flight from **Paramaribo** to **Foengoe**, an island in the **Rio Coppename** in the Reserve, but a cheaper way of getting there is the approximately four-hour drive to **Bitagron** on the Coppename, followed by another four-hour trip up the river to **Foengo Island Lodge**. As well as the widespread high rain forest, the Reserve also has less common types of vegetation such as mountain savanna and liana forest. It is also home to the eight species of monkey found in Surinam, the Yellow-handed Marmoset, the Common Squirrel Monkey, the Weeping Capuchin, the Brown Capuchin, the Red-backed Saki, the White-faced Saki, the Red Howler Monkey and the Black Spider Monkey.

There are many other mammals apart from monkeys which make their home in the Voltzberg region. Agoutis are commonly encountered on the forest floor, as are the acouchis which laymen often mistake for them. The peculiar *Edentata* order of mammals, originally found nowhere but South America, is also represented in the Reserve through the Giant Armadillo, the Nine-banded Armadillo, the Tamandua, the Giant Anteater and two types of sloth. Among

**A closer inspection of a Cannonball tree.**

the larger animals to be found in the region are Collared Peccaries, tapirs, jaguars and ocelots. The forest canopy is home to the brightly-plumed Red Fan, Orange-winged and Mealy Parrots, and Scarlet Macaws, aracaris and toucans with their characteristic fluting call. A number of larger birds also catch the eyes, such as the Black Curassow, Tinamous, the Marail Guan and the Grey-winged Trumpeter. One bird not be overlooked is the Screaming Piha, whose whip-like "pi-pi-yo" cry is one of the most striking sounds of the rain forest, and contrasts with its plain, grey coloring. Of particular interest to the zoologist is the group courtship ritual of the Cock-of-the-Rock, a bird which is not uncommon in the Voltzberg region. The males, which grow up to 35 cm in height, congregate at traditional grounds in the forest and perform their bizarre courtship dance to attract the more dowdy females. Their plumage is a bright golden orange, and is enhanced by a bonnet of feathers round the head; in the dance they show it off to best advantage

by springing into the air, shaking their heads and wings, spreading out their tails, making rattling noises with their beaks, and issuing their harmonious cry while at the same time producing a buzzing sound by vibrating their feathers.

At the time of writing (1991), the political disturbances which have unfortunately been a feature of Suriname since 1987 mean that visits to this region are not to be recommended.

In contrast to the independent states of Guyana and Surinam, French Guiana is an overseas department of France, thus is politically quite stable. Unfortunately in French Guiana, there is no comparable reserve to the Raleighvallen/Voltzberg Reserve yet. Tourists with a taste for adventure and sufficient time for a trip to the interior may care to visit **Saül**, where they will find a network of well-marked paths through the virgin rain forest. With their sparse populations, the beauty and variety of their landscapes and their fascinating flora and fauna, the Guianas have an as yet almost untapped potential for "eco-tourism".

**An Orchid Bee (with pollen on its back) visits a local orchid.**

# TRINIDAD

**Trinidad**, the southernmost island of the Caribbean, supports a unique wildlife link between the Amazonian and the Caribbean biotic regions. Trinidad lies only 12 km from the South American mainland at its closest point. However, 10,000 years ago a great deal of the Earth's water was locked up in ice, and sea levels were much lower than they are now.

Trinidad was at that time connected to the Paria Peninsula of Venezuela. Wildlife of all kinds invaded Trinidad, and many species flourish there to this day. The rugged peaks of the **Boca Islands**, submerged mountains that dot the channel between the two islands, are remnants of that connection.

Trinidad's biogeography still profoundly influenced by South America. The lack of an effective isolating mechanism between the two lands gives Amazonian species a reasonably easy opportunity to colonize the island. Competition between Amazonian and Caribbean species is intense, with new invaders recorded annually as others decrease and vanish.

A powerful agent aiding immigration by Amazonian species is the Rio Orinoco, whose outflow swirls northward like an immense muddy river past eastern Trinidad, depositing debris of all kinds on the beaches. Flood-borne trees from Venezuela serve as life rafts for wildlife, and they often root themselves once ashore.

As in the *tepui* region of Amazonia, elevation in Trinidad dictates the dominant vegetation. With increased elevation comes greater rainfall. The highest and thus wettest areas are the mountains of the **Northern Range** which reach a height of 940 m and are believed to be an extension of the **Andean Cosdillera**. Some peaks receive more than 400 cm of rainfall annually, in sharp contrast to the low lying areas in the west and southwest, areas of rainshadow that receive only 150 cm annually. The difference in available moisture results in lush rain forests carpeting the mountains while thorn-covered scrub struggles to survive in parts of the west and on islands. At low to moderate elevations, there are grasslands, wet savannas, and semideciduous woodlands to be found. These habitats prevail where they have not been destroyed for sugarcane or rice cultivation or for the ever increasing spread of settlements.

The "edge" effects, a phenomenon in which more species are found along interfaces between habitats than within habitats, is partly responsible for the diversity of Trinidad's wildlife. Trinidad lies at the northern edge of Amazonia and at the southern edge of the Caribbean. There are two thus "edges" at work, the first derived from interfaces between diverse habitats in Trinidad, the second derived from the interface between the Amazonian and Caribbean faunal regions. The combined effect yields a greater diversity of species in Trinidad than would normally expected.

For a small island, the types and extent of habitats is remarkable, with montane elfin forest and rain forest, lowland rain forest, savannah, freshwater and saltwater swamps, freshwater impoundments, ocean beaches, and open ocean, as well as a variety of cultivated areas. Perennially available fruits, flowers, and seeds from both Amazonian and Caribbean plants provides sustenance for resident species and for migrant birds from North and South America.

An examination of the wildlife in Trinidad reveals that the birdlife is astonishingly diverse. Amazonian families are strongly represented, along with a large number of Caribbean families and a few exotics. About 25% of the South American species present on Trinidad are found nowhere else in the Caribbean. The presence of Amazonian as well as Caribbean species results in the number of bird species in Trinidad being nearly twice that of any other Caribbean island.

Insect diversity, tremendous in Amazonia, is reduced on most Caribbean islands. With few niches to fill, competition quickly eliminates poten-

This juvenile Tropical Screech Owl has just left its nest for the first time.

tial colonizers that are poorly suited for the environment and which lack adaptivity. Most studies of island biogeography find only about 10% the number of species on islands as on similar mainland areas. Fierce competition among species in the few available habitats renders island depauperate not only of insects for pollination. Species that survive and eventually colonize islands often are larger and heavier than their counterparts on the mainland, a phenomenon resulting from the negative survival value of being easily windblown out to sea.

In contrast to the more usual reduced diversity of insects on Caribbean islands, Trinidad's closely resembles that of Amazonia, being among the richest in the world. For example, some 650 species of butterflies have been cataloged.

Many Trinidadian insects, such as giant katydids and iridescent orchid bees, show unmistakable links to their Amazonian counterparts. Yet certain biogeographical anomalies confound bio-

geographers. For example, populations exist in Trinidad of snail-killing flies known elsewhere only from Central America. Moreover, despite geologically recent connections with the mainland, Trinidad has many endemic species. Such confusing distributional data prevent researchers from generalizing about the origin of groups of organisms.

One striking aspect of Amazonian biology, one more conspicuous on Trinidad than on any other Caribbean island, is the association between birds and a few species of mammals with ants. Ant-following birds such as antbirds, antshrikes, ant-tanagers, antthrushes, antvireos, and antwrens glean arthropods from the forest floor as columns of army and leafcutter ants scour the jungle. The fact that army-type ants are lacking on other Caribbean islands may result from a lack of tracts of rain forest large enough to host viable populations, or perhaps better adapted species immigrate in numbers too small to become established.

Amazonia is clearly the origin of the

**A Rufescent Tiger-Heron fishing at the edge of a lake.**

larger mammals in Trinidad. Large mammals are uncommon anywhere in the Caribbean; in fact, howler monkeys, agoutis, prehensile-tailed porcupines, silky anteaters, and crab-eating raccoons are found nowhere else in the Caribbean except for Trinidad. Unlike insects, plants, and small rodents, large mammals are unlikely to drift on a log, later to be washed ashore on some distant island. Many rodents, however, especially mice and rats, bear close affinity to Caribbean species.

Fogging up the analysis, the reptiles and amphibians in Trinidad are difficult to characterize as either Amazonian or Caribbean. Lizards are abundant throughout the Caribbean, particularly on drier island. Similarly, iguanas, which are chiefly arboreal herbivores, are also distributed throughout the neotropics wherever favorable habitat is found, but are far more abundant in Amazonia. Other widespread groups that defy characterization as either Amazonian or Caribbean include anoles, basilisks, and geckos. Most snakes also are widely distributed, except for the anaconda, which in the Caribbean inhabits only Trinidad.

Trinidad has a higher human population density than most other regions of the tropical South America, and the concomitant disturbance of natural habitats. This disadvantage is more than balanced by the easy accessibility of many habitats, and by the availability of accommodation close to nature catering to the nature tourist.

The two excellent hotels found in the Northern Range are **ASA Wright Nature Centre**, 12 km north of the city of **Arima**, and the **Mount St. Benedict Guest House** close to the monastery above the town of **Tunapuna**. Excellent observation of rain forest wildlife is possible in the estates surrounding these hotels, and in particular along the nearby road from **Arima** to **Blanchisseuse**, which cuts through the forests of the Northern Range. For observation of wildlife of wetland habitats such as Scarlet Ibis or Boat-billed Heron, the place to go is the **Garoni Swamp**, south of **Port Spain**.

This Pauraque, a species of nighthawk, Is nearly invisible as it incubates its eggs.

# ECUADOR

**Ecuador** is one of the smallest countries in South America, but within its 283,520 square km the most varied and extreme habitats exist side by side. The country is bisected by the Andes whose lofty peaks are, geographically and culturally, the backbone of Ecuador. Here the Andes form two parallel ridges, often referred to as the **Avenue of Volcanoes**: over 30 snow-capped cones rise from these ridges, at least eight of which are still active. Between the ridges is a fertile trough divided into several basins which drain alternately east to the Amazon and west to the Pacific.

Ecuador's climatic zones are as complex as its topography. The cold Humboldt current to the west, and the steaming jungle to the east both contribute to the clouds that perpetually condense around the Sierra. The eastern slopes are particularly wet, and long periods of mist and rain may be expected even in the relatively dry months of November/December and May/June. Average temperatures are determined by altitude, and the severe cold at the high altitudes greatly restricts what will grow.

**Ascending the Andes from the Oriente:** In the lowland forest buttress-rooted, columnar trees soar unbranched to a closed canopy some 30 m above. A dense understorey comprising flat-topped minature trees (*Piper* sp., *Coussarea* sp., *Clidema* sp.) and prop-rooted palms (*Socratea* sp.) grasp the thin soil. Filmy liverworts and mosses are largely confined to the lower parts of the smooth-barked trees. Leafy *Polypodium* ferns sprout from their trunks and compete with the many species of climbers which spiral upwards. Here in the dry forest bromeliads and orchids are relatively rare. Numerous monkey species inhabitat this area, including Pygmy Marmosets, Saddle-back Tamarins, Goeldi's Monkey and the Monk Saki.

Upon entering the foothills, the riverine forests merge gradually with the montane species. The subtropical rain forest ranges in altitude from 800 m to 1,800 m and the daily temperature between 18° C and 24° C. It is typically very cloudy with a rainfall amounting to some three meters annually, sustaining a bewildering variety of plant life.

The pale, buttressed trunks of *Higerion*, typical of the lowland are mixed with the *Podocarpus* trees of the cooler slopes. Fruiting fig trees *Ficus* sp. and "*ovilla*" (wild grapes) are abundant, providing good places to observe feeding birds. A mid-storey layer of palms and tree ferns (*Cyathea* sp.) provides a mosaic of variously shaped leaves intercepting the light. At ground level, blood-red begonias and the aptly named pubescent "*Lobias de Novea*" (Lovers lips) flower amongst fallen trees covered with ferns (*Asplenium* and *Polypodium* sp.), mosses (*Lycopodium* sp.) and lichens. The scarlet, barbed flowers of Heliconia, stretching above the dense ground cover are visited by humming-birds. Tracks show that ocelot and margay stalk the forest trails. Rooting marks made by Collared Peccary are common. Fewer primate species are to be found here, but the Night Monkey, Common Woolly Monkey and Dusky Titi may be encountered. A wide range of birds may be seen including the Green Jay, Andean Guan, Golden-plumed Conure, and Grass-green and Golden-crowned Tanagers.

Montane forest extends from 1,800 m to 3,000 m. These steep slopes experience somewhat lower temperatures and are slightly less moist. Due to the instability of the steep, young soils, landslides are a common natural feature of the terrain, although forest clearance and ovegrazing exacerbate the problem. A constant cycle of recolonization, and the slower growth rates in the cold means that the montane forest is typically younger and less complex than the climax rain forest below. But, the vegetation is no less dense, and as the canopy becomes lower and the foliage smaller with increasing altitude, the undergrowth of herbs and shrubs thrive in the increased light. Bamboo (*Chusquea* sp.) is abundant in the higher zones. Characteristic flowering plants are represented by the families *Asteraceae*, *Araliaceae* and *Gunneraceae*. The bright yellow bladder-shaped flowers of *Calceolaria* and the clustered bells of *Bomarea* sp. provide eye-catching displays. Gray-breasted Mountain Toucan and the Rainbow-bearded Thornbill may be spotted here.

From around 3,000 m the rising humid air condenses into an atmospheric mist. The cloud forest is watered more by direct condensation rather than by rainfall, providing ideal conditions for epiphytic lichens, ferns and mosses which clothe every branch and twig of this forest. The tortured slow-growing *Polylepis* tree providing a framework. Dense yellow asters (*Gnoxis* sp.) and the tubular orange flowers of *Tristerix longebracteatus* adorn the canopy above the tangle of flowering creepers and herbs.

Above 3,000 m the vegetation is determined by the terrain as we reach the finely dissected ridges and more open valleys at the top of the *cordillera*. The damp valley floors support grassland and marsh, the steeper gullies sheltering rich cloud forest which thins into stunted elfin forest and heaths on the exposed ridge lines. Dwarf tree ferns and a great variety of bromeliads and heathers are common. The elfin forest is a kaleidoscope of color and texture, and an eye for minute detail is richly rewarded. Hummingbirds frequenting this area include the Collared Inca.

At the most exposed and windswept heights the shrubs give way to open *paramo*, dominated by the dense tussocks of *Stipa*, *Calamagrostis* and *Festuca* grasses. Here daily temperatures range between 6° C and 12° C, the moisture is not so constant and night frosts are not uncommon. From a distance the *paramo* seems simple and uniform, but the tussocks shelter a delicate garden of gentians (*Gentianella* sp.), *Ranunculus*, lupins (*Lupinus* sp.) and the sessile, yellow blooms of *Hypochaeris sessiliflora*. Carunculated and Mountain Caracara skim over this low vegetation for carrion, fruit and insect prey.

On the emergent cones of volcanoes a desolate zone of ash and lava marks the end of the moorland zone. Here there is little vegetation apart from the valiant colonization attempts of lichens. The

**The moss-covered elfin forest at Sangay National Park.**

peaks are usually snow-capped, and glaciers descend to about 4,700 m.

**Sangay National Park:** Situated some 200 km south of **Quito**, the **Sangay National Park** covers some 2,700 square km from the volcanic vents of **Sangay**, **Altar** and **Tungurahua** to the eastern base of the *cordillera*. It includes pristine examples of all the Andean vegetation zones, from sub-tropical rain forest, climbing through montane and cloud forest, stunted elfin forest and desolate open *paramo* to the mountain *tundra*.

There are a variety of routes into the park. Most visitors head for the volcanic peaks, entering from high mountain villages to the west. Sangay, one of the world's most active volcanoes, provides a great challenge to adventurers, which is richly rewarded if the weather is favorable. Small eruptions may occur at 20 minute intervals, and on a clear night, glowing lava and ash provide a spectacular pyrotechnic display. However, as the volcano is almost constantly shrouded in cloud, such vistas may be elusive. Also further ascent may be prohibited by the weather and activity of the volcano; nevertheless, the trek provides a spectacular and varied experience of the high Andean environment.

The first leg of the hike climbs steeply out of the U-shaped **Alao valley**, across **La Trancha pass** at 4,000 m and down into the broad valley of the **Rio Culebrillas**. The local porters can walk the route in five hours; however, the views are worth lingering over. At night the moaning calls of Spectacled Bear emanate from the *Polylepis* forest. The bears occasionally come to the fields to raid maize crops, but are more at home in the cloud forest fringing the *paramo* in the remote areas of the park.

Walkers crossing the pass are likely to be circled by the Andean Condor, effortlessly gliding the updraught from the valleys on a three-meter wingspan as it scour the hillsides for carrion.

The *paramo* of the Alao valley and beyond to the edge of the park is grazed by cattle and managed by frequent burning. This strategy eventually simplifies the grassland and can destabilize the steep

**The Alao Valley entry into the park.**

slopes, but in this remote region the flora is still quite complex. There is little evidence of mammal life here, the occasional paw-print of the Andean Fox, or the skull of its most common prey the Sacha Cuy, wild ancestor of the domestic guinea pig. White-tailed Deer, once common here have now been hunted out, their lands over-run by cattle. Descending from La Trancha, the eerie *Polylepis* dominated cloud forest is entered at about 3,500 m. At its upper limit the hardy bromeliads *puya* and *Tillandsia* are common. Dismantled and scattered remains of this spiky plant testify to a good population of bears, for whom the succulent heart of the plant is a favored food.

The **Culebrillas valley** at 3,000 m is relatively flat and drained by the fast flowing Culebrillas river, surprisingly devoid of fish. Not even the introduced Rainbow Trout (*Salmo gardineri*) can survive in these acidic rivers running off the volcanic slopes. The wide floodplain indicates the intensity of the floods that can affect the area.

The boggy valley floor, although ille-gally grazed by cattle, still provides a favored habitat for Andean and Noble Snipe. Cattlemen have erected a couple of grass-thatched huts and a corral which serves a convenient camp site for hikers and alpinists en route to Sangay volcano. Beyond the camp the valley becomes more enclosed as it climbs towards the mountain. The river is fed by numerous streams that cut steep ravines in the mountain ridges which are clothed in a dense blanket of the giant cabbage-like paraguillas (*Gunnera* sp.) Secluded pools and rapids provide foraging for the White-capped Dipper and the rare Torrent Duck.

Tracks of the Mountain Tapir are commonly found along the river's fringing sandy beaches, and the prints of the diminutive *Pudu* are also occasionally seen. Mammals are rarely observed on the *paramo* during the day since most seek refuge in the fringing cloud forest, only venturing onto the open moors at night or under the protection of the swirling mists Andean Fox and marauding puma are the main predators. Birds of the *paramo* include White-collared Swifts, tapaculos

**Looking into the eyes of a Titi Monkey.**

(family *Rhinocryptidae*), hummingbirds and fringillids. Flocks of Glossy Black Thrushes are common, as are raptors such as the Red-backed Hawk (*Buteo polysoma*) and the Aplomado Falcon.

The route eventually emerges on an exposed ridge of elfin forest, stunted by the low temperatures and high winds that sweep over this area. At **La Playa**, nature finally gives way to the unearthly landscape of the volcano where the subterranean inferno and bitter chill of the atmosphere meet in a constant battle.

**The Napo:** There are two main towns in the Oriente from where it is possible to organise jungle excursions, **Coca** and **Mishuali**. Excursions from these towns used to provide the best opportunities for viewing jungle wildlife, but disturbance from oil exploration and development now means that there is relatively little to be seen in their immediate vicinity, and naturalists are advised to take an excursion, lasting several days, along the **Rio Napo** and its tributaries.

A suitable base – **La Selva jungle lodge** – is situated 60 miles down the Rio Napo from Coca. The river at this point is relatively unpopulated and there are good opportunities for viewing wildlife including the Amazonian River Dolphin Inia on the two-hour boat journey to the lodge. Situated close to palm-fringed brown expanse of **Garzacocha** (**Heron lake**) the lodge is ideal venue for inexperienced naturalists or professionals who want to plan their own itinerary and add to the 425-strong bird list.

Forest trails lead to **Mandicocha** (**Water-hyacinth lake**). From small canoes it is possible to view hoatzins clambering around the marginal vegetation, and to discover the roosts of the Proboscis Bat on fallen trees. Occasionally such rarities as Harpy Eagle, a nocturnal currasow, Long-tailed Potoo (*Nyctibus grandis*) and even the Zig-zag Heron have been observed here. The **Mandiyacu stream** arises from the lake, a canoe trip along its length can provide photographic opportunities with some of the 14 species of monkey present in the area.

Another trail takes in a ridge area, affording views across the canopy. From

*Rupicola Rupicola*, otherwise known as Cock-of-the-Rock.

here Red Howler Monkey and Spider Monkey can often be spotted, as well as fleeting glimpses of mixed groups of tamarins and marmosets which frequent the flowering and fruiting trees. Although well-hidden in the canopy it may be possible to glimpse the ponderous movement of a sloth or Tamandua Aneater.

On the ground the tourist trail is criss-crossed by the tracks of a White-lipped peccary herd. Saucer-sized pad-marks and a pungent aroma disclose the nocturnal passage of a jaguar, following their trail. Squadrons of Scarlet Mawcaws and Mealy Parrots wheel overhead and toucans hop in the dense vegetation. High above the canopy soar Greater Yellow-headed Vultures and King Vultures.

**From the Andes to the Oriente:** The Oriente of Ecuador is still sparsely populated and undeveloped except in the oilfields near the Colombian border, and the rapidly expanding goldfields at **Zamora**, adjacent to the **Podocarpus National Park** in the far south. Elsewhere, roads are few and colonzation is mostly limited to their immediate vicinity and to a handful of ports beyond. Few mountain passes give access to the area; by far the most spectacular is that descending from **Baños** to **Puyo**.

Baños, so named for its hot springs baths, is a popular tourist stop. Its Zoological Gardens house an extensive collection of wildlife from the Andes and Oriente, but many of the animals are poorly kept: not recommended for the soft-hearted. Below Baños, the **Rio Pastaza** enters a breath-taking gorge, and the narrow winding road clings tenuously to the precipe. Above lush forests ascend for over a thousand meters, dissected by numerous waterfalls. The **Agoyan Falls**, some 10 km along the road from Baños, is very spectacular but even their crashing cascade seems dwarfed by the magnificent mountains. After descending a further 900 m the valley widens at **Mera** and extensive agriculture opens up the *selva*.

A circular trip can be made from Quito, heading south as far as **Ambato**, a town within striking distance of the impressive peak of **Chimborazo**, and then east to **Puyo**. From Puyo the road heads north amongst scattered settlements but the land is still predominantly forested. Many rain forest birds can be seen, but mammals have largely been hunted out. The journey now continues along a pockmarked dirt track road whose verges provide a wealth of botanical specimens. Occasional glimpses are gained of lowland birds as they flit about in the fringing tropical vegetation or soar high above.

In the main, agricultural development has only scarred the forest in the immediate vicinity of the road and vast expanses of primary forest are accessible providing good opportunities for viewing birds. The lowland forest is dominated by oil palm with considerable stands of *Cecropia*. The terrain is by no means flat and from vantage points it is possible to look out over the canopy. Brightly flowering yellow and flame-red colored emergent trees proclaim a food source for pollinating birds and the more common species of primates – Squirrel Monkeys and capuchins.

Approximately 48 km north of **Tena**, travelers may visit the limestone caves of **Jumandi** near the **Rio Latas** which houses colonies of Vampire Bats. Farther north one is soon climbing the Andean foothills again, affording memorable views across the expense of lowland rain forest to the east although this is often shrouded by wreaths of low cloud.

From the early missionary town of **Baeza** the road ascends a steep gorge. The eastern *cordillera* is crossed at over 4,000 m affording good views of **Antisana** volcano. Just before the pass is an access to the **Cayabeno** reserved area.

**The Mountain Tapir:** Found in the high pluvial Andean mountain forest and wet *paramo* biomes of Colombia, Ecuador and Peru and possibly western Venezuela, the mountain tapirs have a stout body, (weigh about 200 kg; standing one meter at the shoulder), with a wedge-shaped profile for pushing their way through dense undergrowth. Their diet consists of selectively browsed plant shoots frequently ferns, plantains and *Chusquera* bamboo. Like so many animals found in this region, the tapirs are also under constant threat from habitat modification, illegal cattle grazing and hunting.

**The Passion Flower (*Passiflora coocinea*) is found throughout Amazionia.**

# PERU

**Search for the Source of the Amazon:** The Amazon, by name begins 4,000 km from its mouth in the north-east corner of Peru, where the waters of the **Rio Napo** and **Rio Ucayali** meet, joining the eastern catchments of most of Ecuador and Peru. But the geographical source of the river is defined as the point from which water flows furthest to the mouth. At this junction, the Ucayali is longer than the Napo. But exactly where it can be deemed to rise is still a disputed point.

A short distance upstream, past Peru's major Amazon port of **Iquitos**, is the confluence of the **Ucayali** and **Marañón**. Rising near the peak of **Yerupaja** (6,632 m) in central Peru, the Marañón takes a circuitous route northward before descending east to Iquitos. It often carries more water than the Ucayali and was long argued to be the Amazon's source, but by most current reckoning, the course should continue southward with the Ucayali. As the river is traced to the foothills of the Andes, its identity is confused, and it takes a new name at the confluence of each major tributary, several of which may compete for the distinction of "the source". One is the **Urubamba** which drains the mountains near **Machu Picchu**.

In the sixteenth century Spanish explorer Juan Salinas sought the headwaters of the Marañón and Ucayali. The drainage of the Ucayali extends southward into the **Cordillera del Chile** in the Department of Arequipa. Some believe its source is in this glaciated mountain range, others that it spills from **Laguna Vilafro** on its 7,000 km journey to the Atlantic Ocean.

**Iquitos:** Iquitos is the biggest city of the Amazonian lowlands of northeastern Peru, about 1,000 km by air from the capital Lima, and 1,200 km from tourist destinations like Machu Picchu in southern Peru. Iquitos has grown from a village of only 80 residents in 1814 to a thriving center of jungle exploitation with a present day population of 350,000. The main stimulus for the town's development was the collection of latex from the relatively abundant rubber trees. The rubber barons grew rich at the expense of the native community but at the same time endowed Iquitos with some excellent colonial architecture which remains despite the demise of the rubber industry. Modern day Iquitos is a major port and the center for oil exploration in the Peruvian Amazon.

In the early days the route of access was undeniably difficult: the overland trek from **Lima** took two and a half months through dangerous territory. Rich merchants apparently preferred shipping up the coast of South America, through the Panama canal and the Carribean and thence to Belém and up the Amazon to Iquitos. A road crossing the Andes from Lima to **Pucallpa** on the upper Ucayali was completed in 1943. Pucallpa is a frontier town of some 90,000 people, reliant on timber and oil exploration.

Near Pucallpa is **Yarinacocha**, a lake formed from an ancient meander of the Ucayali. Part of the lake is reserved as a refuge for nature. From Pucallpa it is a two day boat journey to Iquitos, but the Ucayali is sufficiently wide to make bank-side wildlife observation impossible from mid-stream. Because of disturbance from development and oil exploration little wildlife remains in the vicinity of Iquitos for perhaps a 50 km radius. The large mammal species have been hunted out many years previously, although plentiful flocks of birds still commute across this riparian corridor of development.

But Iquitos is still an island in the wilderness, and fruitful natural history excursions are possible. A three-day cruise down the river reaches **Leticia** and **Tabatinga** near "**Tres Fronteras**" – the borders of Peru, Colombia and Brazil. The "jungle tours" involve some hiking, canoe excursions up tributary rivers, night-time caiman spotting and fishing for Red Piranha (*Serrasalmus nattereri*). Good views of birds and occasionally primates including Black Howler Monkey, are afforded along the densely forested river margins. Alternatively boats can be taken from Iquitos to the **Rio Marañón**, or the **Rio Tahuayo** which is also surrounded by dense lush rainforest. In the **Yanayacu** area it is frequently possible to see the *boutu* or Amazon River Dolphin. The very rare Amazonian manatee favors quiet tributaries containing areas of dense vegetation. Infrequently seen, their presence is nevertheless revealed by floating droppings – voluminous quantities of fibrous balls resembling those of the horse.

The river corridors are often the best places to observe wildlife. Caimans and turtles are commonly seen basking on the river banks. Neotropical Cormorants, Roseate Spoonbills and Jabiru Storks fish in the shallow waters.

The forest is at its densest along the river margins. Within, the diffused light reduces the density of the understorey. Long, trailing lianas are plentiful, especially where there are clearings resulting from the death of old trees or removal by Man. These woody vines bind even the tallest trees together in an embrace so tight that even in death they may be prevented from falling to the forest floor. Included here are the strangling figs (*Ficus* sp. and *Clusia* sp.). These start life as epiphytes; birds deposit seeds into cracks in the bark. Aerial roots develop which quickly grow down to the ground eventually forming a fusing mesh that envelopes and kills the host, leaving only the hollow "trunk" of the strangler.

Much of the bird and nammal life is confined to the canopy and its network of vines, allowing only tantalizing glimpses as they move from tree to tree. The animals of the canopy have developed striking adaptations to enable them to exist in this precarious wilderness. The prehensile tail of the Gray Four-eyed Opossum and many of the New World primates provide a convenient extra leg with which to grip slim branches. The peculiar development of the claws of the sloth provide for added security as it browses its way through the dense canopy vegetation.

**Machu Picchu**: **The Historical Sanctuary of Machu Picchu** was established in 1981 and includes the **Archae-**

Nervous stares from a herd of Vicuña.

**ological Park** with the fabulous Inca ruins re-discovered by Hiram Bingham in 1911. Also within the Sanctuary are of surrounding *puna* and cloud forest. A spectacular railway brings visitors from **Cuzco**, ascending the mountain by a series of switchbacks to the peak at about 3,500 m. From the *paramo* the train descends to the **Secret Valley of the Incas**, following the Rios Urubamba and Vilcanota to **Puente Ruinas** at 2,000 m. From here tourists take a bus to the **Inca fortress** from where commanding views of the cloud forest are obtained.

The best way to view nature in this area is to walk the **Inca Trail**, a spectacular three- to five-day hike from near **Ollantaytambo**. The rugged trail passes over a pass at 4,200 m and includes extensive areas of cloud forest.

The area has been affected by agriculture and hunting since the times of the Incas. Even today various conservation problems beset the sanctuary including destruction of the forest by burning and invasion by cattle leading to substantial erosion. Access by 160,000 tourists an-nually, most of them en route to the monument, has led to problems of litter, occasional hunting and general disturbance. Probably as a result of the disturbance, few signs of mammals are apparent, although occasionally the bear-dissected remains of bromeliads (*Puya*) and the tracks of Puma and White-tailed Deer may be seen near the Inca Trail. Numerous species of primate have been reported in the vicinity of Machu Picchu including the Night Monkey, and the Woolly Monkey.

Even in areas where Man has disturbed, secondary growth can provide a rich habitat for a wide variety of bird species. Of the 1,678 species recorded for Peru, some 22 percent (372 species of 49 families) have been recorded in the Sanctuary and its immediate vicinity, including the Andean Condor. The Roadside Hawk, Variable Antshrike, White-winged Black Tyrant and the Green Jay are commonly encountered.

The principal habitat types are *puna* or *pajonale* at altitudes above 3,500 m typified by *Stipa-ichu* grass over which

**griculture as changed ne face of ne Peruvian ndes.**

the Puna Hawk and Mountain Caracara hunt. The Puna Snipe may also be seen. Below this, there is a zone of elfin forest comprising in the main *Clusia*, *Gynoxis* and *Polylepis* amongst a host of smaller shrubs and herbs. This grades into cloud forest particularly in the vicinity of the ancient monument. Tree species such as *Cyathea*, *Alsophila* and *Podocarpus*, *Weinmania* and *Nectandra* are common at the top of this zone. Much of the understorey comprises tangled thickets of *Chusquea* bamboo.

Where the Inca trail passes through the cloud forest zone there are characteristic small trees, Birch forest (*Aliso*; *Alnus acuminata*), terrestrial bromeliads (*Puya ferruginea*) and cactus (*Cereus* and *Opuntia*). The *Polylepis* forest has a distinct and rich avian fauna including hummingbirds such as the Tyrian Metaltail and Great Sapphirewing, the Andean Hillstar, Collared Inca, Bearded Mountaineer and the spectacular Giant Hummingbird and Sword-billed Hummingbird. Various species of Spinetail, are also common. Other interesting bird species that may be encountered here include the Andean Guan, Andean Parakeet. The Golden-headed Quetzal, the Highland Motmot and the Gray-breasted Mountain Toucan are occasionally spotted.

There are a few glacial lakes in the sanctuary and on these it may be possible to observe Puna Teal and in the streams which they feed, Torrent Duck also forage. Other areas of interest close to the sanctuary include **Cuenca del Santa Teresa** and the waters of the Rios Santa Maria and Urubamba, from which the South American River Otter has been reported.

**The Lowlands of Eastern Peru:** East from Cuzco and Machu Picchu lie river systems quite separate from the Ucayali. The **Madre de Dios** is one of the least accessible head-waters of the Amazon, and consequently contains vast areas of undisturbed forests. The watershed of the **Rio Heath** although more accessible is also of considerable ecological significance since it contains the **Heath Pampas National Sanctuary**.

The Alpaca belongs to the llama family.

The Department of Madre de Dios has a total population of only 50,000 most of whom are based in or close to **Puerto Maldonado**, a town on the Rio Madre de Dios not far from the border with Bolivia. The town itself has had a chequered history from the environmental viewpoint. The initial exploration of this area is attributed to Fitzcarraldo who discovered a route from the Ucayali to the **Rio Cashpajali** – a tributary of the Manu and hence giving access to the Madre de Dios. The opening of this route was politically and commercially important in the exploitation of jungle products, particularly rubber, and led to some aggression against the indigenous tribes. With the demise of the Peruvian and Brazilian rubber industry, the attention of the now thriving town turned to other forest products. In the 1920s, hunters entered the area harvesting cats, Giant Otters and caimans for their skins. In more recent years the exploitation of valuable timber including mahogany led to the construction of an airstrip at **Boca Manu**

that is now a convenient route of access for tourists visiting the area. Today gold prospecting and cattle ranching attract thousands of new colonists each year, although spotted catskins are still commonly seen for sale.

The Madre de Dios region is recognized as a World Centre of Plant Diversity by the WWF/IUCN. It is believed that the rain forests, which once covered a much larger proportion of the earth, retreated to a few isolated pockets during the drier climate of the last ice-age. The Madre de Dios was one such refuge, from which plants and animals have subsequently recolonized the Amazon basin. As such, it is one of the oldest, most stable and most complex examples of neotropical forest. Two major protected areas have been created in acknowledgement of the areas value: the **Manu National Park** and the **Tambopata-Candamo Reserve Zone**.

**Tambopata-Candamo Reserved Zone: Tambopata Reserve**, a small reserve of only 55 square km located some 40 km south of **Puerto Maldonado** was

Cloud forest (below) and *Puya* plant near Macchu Picchu (right).

created in 1977. But with the demarcation in 1990 of a huge area of surrounding forest, it has become the hub of the 1.5 million hectares Tambopata-Candamo Reserved Zone. This is Peru's most recently declared "park", protecting the vulnerable basins of the Rios Tambopata and Heath.

Tambopata is a prime location for ornithologists visiting Peru and holds the highest 24-hour bird list in the world: 331 species. The total bird list for this area presently extends to 570 species along with some 91 species of mammals, including the Giant Anteater, Giant Otter and Giant Armadillo. Two of South America's rarest canids are also recorded from the reserve: the Bush Dog and the Small-eared Dog. The tourist facility supports an extensive research program conducted by the scientist guides.

The arrangement of the Reserved Zone is designed to allow for various levels of forest utilization in designated areas. Adjacent to Tambopata a major ethnobotanical project, "AMETRA 2001" was set up to combine modern and traditional approaches to medicine. In part it involves recording traditional knowledge of forest practices and disseminating this information through the local community. Other experiments in sustainable production are also being set up – involving fish, butterfly and capybara farming and Brazil nut collection.

**Manu Biosphere Reserve:** Amounting to some 18,812 square km of mainly humid evergreen tropical forest, laced with countless miles of meandering rivers and glinting lakes, the **Manu National Park** is the largest biosphere reserve zone and the fourth largest national park on the continent of South America. Although the park was established in 1973, it was not until 1977 that UNESCO accepted a proposal from the Peruvian government that it be declared a biosphere reserve, one of 200 such declared areas world-wide. They represent an attempt to preserve pristine examples of all the world's major ecosystems. In the case of Manu, this is the only biosphere reserve that protects an

In flight, the Andean Condor is a majestic sight.

entire unhunted and unlogged watershed. The reserve contains many rare and fascinating species of animals (including over 1,000 species of birds) and plants that have been eliminated from the other areas by Man's influence.

The park is divided into three zones: a core zone comprising 15,328 square km which is strictly preserved in a natural state and in which reside a number of indigenous tribes, an experimental or buffer zone of some 2,570 square km set aside for controlled research and tourism, and the agricultural zone of 914 square km for controlled traditional use. Only six dispersed ranger stations protect the park, but the most effective protection comes from its remote and inaccessible location.

Tourist provision in the park is scant. One company plies the river in motorized dugouts to camp for a few nights at designated camp sites on the sandy beaches deposited at river bends. The only permanent tourist facility in the lowland forest of the Reserve Zone is provided by **Manu Lodge** (elevation 400 m), comfortable and sympathetically constructed from mahogany salvaged from the **Rio Manu**. It is well positioned on the shore of **Cocha Juarez**, some four hours by boat from the mouth of the Rio Manu which can be reached in 45 minutes by light aircraft from Cuzco. However the overland and river route, taking at least two days, is a fascinating journey through all the layers of Andean habitat. The precipitous road is traveled in each direction on alternate days, as it is too narrow to allow vehicles to pass.

**Traveling from the Andes to the Lowland Oriente:** The forested eastern half of the **Peruvian Andes** amounts to some 60% of the countries land area but holds only about five percent of the population. The rivers provide for easy transportation of jungle produce, particularly timber and rubber but also provide a corridor for colonization.

Few roads penetrate the area, the most important of which runs from Lima to **Tingo Maria** and thence Pucallpa giving access to the Rio Ucayali and so the

A collared Gecko from the lowland forest.

mighty Amazon. Another major route runs from Cuzco over the Andes to the watershed of the Madre de Dios river which is accessed through

**Atalaya** or the the missionary settlement at Shintuya. From Cuzco (altitude 3,310 m) an early morning departure is required to reach a suitable overnight camping point within the high cloud forest zone, a journey of some six to seven hours. The paved road soon gives way to a dirt track leading steeply to the Quecha Indian village of **Huacarani** (3,800 m) and vertiginously down into the valley of **Paucartambo**, famous for its many varieties of potatoes.

These habitats are called "*ceja de selva*" (eyebrow of the jungle) and include the high grasslands and cloud forests. Andean Fox and White-tailed Deer may be occasionally encountered in the open grasslands. Herds of llamas and alpacas look up shyly. Overhead Mountain Caracara and Andean Lapwing are commonly seen, and a lucky traveler may spot flocks of Mitred Conure.

Gradually the agriculture thins out towards the **Acjanaco pass** (4,100 m). Here at **Tres Cruces** is an exhilarating view of the Amazon basin and the Manu Park below. Craggy peaks soar over the *puna*, a habitat characterized by tussock grass and pockets of stunted alpine flowers, giving way to relict elfin forest and tangled bamboo thicket. Occasionally the dissected remains of a *Puya* plant denote the nocturnal foraging of the rare Spectacled Bear, a frequent visitor to these exposed lands. Here also pumas (*Felis concolor*) roam preying on the White-tailed Deer and tiny *Pudu* which frequent the mountain grasslands.

The elfin forest gradually grades into mist enshrouded cloud forest at about 3,300 m. These often impenetrable forests cover the Eastern slopes of the Andes. The high humidity resulting from almost perpetual cloud supports a verdant kingdom of dripping epiphytic mosses, lichens, ferns and orchids which grow in profusion here despite the plummeting overnight temperatures. Water seeps into a myriad of icy, crys-

**Colorful lodgings on an Iquitos street corner.**

tal-clear streams and waterfalls. In secluded glades flame-red Andean Cock-of-the-Rock give their spectacular display and Woolly Monkeys are also occasionally sighted. Mixed flocks of colorful tanagers, the Golden-headed Quetzal and Amazonian Umbrellabird are sometimes observed.

At about 1,500 m there is a gradual transition to the vast lowland forests of the Amazon basin. Surprisingly less jungle-like than that we have encountered so far, but warmer and more equable than the cloud forests above. The dense canopy of large leaves inhibits growth at ground level. A great diversity of habitat types is experienced ranging from 10 m high stands of cana brava which bind together the river banks, through lush stands of *Heliconia*, and in the interior – giants like the mahogany (*Cedrela odorata*), and Kapok tree (*Ceiba pentandra*). It is a habitat hung with strangling vines and phylodendrons amongst which mixed troupes of vociferous Squirrel Monkeys, Brown and White-fronted Capuchins

forage. In the high canopy groups of Black Spider Monkeys perform their lazy acrobatics, whilst lower down small groups of Saddle-back and Emperor Tamarins forage for blossom, fruit and occasional insect prey.

The daily temperature varies little during the year with a high of 23–32° C falling slightly to 20–26° C overnight. The two meters of annual rainfall mostly falls from November to April, when access to the park may be restricted. The rest of the year is sufficiently dry, at least in the lowland areas, to inhibit the growth of non-vascular epiphytes and orchids which were so characteristic of the highland areas of the park. For a week or two in the rainy season the Manu rises and floods the forest in its six kilometers wide floodplain. Canoeing and wading are then the only ways of getting around.

The road ends at the **Alto Madre de Dios**. Dugout canoe is still the time honored method of transport, only now it is most likely to be powered by a pair of 50 hp outboards. After traveling for

**Shantytown scene.**

four hours up the main river, we enter the mouth of the Rio Manu. The journey now winds between partially submerged trees left stranded after the floods of past rainy seasons.

The meandering course of the river provides excellent viewing opportunities for herds of russet-brown Capybara, snuffling Collared Peccary and wary Red Brocket Deer. The dipping flight of a toucan demands attention, and the far off screeching of a pair of macaws increases to a deafening cacophony as they pass over the river corridor: asymphony of color if not of sound.

The sinuous meanderings of the river transform the landscape relentlessly and provide a living story of forest succession. At each bend of the 150 m wide river, the forest is undermined by the currents during the seasonal floods at the rate of 5–20 m per year leaving a sheer mud and clay bank, while on the opposite bend new land is laid down in the form of broad beaches of fine sand and silt. Gradually these beaches become invaded by plants taking advantage of the alluvial soils, which are far richer than the weathered upland areas. A succession of vegetation is observed from the fast growing willow-like *Tessaria* which stabilizes the sand. Then tall stands of cana brava (*Gynerium sagittum*) become established. This plant shows a predigious rate of colonization, perhaps some 10 m per year. Finally within these almost impenetratable stands the seeds of tall rain forest trees germinate and over a few years thrust their way towards the light. The fastest growing is a species of Cecropia which forms a canopy 15–18 m over the cana but a little further in this short-lived species gives way to others such as *Erythrinia*, *Guarea* and *Sapium*. Two species *Ficus insipida* and *Cedrela odorata* (mahogany) are capable of outgrowing the others forming a closed canopy at 40 m with an lush understory of shade tolerant long-leaved monocotyledons such as *Heliconia* and ginger. Eventually even the long-lived trees die off to be replaced by others providing a forest of great diversity.

A Common Woolly Monkey.

It has been said that "to walk into the rank vegetation bordering a beach is to walk backward in time" since you are progressing towards older terrain with successive stages of vegetation. As a consequence of these destructive forces the forest bordering the river is of no great antiquity. It is only beyond the flood plain – the zone of *"terre firme"* forest and on the ridges, that the largest and oldest trees are found. The oxbow lakes or *cochas* which remain after the river has changed course are of considerable ecological interest since these provide an abundance of fauna easily observed around the lake margins.

**Wildlife Observation on the Rio Manu and at Manu Lodge:** Much of the lowland wildlife can be seen in the vicinity of the Manu Lodge, including nine species of monkey and over 450 of the 500 species of bird recorded in lowland Manu. In surrounding trees are the pendulous nests of the noisy Yellow-rumped Caciques, Blue-and-Yellow Macaws visit the swaying palms. Black Caiman (*Melanosuchus niger*) prowl the lake on the banks of which the lodge has been built. The dawn is summoned by the gently rising and falling roar of Red Howler Monkeys. Flocks of Pale-winged Trumpeters forage in the undergrowth while guans and curassows haunt the canopy. Particularly noteworthy species are the Black-faced Cotinga, Crested Eagle and the spectacular Harpy Eagle, perhaps the world's most impressive raptor, easily capable of taking an adult monkey from the canopy.

In the trees and on the ground termites and ants play a vital role in the decomposition cycle of the forest. Well-worn trails lead to the vast encampments of army ants. Devoid of vegetation, they often cover several square meters. High up the trunk of a tree a termitarium can be seen. Constructed of chewed wood and soil particles, a drainpipe-like structure connects it to the forest floor. In the ground-litter myriads of these social insects are active in the destruction of decaying timber, carrying particles to temperature-controlled fungus gardens deep inside their elevated fortresses. Here they are

Afternoon tea at Manu Lodge.

protected from the depredations of the Gian Anteater but they remain vulnerable to the arboreal foraging of the Tamandua and the Silky Anteater.

Twenty-seven kilometers of forest trails extend from the lodge from which many animals can be seen at close quarters. Knowledgeable guides with high-powered telescopes ensure a rewarding experience, and can identify the sounds of a plethora of bird species. Having never been persecuted here, many animals are bold and unafraid of humans. Terrestrial animals take advantage of the open trails. Tracks of ocelot and jaguar are commonly found, frequently following along behind groups of tourists. It is estimated that one in seven visitors sees a jaguar.

One of the forest trails terminates at a "*culpa*" (mineral lick). Here, in the very early morning, good views may be obtained of Collared Peccaries, Red Brocket Deer and Lowland Tapir. Occasionally predators attracted by these prey may also be encountered. Rare game birds such as Razor-billed Curassows and Piping Guans may also be seen.

Another lick frequented by macaws and parrots can be reached by river. The birds need a daily dose of the mineral-rich clay to neutralize the toxins in their leaf and seed diet. Here at dawn there may be 600 birds of up to six species jostling for access to the mineral rich clay.

The best way to see the wildlife in the forest is to get into or above the canopy. In Manu, this can be achieved in one of two ways. A sharp ridge provides a number of elevated miradors from which excellent views are obtained. On a clear day one can look across the lowland floodplain to the foothills of the Andes. Flocks of parrots and macaws commute between various fruiting trees. Mixed species flocks are commonly seen in fruiting trees, containing from 25 to over 100 birds of perhaps more than 30 species. Each species occupies a slightly different niche, and since there are few individuals of each species in the flock, competition is minimized. Individual can take advantage of the security of the flock and may even benefit from im- **Tranquil scene at Manu.**

proved efficiency in finding prey disturbed by other species. An even better view of the canopy is afforded from one of the numerous platforms that have been erected in selected trees. The most notable of which is sited in a giant Kapok (*Ceiba pentandra*) stretching above the surrounding canopy to 40 m. Access is provided using *jumars* (rope ascenders). From this aerial vantage point it is possible to view a wide range of primates and birds attracted to the delicate epiphytes and fruiting vines. Plans are in hand to construct a canopy walkway connecting a number of large trees.

One of the most enjoyable ways of seeing the wildlife is to drift noiselessly around the two kilometer long *cocha* in a canoe. The silent approach allows excellent photographic opportunities with Fascinated Tiger-Heron, Sunbittern and raucous, primitive hoatzins. Fleeting glimpses are possible of diminutive Emperor and Saddle-back Tamarins foraging on the fruit-bearing trees that overhang the lake.

**Giant Otter doing the backstroke.**

Nocturnal excursions along the forest trails with a powerful torch provide a different perspective on the fauna of the forest. The booming call of the Bamboo Rat reverberates around the forest competing with the haunting moan of the Kinkajou for attention. The torch beam reflect the neon eyeshine of a Night Monkey disturbed from its foraging activity. On the ground a timid Paca rustles in the leaf litter for oil palm nuts. The warm, humid air is a living soup of insects and the click-click of foraging bats can be clearly heard as they swoop between the forest trees.

The Giant Otter is a particular attraction of the park. Manu is one of the few places where visitors are almost certain to see them at close quarters. Excursions can be made to **Cochas Salvador** and **Otorogo** some three hours up the Manu. Here the otters have been partially habituated during several years of scientific study, and may be carefully approached. The Giant Otter population in the Manu National Park is estimated at over 100, perhaps one-tenth of

that in all South America. To be surrounded by a dozen of these noisy, curious predators is a memorable experience to take home.

**The Brazilian Giant Otter**: This is the world's largest species of otter, males reach a length of 1.5–1.8 m and weight up to 30 kg, females are only slightly smaller. Unlike the smaller Neotropical River Otter (*Lutra longicaudis*), the Giant Otter is gregarious; living and hunting in permanent groups of four to six individuals. Exceptionally, as many as 20 individuals have been seen together. The otters spend long periods of the day hauled out on logs mutually grooming one another and sun-bathing, particularly at mid-day.

The body is superbly designed for motion in water but the gait on land appears cumbersome. The stream-lined head bears long stiff vibrissae which are used to locate their fish prey in the murky waters of the lakes and rivers they frequent. The dense glossy coat is a uniform chocolate-brown apart from the chin, throat and chest which are streaked with cream or buff fur. This unique pattern can be used to identify individuals. The tail is dorso-ventrally flattened and gives rise to the Latin name – *Pteronura*. Sometimes referred to as the "Brazilian wingtail" it is more commonly known as "*lobo del rio*" – the river wolf.

Formerly distributed throughout all the countries of South America, with the exception of Chile, the species is now probably extinct in Uruguay and Argentina, and shows a much reduced range elsewhere.

Observations of the feeding behavior of Giant Otters in Peru, Guyana and Suriname suggest that they prefer to hunt in shallow water. The principal fish prey are piranha and catfish particularly *Hoplias malabaricus* but they will also take caiman (up to two meters in length) and anaconda. The otters cooperate in attacking a caiman, darting at it from different directions. Such large prey are taken comparatively rarely. Fish are commonly eaten in the water while held with the fore-paws. During the seasonal floods, when the river inundates the surrounding lowlands, the fish move into the forest to forage and the otters follow them there.

Normally the otters hunt in a group, diving simultaneously into a shoal of fish. Solitary hunting otters appear to achieve an equivalent capture rate (on average 1.2 fish per hour) to that of individuals in groups. Hunting together need not necessarily imply cooperation, it may just be that, for the individual otter, fish are more easily captured in the confusion brought about by the churning water. It is not known if this is the reason for their gregarious behaviour or whether their sociality confers a defence against predators; young otters are known to be vulnerable to both caiman and anacondas.

The Giant Otter is fiercely territorial, frequently scent-marking its bankside resting places with latrines, areas where they have trampled their faeces and urine into the ground as a pungent "keep-out" signal. Only one pair of otters in the group appear to breed, the female giving birth from four to six cubs. Although the previous years young stay with the group they do not actively participate in rearing the new litter. By the age of six weeks the young may accompany the group, playing boisterously while the adults fish, and wailing pitifully for attention when neglected. They are actively catching prey themselves by nine months of age. Adult Giant Otters have a comprehensive range of vocalizations which they use to great effect to threaten, express alarm, and maintain group contact. They may scream in frustration, snort in alarm or give a wavering scream towards an intruder.

Large, gregarious, noisy and inquisitive, Giant Otters were easily located and hunted from canoes for their luxurious pelt. It was the demand of the fur trade which has led to the present-day endangered status of the otter. Some 24,000 skins were exported from the Peruvian Amazon in the period 1946–73. Skin trading has been banned since the 1970s but it is likely that small numbers of otters are still taken. The entire South American population has been estimated at between 1,000–3,000 individuals.

Spider nabs a victim by casting its 'net'.

236

# BOLIVIAN AMAZONIA

**Bolivia** has been generally associated with the sparse and imposing highland landscapes of the Andes. The snow covered peaks of the **Altiplano** (highland plateau) have provided Bolivia with its signature. However, a close look at a map will show that over half of Bolivia's 1,098,000 square km is in the Amazon Basin. The Amazonian lowlands have played an important, albeit less visible role in the establishments of Bolivia as a nation and still holds great promise for conservation and development – sustainable development, that is.

The **Amazon River Basin** tends to be ecologically oversimplified. The traditional view of Amazonia is one of a flat, tree-covered tropical forest, where great many species thrive in a uniform landscape. Much on the contrary, the habitat diversity of this region is to a large extent responsible for much of the biological diversity it holds. What makes Bolivia's Amazon Basin particularly unique is precisely the fact that it not only contains the classical lowland tropical forests, but also has habitat types as diverse as the montane cloud forests and dry inter-montane valleys of the eastern Andes, the tropical savannas, and the dry scrub forests associated with the central plateau of the Brazilian Shield.

Three protected areas in the Bolivian Amazonia include this wide range of habitats and represent one of the finest examples of the efforts being made in Bolivia to preserve the diversity of life in the planet. These protected areas are the **Amboró National Park**, the **Noel Kempff National Park**, the **Rios Blanco and Negro Wildlife Reserve**. In addition to the biological riches that they hold, these protected areas represent important experiments in the management of protected areas, where private-public partnerships have been taken beyond the formalities of the usually tense and uncomfortable relationship between private non-profit organizations and the public park administration.

These three protected areas are being established, at great cost, to play a significant role in the protection of Bolivia's biological diversity and, at the same time, in the development of the eco-tourism industry. As neighboring countries are plagued by crime and political upheaval, Bolivia remains a peaceful nation that welcomes those who come to learn and see the beauty of the culture, the people and the best of what Nature has to show.

**Amboró National Park:** The Amboró National Park is located in the center of Bolivia in the Department of Santa Cruz, at a point where the Eastern Andes change direction from northwest-southwest to straight north-south. It currently covers approximately 6,930 square km of habitats ranging from humid montane forest in the south to lowland forests in the north. This park was created in 1973 under the name of the **German Busch Wildlife Reserve**. In 1984 the name was changed to the **Amboró National Park** with an extension of 1,800 square km, and in 1990 the park was increased to its current size.

Before the expansion of the Amboró National Park over 540 species of birds and 120 species of mammals had been recorded, among them the endemics Red-fronted Macaw and the Southern Horned Curassow, and the rate Andean Cock-of-the-Rock, Harpy Eagle, Crested Eagle, Spectacled Bear, Giant Otter, jaguar, and many others. Many more species will be included in the park list after the expansion in size since the increased area increases the habitat diversity of Amboró especially at the higher elevations. The two best represented life zones are the cloud forests to the south and the lowland rain forests to the north, described as *lower montane wet subtropical forest*, and *wet subtropical forest*, respectively.

**Amboró** is easily accessible from **Santa Cruz** along two major roads connecting Santa Cruz to **Cochabamba**. The older road, build in the 1950s, goes straight west, up the mountains, and borders the south side of the park. The southern headquarters of the park is located in the city of **Samaipata**, 120 km from Santa Cruz, and it is operated by the *Fundación Amigos de la Naturaleza* (FAN) in partnership with the *Centro de Desarrollo Forestal*, CDF (Forestry Development Center). The southern two-thirds of the park is mountainous and has been significantly less affected by human activity, as compared to the northern lowlands. Although pressure is continuing to increase in the south, there are still frequent sightings of the threatened Spectacled Bear and other critical species.

Along the 180 km of road that borders the south side of the park there are over 40 logging and access roads into the park built before it was enlarged last year. Gradually and with great difficulty, FAN and CDF are beginning to establish a presence in the region and working with the local community to secure the park from invasion and poaching. One of the critical priorities is to develop controlled access to the park so that it can be visited and monitored at the same time. Samaipata has excellent lodging facilities, but adequate visitor access to the park is still limited. Improving the accessibility of visitors to the park is one of the priorities. Access will, however, be controlled and closely monitored to prevent poaching, illegal logging, and penetration by squatters, the major problems afflicting Amboró.

The northern limit of the park follows closely the new Santa Cruze-Cochabamba road, built in the late 1980s. This section of the park is being managed by the Chimoré-Yapacaní Project of the Secretariat General of the Environment, headquartered in the city of **Buenavista**, 120 km off Santa Cruz. The situation in the north side is quite different from that of the south. Being mostly lowlands, and having been exposed to the new road construction, this portion of the park is severely affected by close to 900 families of colonizers that utilize slash and burn practices. Great efforts are being made to prevent more colonists to become established and to control the damage already made, but the very limited resources of the Bolivian government are not sufficient.

A view over Amboró National Park.

**Noel Kempff National Park:** The Noel Kempff National Park is located on the northeastern corner of the Department of Santa Cruz, on the Brazilian border, across the **Rio Iténez** from the Brazilian States of Rondônia and Mato Grosso. It covers a total of 9,140 square km and includes a wide range of habitats. The center of park is dominated by an extensive plateau that rises sharply from the lowlands to over 500 m. The high biological diversity of this area – over 550 recorded bird species – is largely due to the climatic and habitat diversity created by the abrupt topography.

On the central plateau, known as the **Huanchaca** or **Caparuch Range**, one finds perhaps the largest tract of virgin *cerrado* left in the world. The *cerrado* is a sparsely forested scrub savanna found only in Brazil and Bolivia. This ecosystem has become the preferred habitat for soybean plantation in Brazil, what has caused it to disappear at an ever increasing rate. With the unique vegetation, also appear rare, threatened, and endangered species restricted to the *cerrado*, such as the Maned Wolf, Giant Anteater, Giant Armadillo, Marsh Deer, Golden-naped Macaw. To the west humid lowland forests are drained by the large **Rio Bajo Paraguá** and its tributaries teeming with wildlife, including healthy populations of the very rare Giant Otter and Black Caiman.

The **Noel Kempff National Park** was created in 1979 as the **Huanchaca National Park** with 5,410 square km, thanks to the efforts of Noel Kempff, a prominent Bolivian naturalist. Noel Kempff was also the driving force behind the creation of the Amboró National Park, well before conservation was fashionable, and as critical as it is today. Kempff dedicated himself to the study and preservation of the flora and fauna of Bolivia. He designed and established the Botanical and Zoological Gardens in Santa Cruz to bring the people of Santa Cruz closer to their surroundings. There are no pines and zebras in the **Santa Cruz Botanical and Zoological Gardens**. Instead, they are full of *Tajibos* with their brilliantly colored

Sandstone rock face at Amboró (below) and exploring the Caraband Range (right).

flowers, and *Toborochis* with their bulging trunks; there are gentle Night Monkeys, and raucous macaws.

Noel Kempff was killed in September of 1986 when he was visiting the Huanchaca National Park. He and the members of his party landed on the plateau to explore the surrounding vegetation. Unbeknownst to them, the strip was actively used to supply a large cocaine laboratory. What followed resulted in the deaths of all but one of Kempff's party. The surviving member of his expedition made it out of the plateau alive after hiding until a rescue plane spotted the burnt remains of Kempff's aircraft – the plane was set on fire to make it look like an accident – and landed to investigate the incident. It took three days for the police to arrive and to start the investigation. By then it was too late, the cocaine factory had been dismantled and the people running it were nowhere to be found. To this date, the killing of Noel Kempff and his party remains unsolved.

Two months after Kempff's death the park name was changed to the Noel Kempff National Park and the size was increased to 7,060 square km with an additional 2,080 square km of a buffer zone. The management of the park was left in the hands of a group of institutions under the leadership of CORDECRUZ the Santa Cruz Regional Development Corporation. Recently, however, participation of non-government organizations in the management of the park has been increasing. Most notably the acquisition of *Flor de Oro*, a 100 square km inholding, by FAN with the assistance from The Nature Conservancy. *Flor de Oro* will be developed by the FAN as a research and ecotourism station. It is expected that by the middle of 1991, it will be capable of accommodating visitors and researchers.

**Rios Blanco and Negro Wildlife Reserve:** The Rios Blanco and Negro Wildlife Reserve covers an area of 14,000 square km and it is located in the northwestern corner of the Department of Santa Cruz, on the border with the Department of Beni. This reserve was created in 1990

**Arcoiris falls in Noel Kempff National Park.**

specifically to be developed for ecotourism. Management responsibility was granted to the CDF and FAN, which have been working with the business community to open the reserve up for visitation.

The Rios Blanco and Negro Wildlife Reserve is located in the transition zone between the lowland forest and grasslands to the north and dry deciduous and scrub forests of the **Chaco** to the South. Its inaccessibility has kept squatters and poachers out. The best evidence of this is the high densities of tame large mammals in and around **Perseverancia** – the only inhabited site in the reserve which has adequate visitors facilities, accessible only by plane. A few days in Persevreancia are sure to include sightings of tapirs, brocket deers, giant otters, screamers, hoatzins, and several species of macaws.

The success of these three protected areas will, to a large extent, determine the success of many other areas that are only now beginning to become established. The Amboró National Park, the Noel Kempff National Park and the Rios Blanco and Negro Wildlife Reserve, are taking conservation actions to the field at the same time that they consolidate partnerships between the public and private sectors, thus sharing the responsibility to protect Bolivia's biological Diversity with all Bolivians.

The *Fundación Amigos de la Naturaleza* (FAN), a private non-profit conservation organization has been successful at obtaining assistance to support these areas at the same time that it has forged important cooperative arrangements with national and international organizations. FAN and The Nature Conservancy, have been working jointly on the Amboró and Noel Kempff National Parks as the first protected areas in Bolivia to be part of the **Parks in Peril Program**. The acquisition of *flor de Oro* by FAN is a major effort to meet the objectives of this program. It is likely that efforts like this will bring Bolivia to the forefront as one of the leaders in conservation in the last decade of this century.

**Rustic digs at Rios Blanco and Negro Wildlife Reserve.**

# BRAZIL

The largest fraction of the Amazon rain forest lies within the national boundaries of **Brazil**, a huge expanse of three million square km, about eight times the size of Germany or California.

The good news is that only a small part of this area has been "developed" for human use, the bad news, that an even smaller portion falls into the boundaries of national parks, or has any other protected status. Only three parks have been gazetted so far, but, as to touristic use **Pico de Neblina** is nearly inaccessible, and **Jau** and **Tapajós** have neither a headquarter, accommodation, nor regular boat service. Some other areas, like the **Anavilhanas Archipelago** or part of the **Rio Trombetas**, have the status of a biological reserve, and other areas, like **Xingu**, protect the environment of indigenous Indians.

It is difficult to predict the destiny of the region a few decades from now. Most likely, the worst fears will not become reality: today's Brazilian government does not pursue any more a politics of radical development, for example for cattle ranching. The scientific evidence is overwhelming that most areas of Amazonia are unsuitable for intense agriculture. But equally unrealistic would be the conversationist's hope that the whole Amazon could become a biological reserve. Research is going on to identify the areas of greatest biological richness, which might become the cores of future protected areas. Other lines of research attempt to determine the minimal critical habitat that can indefinitely maintain an isolated population of particular species. It is likely that many parks and reserves will eventually become gazetted according to the outcome of this research.

What will happen to the Amazon forest outside the reserves? The answer to this question is speculative. It is likely that the future will see a patchwork of areas that had been – legally or illegally – logged, selectively or by clearcut, or opened by squatters through slash and burn. Other territories will survive in a nearly natural state and used sustainably, for example by the gathering of minor forest products. Locally, agricultural technics of the Indians may expand, that are based on the planting of mixed cultures of useful herbs, shrubs, and trees.

In the absence of developed national parks, traveling in the Brazilian Amazon has to be undertaken by organized boat tours or oriented towards the few accommodations that are embedded in natural surroundings. Consequently, these chapters attempt to make suggestions rather than to give geographically precise directions. Unspoiled nature is still everywhere, it is just a matter of idenitifying a means of transport to reach it.

Since most visitors to the Brazilian Amazon will arrive via cities in southern Brazil, a selection of national parks is included which are either situated on the periphery of **Brasilia** and **Rio de Janeiro**, or which are easily reached by organized tours, namely two of South America's undisputed nature highlights, the **Pantanal** and the **Iguassu waterfall**.

# MANAUS

**The Rubber Metropolis: Manaus** developed from a Portuguese fort which was built in the second half of the seventeenth century on the **Rio Negro**, 18 km upstream from the point where it flows into the Amazon. The town reached its peak at the beginning of this century. Latex from the rubber tree (*Hevea brasiliensis*), a natural substance that at first seemed perfectly worthless brought this jungle outpost fame and fortune.

The *Hevea* tree grows to a height of up to 30 m, and latex, which comprises about 30% rubber, serves to protect it from herbivorous enemies and quickly closes up gashes. Untreated rubber is hard when cold, but becomes soft and sticky when warm because temperature changes cause the very long, unjoined molecules to change position slightly relative to one another. This substance was initially of little importance for the Europeans. In 1839, however, the American Charles Goodyear developed the vulcanization process, which by heating the plastic raw material and adding sulphur transforms it into finished rubber with lasting elastic properties, and this opened up a host of uses. The Amazon basin was for a long time the only area in the world producing rubber, and Manaus was able to dictate prices. The rapidly increasing demand triggered a rubber boom of immense proportions. Attempts to cultivate the *Hevea* in the Amazon were frustrated by outbreaks of South American leaf blight caused by mildew (*Microcyclus ulei*) among the closely planted trees. In 1876, the English botanist Henry Wickham took 70,000 rubber trees to Southeast Asia to form the basis for today's huge plantation industry. Yet it was not until 1915 that Southeast Asia was able to produce rubber in significant quantities and at much lower prices than Manaus, and falling prices on the world market signaled the end of the rubber boom. Five years earlier, Manaus had been at its absolute pinnacle, shipping 38,000 tons of rubber in 1910.

The grandeur of the cast iron which adorns individual public buildings and villas is a lasting reminder of the great age of the rubber barons. Schools, hospitals and offices now occupy the great houses originally built by the gentry, preserving at least the facade of these important parts of the city's architectural heritage. The most imposing symbol of the once accumulated wealth is the 700-seater **Teatro Amazonas**. The design of the building was inspired by the Grande Opera in Paris, and virtually all the materials and fittings were shipped out by ocean-going steamer from Europe – tulip-shaped light fittings from the Venetian glass-works on Murano, marble from Carrara, furniture from Paris and Vienna, cast iron from England, and gilded roof tiles from Alsace – and it was European craftsmen who fashioned the stucco and created the magnificent ceiling paintings.

The citizens of what was in those days the wealthiest city in South America sought to make up for their isolation by making life as comfortable as possible: Manaus had electric street lighting before London, and an electric tram system before Brussels or Boston.

**The Port:** The floating quays which extend far into the Rio Negro are masterpieces of English engineering. Huge iron air tanks make it possible for ships to dock all year round, even though the water level varies by some 15 m between dry and rainy season. All the port facilities, including the elegant **Customs Building**, were shipped out from England in individual pieces at the beginning of the century. Some distance below the now closed Customs Building are the jetties of the Amazon steamers, where these river craft, typically built of wood, take on goods and passengers to transport them all over the Amazon basin.

Large enough to have a superstructure, these ships are usually owned by their captains, and ply regular routes which are announced on large boards on the upper deck. Cabins with bunks are few – and so a hammock is a vital piece of luggage. It is hung up crosswise, providing a place to sleep and rest for its owner, and offers makeshift stowage

**Teatro Amazonas, the famous opera house at Manaus.**

space for his meagre luggage, which is placed in it or beneath it.

The harbor is also where small river boats dock. They bring fruit, fish, cassava flour and jute to the city from the regions nearby. The jungle metropolis is linked to the trans-Amazonian network of roads, many are now sealed, and it has a modern international airport, yet traffic on the river continues to grow. With numerous powerful motor around the beaches of the **Ponta Negra** every weekend – testify to the existence of a modern leisure industry.

**The Free Trade Zone:** Manaus is decaying and developing explosively at the same time. The economic decline which set in after the rubber boom and the population growth and industrial development which started in the 60s have left this city and its one million inhabitants no time to stop and think. Since 1967, when Manaus was declared a free trade zone, foreign firms have not had to pay any import duty, and so goods can be sold in the Amazon metropolis at far better prices than in other parts of Brazil. The once resplendent business quarter near the harbor is degenerating into a bazaar for electrical goods. Manaus is a beacon of civilization, but hardly on its way to achieving modest prosperity. Despite an intensive housing program, a significant proportion of those who have migrated here live in a jumble of wooden huts, with no streets, no drinking water, and no drainage or sewage. The garbage heaps of the slums and markets and the municipal tips with their rotting leftovers attract black vultures (known locally as *urubu*), the most striking members of the city's bird population, in search of food.

**Tourism:** The city's sights take little time: the restored Teatro Amazonas, the harbor quarter, the **Tropical Hotel** with a small zoo of indigenous animals in its own park. For most tourists, however, the real beauty of Manaus begins only as they leave the city, for the most fascinating areas are to be found on trips on the **Rio Negros** and the **Solimões**.

At weekends, the bathing places along the Rio Negro and its tributaries are

**Fruit for sale at the harbor market.**

favorite destinations for the locals. The warm (around 27° C), clear, "black" water (which is actually the color of tea) and the absence of aquatic plants even close to the shore make for delightful swimming. During the dry season, which is called *verâo* (summer) which lasts from May to October, bathers seek out well-known spots such as the beaches at Ponta Negra or the **Cachoeira do Taruma-Grande**, a waterfall on the **Taruma**. Here there are numerous food stalls selling roast fish and other specialities, soft drinks and beer. By Sunday evening, crates of empty beer bottles can be seen stacked around the stalls, clear evidence of the bathers' thirst.

Eighteen kilometers below Manaus, the black waters of the Rio Negros and the beige-colo red Solimões join to form the Amazon. Because of their different temperatures and rates of flow, the two at first remain separate rivers, flowing side by side in the same channel for several kilometers. A number of tour agencies offer half-day trips to see this natural spectacle. In good weather, this *encontra das aguas* and the extensive river landscape around Manaus can also be seen from flights coming in to land at the international airport.

Decades of over-fishing in the waters close to the city have drastically reduced the population of larger fish species. Nevertheless, there is still a very simple way to get an idea of the wealth of fish to be found in the Amazon: all one has to do is visit one of the fish markets (the cast iron market right next to the harbor is a particularly good one), where a great number of species of food fish are on sale.

No visitor to the city should miss the chance to try the *peixada* fish soup (be sure to ask for it *sem espinhas*, i.e. without bones). This is prepared from a favorite Amazonian fish, the tambaqui (*Colossoma bidens*) and the tucumare (*Cichla ocellaris*). The excellent taste of *trambaqui na brasa*, filets of tambaqui cooked on a charcoal grill, is the high point of many barbecues.

**The Institute for Amazon Research (INPA):** Manaus has from its earliest days drawn

Less-than-spacious riverboat interior.

naturalists to recuperate from their exhausting trips into the interior. The establishment of the Institute for Amazon Research, or INPA (*Instituto Nacional de Pesquisas da Amazonia*), in 1952, gave a decisive boost to efforts in Manaus to conduct modern scientific research into the Amazon region. Successful teams swiftly developed in biology and ecology, tropical medicine, agriculture and forestry. In 1962, a link was formed with a tropical ecology team from the Institute of Limnology of the Max-Planck-Gesellschaft in Germany. At the beginning of the 70s, INPA moved to its current home, a site of about 25 hectares at **Estrada do Aleixo**, four kilometers on the outskirts of the town. Here there are spacious laboratories and offices, a large herbarium with well over 100,000 specimens, accommodation for animals, and a good library. Since 1971 the Institute has been publishing its own scientific journal, the *Acta Amazonica*. Numerous INPA monographs devoted to special topics related to the Amazon region are for sale. The forestry division administers two nature reserves (the *Reserva Florestal* "**Adolpho Ducke**" and "**Walter Egler**") on the Manaus – *Itacoatiara* highway and one (*Reserva Florestal Campina*) on the road from Manaus to Caracarai, although these are not open to the public. Considerable international acclaim has gone to the "Biological Dynamics of Forest Fragments" project, which INPA has run since 1976 in conjunction with the World Wildlife Fund and since 1989, the Smithsonian Institution. Some 70 km north of Manaus, islands of jungle surrounded by grazing land and varying in size from one to 1,000 hectares, serve as a natural laboratory in which inventories are made of plants and animals and on their population sizes, to understand how the structure of the jungle and the combination of species are changing in forest fragments of different sizes. The precise reasons for the decline in the number of species and the structural changes of the forest will be investigated in this project planned to last up to 1999.

**Looking across the Rio Negro.**

# THE FISH MARKET

Manaus' fish market is in the middle of perhaps the busiest place in the Amazon. Between the waterfront and the shopping centers of the city's free trade zone is a commotion of stevedores, street traders, shoppers and traffic. Men leave the boats moored to the jetties at a run, carrying bunches of bananas or *pupunha* fruit, sacks of manioc meal or crates of fish. Others hawk on the pavements. In the foul black water around the moorings, vultures fight over offal, and the cooks in the foodstalls wash their plates.

The market is a monument in iron and stained glass – Parisian tastes of the rubber barons, whose extravagance earned Manaus its international infamy. Its the best place to see some of the extraordinary creatures which contribute to the Amazon's fish fauna.

Throughout the year, despite the closed seasons which should restrict the capture of some species, fish such as tambaqui, pirarucu, arowana, tucunare and a great diversity of catfish can be bought, especially in the early morning. Here some of the strangest armored catfish can be found, also the largest specimens of pirarucu: said to be the biggest of all the world's freshwater fish. But while the species on display cannot fail to excite wonder, they also bear testimony to the destructive changes taking place in the Amazon's fishing industry.

Although many of the traditional *caboclo* fishermen (the peasants descended from the Indians inhabiting the floodplains) respect the closed season for certain species, the industrial fishing boats now monopolizing the markets openly flout it. From the beginning of December to the end of February no fresh pirarucu, tambaqui or pau – which looks like a large silver coin – and others, should be traded. But the high profits the biggest boats make and disputes between government agencies monitoring the industry mean that most of the forbidden species can be found on open display in the market.

Means of trapping the fish are also changing rapidly – most *caboclos* still use their skill and local knowledge to trap fish individually. Tambaqui fishermen tie their canoes to fruiting trees and attract their prey by hitting the surface of the water with a weight hanging from a pole, to reproduce the splash of a falling fruit. A traditional hunter of the air-breathing pirarucu, having discovered the fish's lie, waits in his canoe to harpoon his quarry when it surfaces. But the large commercial boats now operating on the Amazon and its tributaries, most which supply the export market, lay down long monofilament nets in the traditional fishing grounds of the *caboclos*. These are responsible for an unsustainable level of harvesting, and could lead to the collapse of some of the Basin's riverine ecosystems. This overexploitation is exacerbated by the tendency of the larger boats to dump their first catch, dead into the water, if they encounter a second shoal of a more lucrative species. On the larger rivers of Mato Grosso, the great piranha shoals are being caught, either to be dried and sold as souvenirs (Japan has recently imported about one million dried piranhas) or still more wastefully, to be converted into fishmeal, for use as animal feed or fertilizer.

The effects of such overexploitation coupled with the destruction in areas of the flooded forest which the fish rely on for their food, are already visible in the markets. The fish on sale are said to be getting smaller. As certain species bear the brunt of the exploitation, in time they are likely to be extinct.

For the people of Manaus, as in other Amazon cities, a dearth of fish would be disastrous. It remains the cheapest form of available animal protein in the Basin, and the means whereby many of the poorer families survive.

**Manaus' fish market is a fascinating place to visit.**

# FLOATING MEADOWS

A widespread phenomenon characteristic of the flood areas of white water rivers, is the "floating meadow". As the water level rises in still or slow-flowing reaches of the river, floating plants can assemble in an almost explosive manner to form patches extending over several square kilometers. Anyone who makes a boat trip during the high water season will see dense growths of these aquatic plants lining the river banks in the stiller reaches of the *várzea*.

The quickest way to reach this floating vegetation from Manaus is to travel along the channel which leads to the **Lago Janauari**. Even more extensive floating carpets of aquatic plants, some of which last all year round, can be found in the quiet waters around **Careiro**, a large island in the Amazon. Here the visitor will still frequently come across expanses of the Amazon water-lily (*Victoria amazonica*). The black waters of the **Rio Negro** (which are crossed to reach these areas from Manaus), are too poor in sustenance to support aquatic plants. But the channel to the Lago Janauari is reached by the nutritional substances of the white water of the Amazon during the high water season.

These floating meadows as the most important element in the production of primary organic material in the low-lying areas of the Amazon basin. Constantly exposed to direct sunlight, leading water temperatures up to 35°C, they afford an environment for developing a massive aquatic ecosystem. The term "floating meadow" derives from the dominant plant forms, the grasses *Paspalum* and *Echinochloa* (which grow up to two meters in height) and in between here are extensive carpets of water-hyacinths, and on the surface of the water – a fine layer of aquatic ferns. It is an astonishing sight: plants like the water lettuce (*Pistia stratiotes*) or the water fern of the genus *Salvinia*, which need the most careful nurturing in home aquaria, cover the surface in a rich carpet.

**Water-lilies decorate this Brazilian lake.**

*Eichhornia crassipes*, with its cluster of hyacinth-like violet flowers, is probably the best known plants which cover the surface of the floating meadows. Its leaf-stalk is characterized by blister-like swellings near the base ("*crassipes*" actually means "thick-footed"), and is filled with a loose tissue of coarse cells which gives it considerable buoyancy. Its roots comprise a loose web of delicate pinnules suspended in the water. This attractive plant and the water lettuce, distributed throughout the tropics to cultivators of ornamental pools from where they spreaded, become serious weeds in lakes, rivers and paddy fields.

In contrast to the deceptive tranquility of the daytime, the floating meadows at night are transformed into a hive of vivid, bustling activity. The beam of a torch pierces the floating carpet of vegetation to reveal a glistening patchwork of dewdrops and pick out the tiny green crystalline eyes of numerous spiders. Spectacled Caimans (*Caima crocodilus*) can be seen as they glide silently through the water. Mosquitoes rise in swarms to attack the visitor. This nocturnal microcosm is dominated by the tumult of colorful tree frogs (*Hyla* and *Sphaenoryhnchus*) which find plenty of food here and seek out this ideal watery biotope for mating and breeding. Grasshoppers cling to the edges of leaves to feed on the abundant vegetation. Along the water's edge flicker the greenish-yellow lights of the glowworm (*Lampyridae*) larvae of the genus *Aspidosoma*, attracting the young operculate snails (*Ampullaria*) which hatch at the height of the flood season.

The water beneath the mass of vegetation provides a safe refuge and a hunting ground for many young fish. The loose roots of the floating meadows teem with insect larvae, planktonic crabs, and innumerable freshwater shrimps. The vegetation of the floating meadows is a source of food for the Amazonian Manatee, the only herbivorous mammal which lives exclusively in freshwater. Unfortunately, these retiring creatures are now threatened with extinction as a result of over-hunting and the influence of increasing traffic on Amazonian waterways.

The Amazonian Manatee is in dire need of protection.

# FOREST LODGES
# AROUND MANAUS

The nature attractions of Manaus are even more limited than the city's cultural highlights: a visit of one of the few very small city parks reveals the presence of some common bird like Kiskadee Flycatchers and Silver-beaked Tanagers, and Short-tailed Swifts hurtle around the neighboring buildings. On a walk down to the harbor one can watch Black Vultures fight for their share of the city's economy. Both species of river dolphins occasionally surface within view of the jetties, and Neotropical Cormorants and Large-billed Terns pass by.

If one has a single day to spend in Manaus, the best way of passing the time may be to visit the **Hotel Tropical**, which is situated on the banks of the Rio Negro about 15 km outside the city. The hotel has nice gardens, a little zoo, and good forest within walking distance.

However, for a stay of several days, the best decision is to book at a local travel agency or stay in one of the forest lodges. There are now nearly ten of them in a 100 km circle around Manaus, and their number is increasing. Most travel agents sell only one particular place, and it is therefore advisable to compare different offers. Illustrated brochures help to make a decision. None of these places is cheap, and one should expect prices between US$50 and 150 per night for one person, and a similar fee for transport by car or boat.

At the top of the price list is the **Pousada dos Guanavenas**, five hours from Manaus, even offers airconditioned rooms and a swimming pool. Cheaper and closer are the Amazon and the **Janauaca Lodge**. On the rustic side, *Selvatur*, the biggest local tour operator, promotes an overnight stay in *tapiris*, thatched roofed wooden bungalows on the banks or floating on one of the Amazon's waterways. The night is spent in a hammock, Amazon-style.

The latest and particular attractive addition to these overnight outings is

**The Solimões (left) and Rio Negro (right) meet to form the Amazon River.**

the jungle tower, **Ariau**. Ariau, promoted by *Rio Amazonas Turismo*, is situated on the banks of a small tributary, 35 km west from Manaus on the opposite side of Rio Negro, and connected with Manaus by daily boat service.

Ariau is built with minimal disturbance of the environment into the flooded forest, and stands on wooden pillars in an area that is actually submerged for several months a year. The lodge consists of three wooden towers that were carefully fit between the treetops. The towers are connected by a phantasy inspiring network of stairs and wooden paths on various levels above the ground. The balconies of 16 comfortable rooms open directly into the canopy, and one has even the choice of a Tarzan's house on top of a 28 m high treetop. A 35 m high observation tower protrudes the treetops and permits observation and photography of the wildlife of the canopy.

Restaurant and bar are screened off from the outside as protection from mosquitoes and from monkeys that de-mand a share of the dinner.

One should be prepared for close encounters with wild animals that have become tame after being fed regularly: a coatimundi may gnaw at an unattended camera, and a Woolly Monkey may wish to welcome a visitor by slinging his prehensile tail around his neck. Squirrel Monkeys, Red Howlers, Brown Capuchins and Black Spiders Monkeys abound in the treetops.

Birdwatching from the towers or in the immediate environment is excellent. Forest birds include Long-billed Woodcreepers, Black Nunbirds, and Cocoa Thrushes. Lesser Yellow-headed Vultures glide over the forest and have replaced the Black Vultures of Manaus. Black-colored Hawks and Yellow-headed Caracaras are some other frequently observed large raptors. On a stroll along the river one may flush Wattled Jacanas, White-necked Herons or Great Egrets, or study some of the region's five kingfisher species, like the crow-sized Ringed or the sparrow-sized Pigmy Kingfisher.

**Wild coati finds a helping hand at Ariau Lodge.**

# RIVERBOAT EXPEDITIONS: THE ANAVILHANAS ARCHIPELAGO

Without question, the most versatile and comfortable way to explore the Amazon is by a passenger boat, a privately rented boat, of course, not the ships that are operated for public transport. It is home and observation platform at the same time, it permits the transport of heavy equipment and of a large reserve of food and drinks, and the itinerary is established following the preferences of the passengers – be it birdwatching, collecting of plants, fishing, or simply swimming. Stops are made at idyllic achorage rather than close to a town, and the accompanying canoe permits the exploration of shallow tributaries or floating meadows.

From a beginning with some few small boats in the 1970s, a fleet of several dozen ships is now established in Manaus. Many of these boats are operated by *Selvatur*, the leading local agency, but there is a growing number of smaller enterprises like *Amazon Expeditions* and *Amazon Nut Safaris*. The owners, like Moacir Fortes of *Amazon Expeditions*, are often the captain and nature guide at the same time, and may excel in both skills.

Traditionally, river expeditions were done with 6 to 8 m long canoes with outboard motor, and the night was spent in a hammock or tent on the beach. Canoe trips are still very much alive, but the trend goes to larger riverboats. These are 15 to 25 m long and built locally from wood in the traditional style of riverboats. They have cabins for five to sixteen passengers, and additional space for hammocks. The cabins are small, with berths and showers, and are kept reasonably clean. A fan is sufficient to find sleep during the surprisingly cool nights. In addition to the captain and his helpers, the tour is accompanied by a cook, who may turn the catch of the day,

**High tide during the flood season.**

Black Piranha, Rainbow Bass, or Peacock Bass, into an excellent dinner. Soft drinks and beer come refrigerated. The boat may even have a small library with scientific books on the Amazon.

Boat rentals, which include transport, accommodation and food for all passengers, range from US$400–1,200 per day, depending on the size of the group. These prices sound more expensive than they really are: if the group size is right, US$1,000 per person may be the fare for a two week trip covering a 1,500 km distance.

These trips can be taken at any time of the year, but boats tend to be booked out long in advance for the high season, which lasts from March to August: low precipitation and good weather coincide with high water, and even small waterways become navigable. The forest is flooded all the way up to the canopy, and monkeys, parrots and toucans can be observed at eye level from the upper deck.

One's imagination is the only limit to the objectives of any particular tour, if there is consensus among the passengers. You may wish to go up the **Solimões** to find the rare Red Uakari on river islands around **Tefé**, and to stop on the way on the **Janauaca Lake**, an area rich in Giant Amazon Water-lilies, to search for birds like hoatzins or Horned Screamers. Alternatively, one can travel several hundred kilometers up the **Rio Negro** and turn into its tributaries.

A small waterway from the south, the **Rio Caures** is a good area to search for the Brown Uakari. The **Rio Branco**, a large tributary from the north, is also navigable, and a 250 km trip, first by boat, then by canoe, leads to the **Rio Catrimani**, a particularly unspoiled part of Amazonia, with a chance for observation of large flocks of at least five species of macaws, or of the rare Harpy Eagle.

The preferred destination of a tour of four to eight days is **The Anavilhanas Archipelago** (*Archipelago das Anavilhanas*) in the Rio Negro west of Manaus. The first islands are reached after 60 km, and the archipelago extends another

**Tour boats beach at Praia Grande.**

80 km further upstream. The Rio Negro at its greatest has a width of about 30 km – such a vast expanse of water that the forest on the southwestern bank of the river becomes nearly invisible.

Completely different is the mosaic of small landscapes of the Anavilhanas: The river forks into legions of often only 50 m broad channels, which are separated from one another by spindle-shaped islands, many of them also only 50 m broad, but with a length of several kilometers. The channels are often ten meters deep, and the islands are bordered by steep banks astonishing differences in height, if one considers the extremely slow movement of the river. During the low water season, it takes a little climb to enter the *igapó* forest on the small plateaus of these islands.

The area is gazetted as a biological reserve, which has at least the consequence of keeping squatters away. Small populations of large mammals like jaguar, tapir, capybara, and manatee still persist, but poaching, particularly of the latter species, is not under control.

Anavilhana is a birdwatcher's paradise. From a canoe that slowly follows the river banks, one will come across a great number of large water birds, anhingas, Green Ibises, Great Egret, Great Blue and Boat-billed Heron, just to name a few. There are good chances to find sunbitterns and sungrebes, or a flock of hoatzins that rest in a treetop. Interspersed in the network of narrow channels and spindle-shaped islands are some large, shallow lakes. Astonishingly, the big, black Muscovy Ducks are the only waterfowl which is abundant in these areas.

For the observation of parrots, the only right thing to do is to get up with the sunrise, when large flocks of Tui Parakeets, Orange-winged and Festive Parrots move from their nightly roosts to their feeding grounds. There are also good numbers of Blue-and-Yellow Macaws, and with some luck one of the other macaw species may be spotted. The early morning is also the best time to spot toucans or an Amazonian Umbrellabird, which may quickly hur-

tle across the channels. Large raptors like Yellow-headed and Red-throated Caracara follow rather than pass the channels and forage in the open land at the interface between forest and river.

In the quietness of the heat of the day one may well take a nap in a hammock, or use one of the occasionally interspersed white sandy beaches to go swimming. However, inside the forest birdlife does not become as quiet as in the open. A dive through the brush at the margins into the interior of the forest leads into a generally quite open vegetation and into pleasantly cooler temperatures. It is easy to walk about, and to become overwhelmed by the area's richness in small forest bird species. It is not difficult to identify some colorful species like Yellow-tufted and Cream-colored Woodpecker, or the Green Honeycreeper. But the seemingly dozens of species of fairly uniform colored woodcreepers and antbirds will certainly test the limits of any die-hard birder.

As darkness falls, bats and birds of the night like Blackish Nightjars and Band-tailed Nighthawks flutter around the boat. A startling observation in the perfectly dark night of the Amazon: Manaus, though nearly 100 km away, can be located as a golden glimmer of the eastern horizon.

A trip to Anavilhanas frequently encludes a detour into the **Rio Cuieras**. On a length of nearly 50 km along its course, the forest consists mostly of the trunks of dead trees. Man was not the culprit: a few years ago high water levels persisted several months beyond the normal end of the high water season. The stress exceeded the adaptation of these plant species to the flooded forest with fatal outcome for most trees. Another detour leads to **Novo Airao**, a small village on the southwestern bank of the Rio Negro, where Amazonian river boats (most likely including the one on which you travel) are built from hardwood trees of the surrounding forest. Further downstream, one may like to take a bath at the beach **Praia Grande**, whose white sand would be the pride of any seaside resort.

The Red Howler Monkey (**left**) and the Squirrel Monkey (**below**) inhabit the forests of the lower Rio Negro.

# PICO DE NEBLINA NATIONAL PARK

**Pico de Neblina** is a national park of superlatives: It contains the highest mountain of Brazil and the largest national park of Brazil. Taken together with a national park across the Venezuelan border, it forms one of the largest tracts of protected land on earth. It has the biggest number of endemisms in the Brazilian Amazon. And it is one of the least accessible areas of the Amazon. To be more accurate: for touristic purposes, its central part is inaccessible, unless one enters the area by helicopter.

The park stretches from the Venezuelan border south to the Rio Negro which forms its southern boundary, roughly between Uaupés and Tapurucuara (about 700 km upstream from Manaus). In Brazil, 22,000 square km are gazetted as park, and the adjoining **Parque Nacional Serrania de Neblina** on the Venezuelan side has an area of 13,600 km.

The biggest part of the Brazilian park is in the lowlands, but the northern sector rises to the **Pico de Neblina** and the **Pico 31 de Marco**, with 3,014 m and 2,992 m repectively, Brazil's two highest mountains, positioned right on the Brazilian/Venezuelan border. Geologically they consist of plutonic rocks like orthoquartzit and gneiss, which are typical for the Guayana shield like the Roraima tepuis, which is about 700 km to the northeast from Pico de Neblina. The mountainous areas also receive with about 3,000 mm annual rainfall, the highest precipitation in Brazil, and no part of the year is particularly dry.

The lowland part holds the animal species that are typical for this habitat throughout the Amazon. Due to the remoteness of the area, good populations of some larger mammals and birds that are rare elsewhere like Brazilian Tapir, jaguar, or macaws survive here. Even the Giant Otter, an endangered species in many parts of the Amazon, still survives here. An attraction for birdwatchers is the Guianan Cock-of-

**Larva of an Owl Butterfly (*Caligo*).**

the-Rock, which is restricted to districts close to the northern border of Brazil.

The real biological interest in the park stems from the large number of endemisms in the montane vegetation, which may exceed that of other tabular mountains of the Guayana shield like Roraima. Botanical expeditions that explored over the last decades the upper ranges of these mountains estimate that more than 50 % of the plants are species that are new to science.

Access to the park is possible by canoe on the **Rio Cauaburi**, a widely meandering river that is the principal drainage of the mountain range to the south into the Rio Negro. A 200 km trip along this river comes within 20 km of the mountains. A closer approach, however, requires a full-fledged expedition.

The park, in time may be accessible by road: if things go according to the planning of the *Perimetral Norte*, Brazil's controversial highway BR 210. This road would cut right through the center of the national park, and connect it with the city of **Boa Vista**, 600 km to the northeast. The *Perimetral Norte* is the northern counterpart to the *Trans-amazon Highway* which would provide access, to undeveloped parts of the Amazon from the Atlantic coast and the west of Brazil. The plan was initiated by Brazil's former military governments – militaric stabilization of Brazil's northern frontier as one of the motives to build this road – a poor reason in a continent whose countries have a long peaceful tradition. The past years have seen educational processes which may provide arguments to stop a further development of this road. The *Transamazonica* is basically a failure – only parts are open to traffic, and generally impassable during the rainy season. The threat to the existence of the Yanomami Indians due to hostilities of goldminers, and indiscriminate forest destruction in the states of Rondônia and Acre show problems that resulted from access to the wilderness if the authorities are not able or willing to enforce law and order. Thus, hoping that Brazil's new democracy will lead to a revision of these projects.

**Lesser Anteater waits for a meal to appear.**

# THE TRAGEDY OF RONDÔNIA

The development of the state of Rondônia, in the west of the Brazilian Amazon, has been described as the world's greatest environmental tragedy. Until the late 1960s the state was effectively pristine, and contained some of the most diverse ecosystems on the planet. Today more than 23% of the virgin lands there have been cleared, and some projections suggest that all of Rondônia's primary forests could be gone by the end of the century.

The destruction began when impoverished peasants left their lands in the center and south of Brazil and followed a new road, the BR364, built by the government in 1967 to provide access to the western Amazon. In the early 1970s the government founded settlements intended to accommodate them. When news of this distribution of land and housing, coupled with rumors of fertile soils, spread among the peasant communities of the south, colonists began to flood towards the new frontier: in one decade the population of Rondônia rose from 110,000 to 500,000.

The government colonization agency established more settlements in response, but these could absorb only a small number of the new arrivals. Others took land for themselves, clearing the forest to secure their rights to ownership. Ranchers from elsewhere in Brazil arrived and began to establish properties by pushing the small farmers out. The government disbursed land to them too, often – as the ranchers applied for adjoining plots with false names – in tracts of tens of thousands of hectares. The ranchers cleared their land to raise its speculative value, earn government incentives and prevent others from taking it.

Amid this rush for land, and a failure on the part of the government to take account of the conditions in which the colonists were being settled, the infertility of Rondônia's soils was ignored. In contrast to the rumors prevailing elsewhere in Brazil, only 9% of the land in the state could be considered suitable for agriculture. As much of this fertile territory fell into the hands of the large landowners – many of whom simply cleared it and left it unfarmed – most of the peasants arriving in Rondônia settled on lands which could not sustain their crops.

Most of the land in Rondônia was owned or used before the settlement began, by both Indians and peasants collecting rubber and other forest products. The government settlement schemes tended to ignore the indigenous inhabitants of the lands they were distributing, and the settlers seeking to establish properties for themselves drove out and in many cases murdered the original owners. The colonists introduced new diseases to the Indian communities, many of which were destroyed by epidemics. Between 1971 and 1974, for example, the population of the Suruí Indians is believed to have declined by half. The Uru Eu Wau Wau tribe, some of whose members have had no peaceful contact with the outside world, may at the moment be suffering a tragedy of a similar scale, as failed colonists flood into their territory to mine tin.

Settlers arriving in Rondônia during the 1980s found that they had to buy land from other colonists. They then had to survive on holdings so far from the roads that they had no access to markets, schools and hospitals; or find work as waged laborers. For them, as for the established settlers, life was, and remains, extremely hard. The infertility of the soil and the abundant pests and weeds of the Amazon mean that crop yields are low and the land is quickly exhausted. Middlemen monopolizing the marketing networks force the settlers to sell their produce at low

**Nature destruction in Rondônia.**

prices, and government trading agencies may take so long to supply the money for the crops they buy that inflation renders it worthless. Agricultural laborers can make more money in Rondônia than small landowners.

The circumstances of both the settlers and the original inhabitants of the state deteriorated markedly in the 1980s, as the Brazilian government, with the help of the World Bank, launched a project designed to accelerate the development of Rondônia. The government pledged to protect the Indians and the environment from the effects of its new program. All of these plans were to end in disaster.

By 1988, 1.5 million people were believed to have arrived in Rondônia. The road networks absorbed over half the program's budget, while only 2.5% was disbursed for protection and aid for the Indians, and conservation received only 1.4%. Indian reserves were invaded, a new road, the BR429, was built through the fragile ecosystems between the main highway and the border with Bolivia, fertile land continued to concentrate in the hands of big ranchers.

Most of the colonists were unable to invest in the perennial crops – like rubber and cacao – the project was supposed to be encouraging them to farm. These crops required capital they did not have, and a guarantee that the land would remain in their possession. As the peasant settlers feared expulsion from their lands, they farmed as if at any moment they would have to leave, planting crops which matured quickly but exhausted the soil, as opposed to trees whose cultivation might have been sustainable, but which would have been of benefit only to the land invaders, rather than the people who planted them. As the colonists failed financially, most of the agricultural land fell into the hands of large ranchers: by 1985, 66% of the land was owned by 1.9% of the farmers.

The World Bank is now considering the funding of a new government project, to rectify the mistakes of the old ones. Already the latest proposal has attracted criticism, as it appears to be imposing a development model conceived by planners in Washington and Brasilia, rather than the people who are to be affected.

Despite all the bad planning, corruption and insensitivity which have accompanied development in Rondônia, some of the settlers have benefitted from their migration. **Settlers found Rondônia was not the land of milk and honey.** For some the acquisition of land, however infertile, was an improvement on the landless and laborless circumstances they suffered in the south of the country. Some peasants have succeeded, with the help of the unions they have formed, in defending their properties from the ranchers, and in planting crops which are economically viable and do not destroy the lands they farm. In the oldest of the government-sponsored settlements, at Ouro Preto d'Oueste, the peasants are experimenting with the production of honey, vegetables and treecrops, with some success.

Perhaps the greatest current threat associated with the Rondônian frontier is that of timber cutting, for it extends exploitation into areas which would otherwise have remained untouched. The sawmills now lining the BR364 and many of its feeder roads, having exhausted the high value timbers elsewhere in the state, are invading the reserves set aside for Indians or for wildlife.

As peasants are forced by poverty or ranchers to leave the Rondônian frontier, they move northwards, in search of new lands in the heart of the Amazon. The BR364 has been extended deep into the state of Acre, and many of Rondônia's failed colonists, as well as its expanding landlords, have followed it, coming into conflict with Indians and rubber tappers as they try to take possession of their lands. Rondônia has become the gateway to the western Amazon, and the tragedies taking place there are likely to be repeated elsewhere.

# TAPAJÓS
# NATIONAL PARK

**Tapajós National Park** is one of the few areas in Brazilian Amazonia with formal legal protection. Although no facilities exist in the park nor is there regular access to it, the scenic beauty of the park and the proximity to the city of **Santarém** make it likely that this park may become a tourist attraction in years to come.

Santarém is, with about 200,000 inhabitants, the biggest settlement between Manaus and Belém, and connected with these two cities by daily flights. The city is located at the southern bank of the Amazon, east of the mouth of the **Rio Tapajós**. There are several good hotels in Santerém, and several local travel agents offer boat tours on the Amazon and the Tapajós, from a few hours to four days duration. However, no regular tours to the park are offered yet.

Tapajós National Park, also referred to as *Parque Nacional da Amazonia*, is reached about 300 km from Santarém. One can either travel on the road that detours from the *Transamazonian Highway* to Santarém, and then continue on this highway towards **Itaituba**, or by boat, which reaches the park a few kilometers after the first waterfall, **Cachoeira Maranhao Grande**.

The park has a total size of about 10,000 square km, and most of a it, a strictly reserved zone, extends from the river to the northwest. Tapajós holds a full complement of large vertebrate Amazonian fauna, including endangered species like the Giant Otter, and provides an excellent opportunity for the observation of mammals and birds.

The mouth of the Rio Tapajós, about 700 km from the Atlantic, is one of the furthest points on the Amazon that is still reached by the oceanic tides, which cannot run further up the Amazon beyond the narrows close to Obidos or up the Rio Tapajós because of its cataracts. The lower part of the Rio Tapajós, however, is a particular broad and slow-running part of the river system. This expanse of water, up to 30 km wide, is believed to be an inundated valley, which had been created during the Ice Age. Due to the enormous amount of water that was bound in form of ice in polar regions, the sea level was then up to 180 m lower than today. And the resulting swift current of the lower Amazon and Rio Tapajós eroded deep valleys, which were below sea level as the sea rose with the melting of the ice.

Most rivers of the Amazon basin are either classified as black water or white water rivers. Black water rivers drain massive lowlands, thereby extracting the brown colored humic acids from the forest soil. White water rivers originate from the run-off of the Andes and owe their color and name to the suspended soil particles. The Rio Tapajós, like the Rio Xingu, is put into a third category of clear rivers. These rivers drain the extensive highlands of the Brazilian shield, whose ancient rocks resist quick erosion. Lack of suspended particles or humic acids makes the Tapajós' water beautifully transparent.

**Left**, wood decayed fungi. **Right**, Yellow-rumped Cacique perched above a colony of nests.

# THE THREATENED TURTLES OF RIO TROMBETAS

During the past three centuries, the Giant River Turtle (*Podocnemis expansa*), whose shell can be as long as 90 cm, has been intensely persecuted for its meat and eggs; and as a result, its populations have suffered a catastrophic decline. Historical records show that during the 19th century, in the Amazon basin alone, millions of turtles nested annually, and as many as 48 million eggs laid by these turtles were harvested by humans each year. In the mid-1970s a survey of 14 Amazonian rivers in Brazil conducted by the Brazilian Institute of Forest Development (IBDF) showed that there remained fewer than 15,000 females nesting annually in Brazil.

The most important nesting area identified by the IBDF survey was at sand banks (or "*tabuleiros*") located on the **Rio Trombetas**. There, during the 1970s, an average of 5,500 females reportedly nested each year. These sand banks are inundated most of the year, but during the dry season, in the months of September through January they are exposed. When the dry season commences, the reproductive adults leave their feeding areas in the lakes and flooded forests, where they feed largely on fruits, and travel to the *tabuleiros* where nesting will occur. They remain offshore for about a month prior to nesting. Nesting on the Trombetas takes place during September, October or November and often lasts about two weeks. The turtles usually nest at night in groups of tens or hundreds of individuals which all emerge onto the beach at approximately the same point and lay their eggs within a small area of beach. Often they choose a site where the elevation is highest. This will protect the eggs from flooding should the river rise before the eggs have finished incubating. Females lay about 90 eggs per clutch. The eggs take about 44–55 days to hatch.

Nesting *Podocnemis expansa* are ex-

**Turtle trio at Rio Trombetas.**

tremely shy and are easily disturbed by light, noise, human activity, and by pollution of the water. For this reason, visitors are not allowed on the beach when the turtles are nesting, or even in the vicinity during the breeding season. Boat traffic on the river can discourage the turtles from laying eggs. During 1984, little or no nesting occurred along the Trombetas, possibly because of heavy boat traffic on the river during the rapid growth of a town 50 km upstream.

Although the species is protected by law, the meat fetches a high price on the black market – each can be sold for more than US$100. The turtles are illegally captured all year round but most vulnerable when they are nesting. Incidents that occurred during the 1985 season dramatically illustrate the lengths people will go to harvest the animals. Shortly after nesting began in November 1985, a group of 30–50 armed men from the impoverished local *caboclo* population drove the IBDF personnel away and took control of the *tabuleiro*. They held the *tabuleiro* for two days,

and captured about 100 to 300 animals. The *tabuleiro* was finally liberated by a contingent of Federal Police armed with submachine guns. The turtles nested in relative peace after that. Although hundreds of egg clutches had been laid by turtles during the course of the season, only one egg clutch eventually hatched out. The remainder had been clandestinely excavated by poachers.

The level of protection afforded the turtles improved greatly in subsequent years, nevertheless the nesting population has continued to decline. During 1989, approximately 850 females nested. Although the poaching problem on the nesting beach has been brought under control, the remaining turtles are seriously threatened by habitat disruption. Construction of a hydroelectric dam has been proposed upriver of the nesting site. Should the dam be built, the turtle population would be threatened by an increase in boat traffic, possible erosion of the *tabuleiros* induced by unnatural oscillations in water levels; and water pollution.

The turtles come ashore to lay their eggs.

# THE JARI PROJECT

The Jari Project, one of the largest development projects in Amazonia, is located on the Rio Jari. The Project, second in scope only to the Carajás mining project, lies 500 km west northwest of Belém and extends from the main Amazon River 120 km up the Rio Jari. The Project area covers a plateau extending 160 km east to west.

The Project was the dream of the multimillionaire Daniel K. Ludwig, who in the early 1960s, forecast a shortage of paper in the future. As a result, Ludwig bought a tract of land of 15,000 square km for three million dollars. He planned to plant fast-growing timber on the upland and rice in the floodplains.

**Towns and Populations:** Approximately 30,000 employees were required to run a project the size of Jari. Most were housed in **Monte Dourado**, a completely new city on the banks of the Rio Jari located some 22 km from Munguba, the site of the industrial complex. Monte Dourado has all modern facilities – 55 MW power plant, schools, a hospital and clinics, roads, police, supermarkets, social clubs, sports teams, etc.

**Goals:** The Jari Project was based on the use of gmelina (*Gmelina arborea*), a fast-growing Asiatic timber that produces up to 160 cubic meters of wood pulp per hectare on a seven year rotation. Eventually, the Caribbean pine (*Pinus caribea*) was also introduced and recently *Eucalyptus deglupta* was added as the third major timber tree. Experiments with other species are underway. The Project also planned to produce rice in the floodplains to feed the growing population.

**Industry:** Jari is the best known for its pulp mill which was floated over from Japan. The pulp mill were designed in Finland and constructed in Japan at a cost of $270 million. It was built on barges and towed to Amazonia over the Indian and Atlantic Oceans and placed on pilings at their present sites using the flood season and additional dams to float them to their place. The power plant is fueled by wood and black liquor, a by-product of pulp production.

**Forests:** At present, 1,100 square km of forest are planted, mainly with gmelina. The plantations are being increased at the rate of about 50 square km per year until the goal of is reached. This is well within Brazil's regulation stipulating that development projects leave half of the forest intact.

**Floating bar offers liquid sustenance to tired Amazon travelers.**

**Livestock:** Jari is also a major producer of livestock and is now concentrating on the uses of the water buffalo in the lowland floodplain area. These animals are ideally suited to the region and plans call for building up the present herd of 6,000 head to 35,000. Although the present food is flown in from Belém, plans for producing food at Jari are underway, such as rice, that is produced in the floodplain area of the main Amazon River. Rice fields were made by clearing areas of the flooded forest and savanna, followed by leveling and diking.

**Kaolin:** After the Project was well underway, an enormous deposit of kaolin was discovered just across the Rio Jari from Munguba. Kaolin is a fine white clay used for paper surfaces and in porcelain, paint and other products. As it is processed, kaolin is mixed in an open pit and pumped as a slurry through a pipeline that goes under the river to the Munguba processing plant. At present, 200,000 tons are shipped to Europe each year.

**The Current Status of Jari:** Unfortunately, the trees at Jari grew much slower than predicted and the Project ran into many political difficulties in Brazil. The Brazilian authorities refused to support the infrastructure of the Jari towns, and in December 1981,

Ludwig decided to sell out his interest in the wood and pulp project.

Therefore, Ludwig's dream came to a bitter end and caused him considerable financial loss, casting doubt on the viability of such large silvicultural plantations in the Amazon region.

As part of a cost-cutting program initiated before Ludwig sold out, the number of Jari employees was reduced from 7,000 to 3,000. At the time of its sale in early 1982, however, Jari was posting losses at the rate of $100 million per year. These losses were caused by costs overruns, excess turnover of management and disputes with the government.

After Ludwig, the Project was taken over by a consortium of 27 Brazilian companies led by the mining entrepreneur Augusto Trajaro de Azevedo Antunes, who put up $40 million of the $100 million cost. Ludwig invested over one billion in Jari. It was sold for $280 million ($180 million in debt assumed by the consortium and $100 million payable over three years). The debt on the $200 million loan for the pulp mill was assumed by the Brazilian government through the semi-official Banco do Brasil. The government also agreed to provide transportation and communication facilities to serve the Project.

**Large-scale logging operations.**

# BELÉM

**Belém** is the biggest city of the Amazon delta, and, second only to Manaus, the most important access to Brazilian Amazonia. It has daily connections by plane with Rio de Janeiro and Brasilia. The only road of the eastern and central Amazon that is reliably open to traffic, permits regular bus service from Brazil's south. For nature tourism, Belém serves as a gateway to **Marajó Island**. Belém has not yet become like Manaus which is a regular point of departure of expeditons by passenger boat, but such a development is likely in the future.

Belém was founded by the Portuguese as early as 1616 to initiate the colonization of an Amazonian empire. With its importance as the leading port of the region, it has grown to today's population of about one million inhabitants. The city is built on a promontory, 120 km from the Atlantic Ocean, between the **Baia do Guajara** to the northwest and the **Rio Guamá** to the south. This river comes from the east and flows around Belém into this bay. This bay with a breadth of 40 km, which separates Belém from Marajó island, is still part of the gigantic Amazon delta and formed by the confluence of the **Rio Tocantins** and the **Rio Pará**, a southern arm of the Amazon.

Like Manaus, Belém cannot be characterized as a particularly clean and attractive city, but a combination of several city parks, the zoobotanical garden of the **Goeldi museum**, and sidewalks and markets along the bay invite to stay in this city for two or three days.

Most of the hotels are downtown, more of less in walking distance from bay, port and fish market. The **Ver-o-Peso market**, is an interesting starting point for a walk through the town. The market hall, that dates back to 1688, when it served as a custom checkpoint, and surrounding stalls serve as an excellent introduction to the fish fauna of the delta of the Amazon. After cleaning of the fish, the innards are thrown on the

**Belém port, near the Ver-o-Peso market.**

rivers' mudflats, a haven for Black Vultures. It may be a bit strange to find beauty in this scene, but it is somehow reassuring that so many large birds can get by close to human habitation. Belém is actually only their feeding territory. Black Vultures breed on the ground of closed forests, not in cities. Their population density as a breeding bird is considered to be quite low, and consequently, many of the thousands of vultures that feed around Belém are likely to breed several kilometers away. Only the Black Vultures come into town to forage; the Lesser and Greater Yellow-headed Vultures and King Vultures, are only occasionally seen circling high in the sky.

A few species of birds, that naturally occur on forest margins, have adjusted to life in the city. Palm Tanagers sing from the roof of the Ver-o-Peso market, and on a stroll through some nearby parks like the **Praça D. Pedro II**, one sees Pale-breasted Thrushes, Blue-grey and Silver-beaked Tanagers. The latter ones may compete with various Kis-kadee Flycatchers for being the birds that have gained most from Man's transformation of Amazonian landscape, since these species are seen virtually everywhere in the region in gardens, parks, scrub, and secondary forests.

Continuing to the southwest, the stroll through town leads to the **Catedral da Sé** and the nearby **Forte do Castelo**, the first building that was erected after the founding of Belém. The park between cathedral and fortress, the **Praça F.R.C. Brandão**, is a particularly quiet pocket of town, a place to relax. Many of the impressive trees of this and other parks are mangos, a species that actually originated in tropical Asia. (Beware of falling objects!) The children of Belém throw sticks and stones into the trees to harvest the ripe fruits.

A safer place to relax is a restaurant in the fortress: while enjoying some excellent lunch or dinner of fried fish, one can watch the live river dolphins surface in front of the restaurant's veranda. One should not mistake the view over the huge expanse of water from this veranda to a

**Park facing the Cathedral da Sé.**

remote mangrove as a view of Marajó Island across the Amazon. Marajó is to the north, beyond the horizon, and this is only the relatively small mouth of the Rio Guamá into the Guajara bay. The adventurer, who wants to explore this region by public transport will find the river boats jetties along the road **Siqueira Mendes**, right behind the fortress.

Among naturalists, Belém is famous for the **Museu Paraense Emilio Goeldi**, referred to as "**Goeldi museum**". The museum is a center of ecological research in the Amazon. It was founded under the leadership of the naturalist Ferreira Penna in 1866 as the Pará State Museum of Natural History and Ethnology. It owes its present name to the Swiss zoologist Emil August Goeldi, who was around the turn of the century particularly influential in promoting public programs of the museum. The museum has research departments that include anthropology, archeology, botany, zoology, geology and geography. Today, most of the research facilities are located remote from the traditional site of

the museum, at the periphery of Belém, associated with the university.

Most visitors to Belém will not be interested in these research facilities but rather in the museum's traditional site, the zoobotanical garden, which is close to the center of town, about three kilometers to the east of the Ver-o-Peso market. The park, a five hectare block of the city, holds about 1,000 plant species, from gigantic emergent forest trees to tiny orchids and the water-lillies (*Victoria amazonica* sp.).

Mammals like sloths and agoutis roam freely in the park, which is also home to some forest birds, like the Canary-winged Parrot. The biggest attraction are the captive animals, a unique opportunity for close observation of well-kept animals of the Amazon, that are rarely seen in the wild. An enclosure to the right of the entrance gives a chance to pet a capybara or a Brazilian Tapir. A pool further ahead to the left is home to an Amazonian Manatee. Other mammals on exhibit are jaguars, ocelots, several monkeys, and anteaters. Among the birds are Harpy Eagles, Golden Parrots, macaws, guans and curassows. The compound for the Black Caimans holds a particularly large individual, and a nearby pond a number of Giant Amazon Turtles. Two hundred and fifty species of fish are kept in the aquarium, nearly 10% of those found in the Amazon river system.

An even larger place to explore nature in the middle of Belém is the **Bosque Rodrigues Alves**, 16 hectares of protected forest with about 2,500 Amazonian trees. This park is on the **Avenida Almirante Barroso**, four kilometers further east from the Goeldi museum.

Belém may become a bit tiring after two or three days. The best escape is one of beaches of the city, from **Outeiro**, only 35 km away, to **Mosqueiro**, 86 km from Belém. This latter one, an island, accessible by bridge, has numerous hotels, and is quiet during the week, but crowded during the weekend. Mosqueiro and other beach resorts have a hinterland with scrub and secondary forest that is richer in wildlife than one may suspect at first glance.

**Touring a forest reserve.**

# TOMÉ-ACÚ

Of all the unlikely colonists in the rain forests of Amazonia, it is the Japanese settlers living five hours to the south of Belém who have developed some of the most exciting new farming techniques. In 1929, in order to relieve domestic unemployment and population pressure, Japanese farmers were encouraged to move to the Amazon by their government.

In the new colony of Tomé-Acú, they tried at first to cultivate rice, much as they had done at home. Like the crops cultivated by many other newcomers to Amazonia, this failed, smitten by poor soils and pests. They turned to cocoa production, however were foiled by fungus. Finally, the colony, racked by economic failure and disease, began to depopulate.

But in 1942, when Brazil declared allegiance to the Allies in the Second World War, the many Japanese in the country were rounded up. Tomé-Acú became one of the internment centers. Immediately after the war, the replenished community began to plant pepper, which until then had never been grown in the Amazon. It was immediately successful, and by 1960 the Japanese colonists had not only replaced Brazil's pepper imports, but were responsible for five percent of world trade.

In 1961 they were hit simultaneously by a plant disease and low crop prices. Once more, many farmers were bankrupt and forced to leave. But those who remained recognized the need to diversify their production, to avoid the dangers of relying on just one crop.

They began to experiment with trees and shrubs both native to the Amazon and from elsewhere, and soon found that they could create markets for products which had seldom before been grown commercially in Amazonia. By planting treecrops of different heights on the same patch – mimicking in other words the structure of the rain forest – they discovered that they could not only reduce the area that they needed to clear, but also conserve nutrients and rainwater. This enables each farmer to confine his activities to a single plot, unlike most of the other Amazonian colonists, who are characteristically forced to abandon their land when the soil became exhausted after just a few years.

The result of this diversification is that the Japanese farmers are now selling at least 55 crop products. This means that their economy is more robust than that of the other settlers, as they are not seriously affected by the failure of one crop. The farmers process and market their products together, through a cooperative which ensures that outside middlemen cannot impose low prices.

The strength of the cooperative is such that the farmers cannot be dislodged from their lands by ranchers. While other colonists are forced by the threat of expulsion to grow crops which mature quickly but exhaust the soil, the Japanese can invest in trees which take thirty years to yield.

Like the longer-established residents of Amazonia, but unlike the recent colonists, they are anxious to preserve their resources for the benefit of their heirs. But while some of the agricultural principals they have developed may be of benefit to other Amazonians, the fertilizers, pesticides and intensive labor they employ are likely to render their system too expensive for others to copy faithfully.

Tomé-Acú and the surrounding villages bear testimony to the success of the Japanese farmers; in the banks, the offices and some of the hotels. Japanese is the first language spoken, and in the settlement of Quatro Bocas, within the heart of Amazonia, are two of the best Japanese restaurants outside Japan.

**Proud pepper planter poses for a picture.**

# MARAJÓ ISLAND

The mouth of the Amazon River extends from Cape North over 300 km to **Curuca Island**, about the same distance that separates London and Paris. The river discharges through two large arms, the main channel to the north of **Marajó Island** and the **Rio Pará** to the south. The estuary embraces several large islands and is in a constant state of transformation, causing navigation charts to become outdated before they are even printed. One feature of this great river is that it has no delta of accumulated mud which extends into the sea like the Mississippi or the Nile, but the amount of sediment carried in its waters stains the ocean for hundreds of kilometers and turns sandy beaches into mudflats in French Guiana.

The main city on the north arm of the Amazon mouth is **Macapá**. The historic fort of Macapá on the promontory in front of the city attests to the defensive role of this settlement in securing the Amazon for Brazil. Today Macapá is a modern city, in spite of its lack of road connections. A three-day boat trip or 45 minute plane flight separates Macapá from **Belém**, the principal city on the southern arm of the Amazon.

The Amazon estuary is the site of a tidal bore, locally called the "*pororoca*", which has received more than its fair share of publicity. This wave, which goes upriver at a speed of 10 to 15 km per hour, results from the strong spring tides overcoming the river's current in shallow waterways of no more than four meters in depth. When the bore occurs, which is always at the lowest point of the tides, a roaring sound is heard as far away as five kilometers, followed by the appearance of a wave about one to two meters in height. The phenomenon is most common in January to June on the coast near **Maracá Island** and on the **Rio Araguari**.

Marajó Island, in the Amazon delta, is the largest river island in the world, with over 48,000 square km in total area

**Preceding pages: rounding up the herd at Marajó. Below, Scarlet Ibis and Wood Stork flocks gather over the floodplain.**

(about as large as Switzerland). It is really a complex archipelago. Soils under Marajó are river sediments going down to a depth of over 2,000 m and those on the western portion of the island are recent deposits. Throughout the island, high soil fertility persists. The island's eastern half consists of low-lying natural grasslands that can be underwater for as long as four months each year. The western portion is forested with some of densest and the most handsome vegetation in the whole Amazon. Mainly cattle and the Asian Water Buffalo are found on the large ranches that dominate the region near **Soure**, **Salvaterra**, and **Cachoeira do Arari**. The economic mainstay of the forested part of the island is now timber extraction, rubber extraction no longer providing an adequate income for the tapper and his family.

Ever since colonial times, Marajó, and the associated islands of **Caviana** and **Mexicana**, have been known for Marajorara ceramics, distinctively designed pottery, associated with Indian cultures that disappeared before European contact. The finest collections of these superior ceramic works are in the **Goeldi museum** in Belém, and the **Marajó museum** in the picturesque town of Cachoeira de Arari. Vases, pots, plates, bowls, and burial urns with the same painted figures and geometric designs are still produced, as replicas, on **Marajó** and in **Icoaraci**, near Belém, but the mystery of the pre-colombian inhabitants of Marajó is as yet not completely solved.

Water forms the background for life in Marajó. River transport is all important, and the people are almost born with the canoes and boats that are their cars and buses. A river trip to **Breves** or one of the interior towns of Marajó takes one along narrow tidal canals, often lined with houses. The "**Straight of Breves**," a shortcut from the southern mouth of the Amazon to the main Amazon channel, is so narrow that ships will often brush the trees on the banks. Navigation is tricky here and some blind corners require 120° turns in a few hun-

**The Southern Lapwing is one of the few shorebirds of the Amazon.**

dred meters. Passengers are always greeted by the inhabitants of the Breves channels with call of "*Cunardo*." This harkens back to the first years of this century when English ships under the *Cunard Line* where regular traffic in these waters. Even quite young children take to canoes when a ship passes, to ride the ship's wake like a surfer on an ocean beach and solicit tokens from passengers.

Regular passenger boats run between Belém and Soure and other towns, such as **Ponte de Pedras**, **São Sabastiao da Boa Vista**, **Breves**, and **Gurupá**. Destinations on the northern coast or the interior of Marajó are more difficult to reach.

**Arari Lake**, in the interior of Marajó, is only four to seven meters deep in the rainy season and dries to pond-like dimensions in the dry season. This is one of the principal fishing grounds for the Belém market, but over-fishing has caused a serious decline in the stocks of the Peacock Bass (*Cichla ocellaris*) and the pirarucu (*Arapaima gigas*). A brisk trade in ornamental fish for the export market, almost on the scale of Manaus, begins in the small streams near Soure and Ponte de Pedras. How long this uncontrolled fishing can continue is still unknown. Commercial fishing in **Marajó Bay,** actually the southern mouth of the Amazon River, concentrates on river catfish, such as the Piramutaba (*Brachyplatystoma vaillantii*), and the fishing fleet is both large and well-equipped with trawlers and seines. On the northern coast of Marajó, fishermen still use fish traps, called "corals" locally, that catch fish and Amazon River Manatee stranded at low tide. Cattle ranchers complain that Marajó fishermen are also adept at getting a cow on their line when no one is watching.

Birdlife abounds on Marajó, and several lodges have installed special blinds for close-up viewing of water fowl and towers for tree-top birds. One of the finest lodges is the **Fazenda Bom Jardim**, near Soure, where much of the original faunal research for the island has been conducted. The southern mar-

**The Brazlian Tapir is an adequate swimmer.**

gin of Marajó near Belém has mangrove forests and extensive mudflats covered with the six meter high aroid (*Montrichardia arborescens*) with giant heart-shaped leaves. This plant cover serves as room and board for the hoatzin, a relative of the cuckoos. The hoatzin has attracted attention because its young have claws on the wings which enable them to scale vegetation.

Mangrove forest is an experience in itself. The Amazon estuary has mangrove around **Caviana**, **Mexicana**, **Janauca Islands**, and on the western margins of Marajó. In the southern arm of the Amazon estuary, mangroves penetrate as far as the saline waters does. Isolated mangrove trees, such as *Avicennia* and *Rhizophora* can be found as far upriver as Breves. These mangroves provide the crabs which are so avidly eaten in Belém. Ocean shrimp – thousands of tons caught by trawlers off the Amapá coast – are believed to pass their larval stages in mangroves too.

Forest is the source of income for most Marajó residents. In the past, the region was known for the high quality of its natural rubber. Today, however, the low price paid for wild-tapped rubber has nearly driven the rubber Marajó tapper to extinction. Lumber extraction on Marajó is centered near Breves where dozens of saw mills are located. Only a few tree species, however, are sawn, resulting in short-sighted overexploitation, without reforestation, of the Kapok Tree (*ceiba pentandra*) used in plywood, and *Virola* exported for fine millwork.

Marajó dwellers consume the purple juice of the Acaí Palm fruit (*Euterpe oleracea*) by the liter each day. A symbol of the Amazon River delta, this palm grows in pure stands on the banks of streams and sparsely in upland forest. The trunk is almost too slender for the height of the crown, but, with little prodding, local folks will demonstrate how easy it is to shinny up the palm and pick the berry-like fruits. To aid them in their daily practice in climbing palms, these folks use a vine strap called a "*peconha*" to hold their feet to the tree.

**The Capybara, a pig-sized rodent, is at home in the water.**

# BRASILIA NATIONAL PARK

Evergreen tropical rain forest can only grow in regions with high precipitation, normally more than 2,000 mm, more or less equally distributed throughout the year. If there are several consecutive months without any rain, trees can only cope with evaporation by loosing their leaves. This type of climate is found at the southern rim of Amazonia, approximately 1,000 km south of the Amazon river. With increasing length of the draught, the rain forest gradually blends into a different plant community – tropical deciduous forests. South and southeast of this region the climate becomes even drier, and does not permit the growth of any type of forest. Annual precipitation as low as 1,500 mm, occurring across wide areas of interior Brazil, give rise to a distinctive type of dense grass cover with scattered gnarled trees, a type of savanna that is called *cerrado*. Some areas in the northeast of Brazil are even drier. Here grows a vegetation called *caatinga*, an arid scrub and low woodland with little grass cover.

With the changes in vegetation most of the wildlife changes as well: some mammals and birds, like the Maned Wolf and the ostrich-like rhea, are adjusted and restricted to this kind of open country. Others, like the Giant Anteater, which is very rare in the rain forest, become more common and easier to observe in this open grassland. However, most species of the arboreal monkeys are absent.

The *cerrado* is, in contrast to rain forest, often of high value for agriculture and cattle farming. Consequently, much of the original vegetation in Brazil is today extensively modified by Man's activities.

In contrast to rain forest, *cerrado* is also easily accessible to Man, and consequently hunting pressure is heavy. The beautiful Spix's Macaw from the northeast of Brazil is now extinct in the wild, and some of the most characteris-

**Dry season in the *caatinga*.**

tic species of the *cerrado* and the *caatinga*, particularly edible mammals like the Giant Armadillo, are now quite scarce and localized.

Nearly all flights from São Paulo or Rio de Janeiro to the Amazon stop in Brazil's capital **Brasilia**, and the landing approach gives a good chance to study little disturbed *cerrado* from a bird's eye view. A typical bird of the *cerrado*, the Crested Caracara, is normally seen right next to the runway. A stopover in Brasilia provides a unique chance to visit an outstanding example of protected *cerrado*, **Brasilia National Park**.

The park, created as early as 1961, is within view of the city, namely eight kilometers to the northwest, and has the respectable size of 300 square km. Areas of the park fall into six different categories of protection. Following Tijuca and Iguassu, the park has the third highest number of visitors of all Brazilian national parks. This is mostly so since it caters in particular to the crowds of local visitors who are inter-ested to enjoy their weekends in the zone of intensive use with picnic areas and artificial swimming pools. There is no accommodation, however, and one has to spend the night in the city.

The naturalist interested in the park's "*zonas primitivas*" has to rely mostly on himself, an easy task though in this open landscape. In the sunny and dry climate with temperatures sometimes exceeding 35° C, it will be advisable, though, to bring drinking water on a walk and to wear a hat for protection against the sun. One may also need protection against the cold: in the dry season, temperatures can approach the freezing point.

Only a small population of the Maned Wolf exists, while the Giant Armadillo persists at fairly high densities in the park. The Brazilian Tapir has increased in number since declaration of the park. Fairly frequently encountered are the Tufted-ear Marmoset, the capybara, peccaris, and the Giant Anteater. Attractive large birds include the rhea, the seriema, the Red-winged Tinamou and the Spotted Nothura.

**The Greater Rhea is a South American version of the ostrich.**

# PANTANAL

The **Pantanal** is a result of a unique blend of species and communities from Amazonia, the dry savannas of central Brazil known as *cerrado*, and the *chaco* scrublands from Bolivia and Paraguay. It is described by Carlos Toledo Rizzini, Adelmar Coimbra Filho and Antônio Houaiss in their book *Brazilian Ecosystems* as a region undergoing transformation. It is relatively new and unstable from a geological point of view and is dominated by a complex mixture of plants and communities. The flora is thus a mixture of various elements: Bolivian and Paraguayan xerophytes, savanna species from central Brazil, species from eastern Brazil and Amazonian forests, and hydrophytes which have a wide distribution in the neotropics.

But, what gives the Pantanal its name (*pantano* = swamp) is the fact that when the rainy season comes, between December and March, over 150,000 square km become a flat, flooded lowland with scattered islands, three times the size of Costa Rica – the largest wetlands in the world. The rest of the time the Pantanal is dry, or in different stages of flooding and draining.

The rains start coming down in late November but they do not fall uniformly throughout the Pantanal. Near **Cuiabá**, in the northern Pantanal, the average annual precipitation is 1,388 mm compared to 1,246 mm in the city of **Corumbá**, and even less as one moves further south where precipitation does not exceed that of the dry central plateau of Brazil.

Thus the floods can be attributed primarily to the overflow of the shallow river channels as the water rises. At times, the waters come up so rapidly that herons, egrets, and jabirus have been reported to move in front of the raising waters picking up snakes, small mammals, and other animals as they try to escape the flood. This is reminiscent of what antbirds and other ant-following birds do as the swarm of army ants

moves through the Amazonian lowlands forests flushing small animals and insects from the leaf litter.

At the peak of the flood, the only thing that stays above water are the elongated *cordilheiras* which are just 1.5 to 4 m above the surrounding floodplain. The smaller *cordilheiras* are known as *capões* and look like islands of vegetation. Periodically, however, all but the highest *cordilheiras* get flooded, as was the case in 1988 when the waters rose two meters above the level of previous years. The topography of the Pantanal is so flat that a variation of as much as a few centimeters in water level during the wet season means many hundreds or thousands of hectares being flooded or staying dry that year. Just consider that the Rio Paraguay drops only 30 cm in its 1,300 km journey across the Pantanal.

Between the *cordilheiras* and *capões* there are a number of types of bodies of water that play an important role in the dynamics of the Pantanal. The *bahias* are lakes of varying sizes, from a few hundred meters to several kilometers in diameter, that form between the *cordilheiras*. These *bahias* are connected to the river or to other *bahias* by seasonal streams known as *corixos*. When the *bahias* become isolated from the cyclical flood patterns they become saline due to the evaporation of the water during the dry season and the filling with rainwater in the wet season. These isolated *bahias* are known as *salinas* when they have water, and as *barreiros* when they dry up. These *barreiros* are an important source of salt for the wildlife of the Pantanal and for cattle.

It is in this patchwork of ponds, lakes, streams, rivers, gallery forests, and forested islands that a naturalist finds one of the most exciting ecosystems of the world. The Pantanal teams with life as Amazonia's *cerrado* and *chaco* blend to create a "hybrid" ecosystem, where 600 species of fish, over 650 species of birds, approximately 200 species of mammals, thousands of species of plants, and an untold number of terrestrial and aquatic invertebrate species find their home. Each and every one of these species is inextricably linked to a delicate and all-determining water cycle.

The *piracema* (the spawning season) takes place, for most species, between February and April, as the waters begin to recede, when fish migrate from the deeper river channels to the shallower *bahias*. It is during this period that many aquatic plants and animals breed. The apple snails, for instance, lay their eggs in brittle clusters attached to vertical twigs and stems. As the water rises, the larvae mature and hatch into the water where they complete their development.

As water recedes and the aquatic life becomes concentrated, the terrestrial animals, faced with an abundance of concentrated resources start breeding. The most visible example of this is seen in wading birds, where over ten species of herons and egrets, three species of storks, and six species of ibises and spoonbills, start breeding as soon as the water levels drop sufficiently to make fish and other aquatic prey easier to catch. Apple snails fall easy prey to

Snail (Everglades) Kites. And Wood Storks methodically crisscross the shallow *bahias* with their beaks in the water, snapping them shut and catching unsuspecting fish as they swim between the birds' half-opened bills.

Mammals are more visible in the Pantanal than in any neotropical rain forest. Capybaras along the water's edge are a very common sight. These large rodents spend their days in family groups feeding on aquatic vegetation. In the Venezuelan *llanos*, for instance, they are being raised for meat production. In the Pantanal, this idea has not caught on, and the capybaras are left alone to be used as an important resource for the growing tourism industry.

Marsh Deer abound and are generally also left alone by the *pantaneiros*, the men of the Pantanal, who prefer beef and the meat of the feral pigs to the leaner deer.

Feral pigs, known in the Pantanal as *porco monteiro*, have been roaming the Pantanal long enough to exhibit the signs of natural selection. They have longer

**Swamp Deer of the Pantanal.**

legs and can be easily told apart from peccaries by the diversity of colors, a trait that they have retained from their domestic ancestors. The *pantaneiros* of most parts, actively manage the feral pig populations and have turned them into an important source of meat for the human population. When feral pigs have been spotted, the *pantaneiros* rope and catch young adult males which are castrated, one of their ears notched, and released back into the wild. The *pantaneiros* are then on the look-out for those marked individuals which are eventually recaptured and slaughtered after they had put on enough weight. The use of feral pigs is one of the best examples of low intensity resource management that characterizes traditional life in the Pantanal.

Among the reptiles of the Pantanal, Black Caimans are rarely found nowadays because of the unabated poaching that is, unfortunately, so typical of the Pantanal. Black Caimans have been hunted so intensively that the abandoned poachers' camps are known locally as *cemitérios* – cemeteries – because of the large quantities of animal remains left behind. Fortunately, other types of caiman are still quite common. A walk at night to the water's edge can bring the hair-raising sight of dozens of pairs of little eyes reflecting the flashlight's beam. But, there are few sights that match that of a *sucurí* – anaconda – submerged in the clear water, with up to seven meters of powerful coils moving effortlessly in the *bahias* and *corixos* in search of prey.

The Pantanal has been occupied by humans for over two centuries. The *pantaneiros* are proud to say that they have struck a deal with Nature, whereby their low intensity cattle ranching practices and management of feral pig populations allow the Pantanal to remain rich and diverse. The threat, they say, comes from the large soybean farms that clearcut large areas and the agroindustrial complexes that pollute the waters. Although the expansion of the agricultural frontier and the heavy mercury pollution resulting from allu-

**The Red-billed Scythebill is aptly-named.**

vial gold mining, are major threats, the *pantaneiros'* claim of their sustained coexistence with nature is being contested. Cattle enclosures have been observed to differ significantly from the grazed surrounding, suggesting that grazing does have an effect, at least in vegetation structure. What effect feral pigs have on the diversity of plant and animal life is at present not clear.

A more important threat, however, may be the gradual shift in ranching practices, from the production of *boi magro* – thin cattle – to more intensive ranching, with more productive and demanding breeds. The *boi magro* practice involves very extensive ranching of Pantanal-adapted breeds left to feed on the native grasses, forbs and woody shrubs. When the cattle reach a certain age, they are herded, once a year, for over 50 to 70 days, or trucked to ranches near major cities where they are fattened before being slaughtered or shipped elsewhere for processing.

The new practices involve extensive clearing of *cordilheiras* to plant introduced grasses, in an effort to increase yield. The loss of the *cordilheiras* is having a major impact on the habitat and species diversity in the region because they harbor a unique ecological complement that include species with very restricted distribution and narrow ecological requirements.

It is difficult to rate the environmental threats facing the Pantanal with regards to their impact on the biological diversity, the quality of life of the *pantaneiros*, or other parameter. Each and every one of them affects the integrity of this ecosystem in different ways and we are still learning how all these threats affect specific ecological processes.

Gold mining, however, has become a particularly serious problem. It has gone much beyond affecting the Pantanal alone and is now reaching human populations inside and outside the region. In addition to the extensive habitat destruction caused by the mining process itself, there is a tremendous amount of mercury that finds its way to the bottom of the river channels. Heavier

**Spectacled Caiman lies quietly in wait.**

than water and not soluble in it, mercury concentrations are much higher in the mud than in the water column, consequently it is being picked up at higher concentrations by the bottom-feeding catfish that are a major source of protein for families in the Pantanal and in the nearby cities.

Recent studies have shown that mercury has also been detected in cattle at higher concentrations than expected. The more intensive gold mining operations are restricted to the vicinity of Poconé in the State of Mato Grosso, leaving fortunately most of the Pantanal free of the direct effect of this environmental catastrophe, but not free from the threat of mercury.

The abundance of large animals makes the Pantanal a hunters' paradise. Avid hunters from the city are not very discriminating. Some have been known to shoot everything in sight. Poachers, on the other hand, are a very different matter. They go after a few valuable species, like jaguars, otters, and Black Caimans, for their furs and skins. The number of caimans taken by poachers range from one million to 1.5 million per year! In some parts of the Pantanal, they have been associated with the drug cartels and are considered very dangerous by the police force responsible for controlling their illegal activities.

In spite of the difficulties resulting from the extensive areas that require patrolling, the lack of equipment, logistic support, and proper training, there may be hope in sight. Actions taken in the Pantanal by private landowners and government agencies, coupled with very vocal and intensive public awareness campaigns may significantly help reduce the magnitude of the problem from the supply side as well as curb the almost insatiable demand for the products of poaching by industrialized nations.

There is another seemingly harmless activity – tourism which is having a major effect in some parts of the Pantanal. Tourism has been heralded all over the world as the alternative to the non-extractive use of the tropical riches. Tourism, it has been said, can provide

*Crested Caracaras survey the area.*

sufficient income to support local communities, at the same time it provides incentives to maintain the biological systems in their original state. Unfortunately, this is not so everywhere at all the time. There are still too many tourists who are all too eager to see an entire rookery take wing and observe the majestic herons, egrets, and storks fill the sky. There are also too many tour operators that are all too eager to please their misguided clients and scare off breeding colonies of these birds by firing guns into the air. Little thought is given to the hundreds of young that fall off their nests to their deaths, or to the severe disruption that these events represent to the breeding of the "resource" that makes the Pantanal so spectacular. Actions like this kill the goose that lays the golden eggs.

Nevertheless, tourism is, in fact, a key solution to the loss of biological diversity and the increasing poverty of rural areas. Although tourists are ultimately responsible for shaping the behavior of the "overeager-to-please" tour operators that they hire, nature tourism is a viable alternative to more destructive economic activities.

Currently there are a number of efforts underway to protect the Pantanal and all the biological diversity it holds. First and foremost, the new Brazilian constitution lists the Pantanal, along with the Atlantic forest, the Amazon, and coastal ecosystems as conservation priorities. The commitment to protect these ecosystems has been followed by action. The Brazilian government has obtained an unprecedented US$117 million World Bank loan to be matched by US$50 million in Brazilian funds to implement the first three years of the Brazilian National Environmental Program. One of the key ecosystems of this program is the Pantanal. Bolivia and Paraguay have started by joining Brazil and forming an international working group to address the conservation needs of the Pantanal.

Organisations like The Nature Conservancy (TNC) and the World Wildlife Fund (WWF) have been active in the Pantanal for many years and have made significant progress strengthening the local conservation capacity. In 1989 TNC and WWF signed an agreement with Mato Grosso's State Environmental Foundation (*Fundacão Estadual do Meio Ambiente* – FEMA) to create the first Conservation Data Center (CDC) in Brazil using technology developed by TNC in cooperation with over 60 CDCs in the Americas. The Mato Grosso CDC has been critical in providing information for the design and implementation of a protected areas system in their state under the National Environmental Program.

A number of Brazilian organizations, like SODEPAN in the Mato Grosso do Sul and *Ecotrópica* in Mato Grosso, have been active in the pursuit of creative ways to conserve and sustainably utilize the natural resources of the Pantanal. And, universities in both states, their respective state environmental agencies, the Brazilian Institute of the Environment (*Instituto Brasileiro do Meio Ambiente* – IBAMA), and the Brazilian Research Agency (*Empresa Brasileira de Pesquisa Agropecuária* – EMBRAPA) have been working to address the conservation needs of the Pantanal for years and now, with the upcoming influx of funds from the National Environmental Program, face days of improved working conditions and a much awaited expansion in the scope of their programs.

In Bolivia, the Pantanal is now also beginning to become a conservation priority. In the private sector the *Fundación Amigos de la Naturaleza* (FAN) has been evaluating their capacity to work on projects aimed at preserving the biological diversity of the Bolivian Pantanal. They have the support of the General Secretariat of the Environment (*Secretaría General del Medio Ambiente* – SEGMA) and the National Fund for the Environment (*Fondo Nacional para el Medio Ambiente* – FONAMA).

In Paraguay, the *Fundación Moises Bertoni para la Conservación de la Naturaleze* in also pursuing similar objectives with the support of their government.

**Guira Cuckoos look eternally surprised.**

# THE ATLANTIC FOREST

**The Atlantic forest** of Brazil covered originally over one million square km along a narrow band, 4,200 km long, hugging the coastline from the state of **Rio Grande do Norte** to the state of **Rio Grande do Sul**. Its long and narrow shape makes the Atlantic forest a very heterogeneous ecosystem. To the north, it is influenced by the drier, hotter climates characteristic of Brazil's *nordeste* and to the south by the cooler climates of the temperate forests. There is also a marked effect of the distance from the ocean which is reflected in the structure and composition of the vegetation. These north-south and east-west gradients interact strongly adding significantly to the richness and diversity of the Atlantic forest.

But there is a third axis that also contributes to the heterogeneity of the Atlantic forest: elevation. The coastline is flanked by a series of mountain ranges – **Serra do Mar**, **Serra da Mantiqueira**, and others – that provide marked elevational gradients and create a complex landscape, rich in isolated valleys and ridges with different microclimates and corresponding changes of the biological composition.

Any brief characterization of the Atlantic forest is inadequate and must be regarded as an oversimplification. Nevertheless, it is still possible to make a thumbnail sketch of what this ecological "hot spot" is all about. One must always keep in mind, however, that the Atlantic forest is in reality the host to many vegetation types in the broadest sense of the word.

Moving from the coast inland there are mangroves, *restingas*, rain forests, high-elevation grasslands at the top of the highest mountains, and in the most temperate areas to the south there are – or at least were – large stands of *Araucaria* pines. Mangroves, although not unique to the Atlantic forest, are an integral part of this ecosystem. This habitat type is associated, wherever it is found, with coastal areas where rivers and oceans mix their waters creating a brackish environment. The resulting system tends to be oxygen-poor and exposed to the ebb and flow of the tides twice a day. These waters are known for their high productivity as well as for the fragility of the ecological communities which they support. Many species of commercial fish breed, and complete the early stages of their development under the the mangrove roots which are also an important substrate for other commercially important species such as oysters.

In the **Lagamar**, a region of 32,000 square km between the states of São Paulo and Paraná, one finds, in addition to the 20% of the remaining Atlantic forest, one of the most extensive mangroves in the coasts of Brazil.

Mangroves had virtually disappeared from the shores of **Comprida Island** and the vicinity of the city of **Iguape** as a result of a channel being opened in the first decades of the century to shorten the distance that ships had to travel from the coast up the **Rio Ribeira**. The channel immediately changed the water quality of the **Mar Pequeno**, the estuary that separates the Comprida Island from the mainland, killing all the mangrove stands and reducing fish catches significantly for many years. But the resilience of the mangroves can be demonstrated by their quick recovery. Together with the mangrove, the local fishing industry also recovered, after the channel was closed in the late 1970s. Now, Comprida Island has regained its healthy mangrove stands and a reviving fishing industry.

Inland from the mangroves, one finds a very different habitat – the *restinga*. This mix of shrubs and short forest is the result of a successional process that takes place as the coastal dunes are stabilized. This complex includes dunes, interdunal ponds, and extensive wetlands. Some of the best examples of them are the large wetland systems of **Rio Grande do Sul** (**Lagoa dos Patos** and **Lagoa dos Peixes**), and the easily accessible *restingas* of **Rio's Barra da Tijuca**.

**Preceding pages:** Atlantic rain forest and 'God Finger Peak' at Serra dos Orgãos National Park. **Left,** tucked-in tree frog.

One important characteristic of all dune systems in the world that is also apparent in the *restingas*, is the prevalence of vegetative growth as a method for invading new and unstable dunes. This process is slow and can be easily disrupted. Consequently, any damage made to the *restingas* by intensive visitation and off-road vehicles results in setting back plant succession in a significant way. These disturbances can initiate a process that reverses succession, taking stabilized dunes with vegetation cover, back to the stage of shifting dunes – dunes that move due to wind action.

Further inland from mangroves and *restingas* are the rain forests that have become the signature of the Atlantic forest. Lush vegetation covers the rocky mountains that follow the coast and touch the waters as in **Rio de Janeiro**, or that are well inland as in **Minas Gerais**. The vegetation of those mountains varies significantly with elevation. The lower strata, from sea level to 800 m is lush and has a canopy that can reach 25 m in the wettest sites with emergent trees that can easily exceed those heights. In the drier ridges the canopy drops to as low as 12 m. Further up, between 800 and 1,700 m, is the montane forest with their epiphyte covered trees. In the **Serra dos Orgãos National Park**, close to 800 species of epiphytes have been collected and it is expected that the number of species present can even go as high as 1,000.

At the top of the highest mountain ranges are the *campos de altitude* – the high elevation grassland which are like islands of grassland in a sea of forest. This remnants of a habitat that was more common during the last glaciation period have a species composition that sets them apart from everything around them. The level of endemism – species that occur only there and nowhere else – by virtue of its isolation from similar areas and its history is astonishing. Near Rio de Janeiro, within the **Desengano State Park** and the **Serra dos Orgãos National Park,** are excellent examples of all these habitat types.

In the more temperate and humid areas of the southern states, there were once large single-species stands of conifers like the *Araucaria* and *Podocarpus*. However, these stands have been reduced to isolated patches by the intense harvesting that they have been subjected. A number of parrot species, endemic to the Atlantic forest, have been associated with this habitat type such as the Red-spectacled Parrot, Red-tailed Parrot, Vinaceous Parrot, and the now considered extinct Glaucous Macaw. Other bird species like the Azure Jay have been known to be specialized seed dispersers of *Araucaria*. The decline in this conifer is thought by scientists to be largely responsible for the reduction in the jay populations.

This complex of habitats and ecological associations harbors one of the most diverse plant and animal life in the planet. Because of its high diversity and endemism, the Atlantic forest is considered one of top biological hot spots on Earth. In spite of having been reduced to less than 14% of its original size, it still contains an estimated 10,000 species of plants which include 53% of the tree species, 64% of the palms, and 74% of the bromeliads are endemic.

Animal life of the Atlantic forest is equally diverse and endemic. Out of the 130 species of mammals, 51 occur nowhere else in the world. Primates are a group of mammals with one of the highest levels of endemism in the Atlantic forest: 10 species of marmosets, three species of lion tamarins as well as the Woolly Spider Monkey, are endemic. Among the birdlife, there are 30 genera and 160 endemic species found in this ecosystem.

Most of this biological diversity still remains after intensive exploitation for close to five centuries. Since the year 1500, when Brazil was discovered by the Portuguese, the Atlantic forest has gone from occupying 12% of Brazil's territory to approximately 1.6%. This forest has been going through many phases of resource use, for example: logging which still goes on today with disproportionate intensity considering

**The Green-headed Tanager (left) and the Red-necked Tanager (below), two of more than 100 members of the same genus.**

the perspective of sustainability of this activity; sugarcane, coffee, and industrial development..

In the process, close to 60% of Brazil's population – over 80 million people – has been concentrated in the region that was once the original Atlantic forest, maintaining tremendous pressure on the little that is left of this ecosystem.

The responsibility for the current condition of the Brazilian Atlantic forest must be shared by all. Poverty drives most of the inhabitants of the forest to over-exploit the resource base, as it happens everywhere in the world. And greed drives businesses to engage in activities that destroy the last remnants of this imperiled ecosystem for the sake of short-term profits.

There are, however, communities that are using sustainable practices to increase the productivity of the system with a minimum environmental impact. Local fishermen in the Lagamar for instance, are beginning to use low technology methods to increase oyster yields by providing artificial substrates for the attachment of oyster larvae. These methods are helping the fishermen gradually abandon the non-sustainable practice of cutting mangrove roots to collect the oysters.

Other important changes in resource use practices are also being implemented by some businesses and farmers. However, in spite of all the efforts being made, the Atlantic forest continues to disappear at an ever increasing rate. This is one of the cases in which strong protection is still needed because there is simply not sufficient time to change old habits based on the misconception that the riches of the forests and the seas are inexhaustible.

On the conservation side, there are major efforts being made to find integrated ways to deal with the critical problems of poverty and habitat loss. Brazilian government and non-government organizations in São Paulo and Paraná, with assistance from WWF, The Nature Conservancy and other international organizations have focused on the development of a major ecosystem

**A Black-fronted Piping Guan.**

302

conservation program for the Lagamar where 20% of the remaining Atlantic forest can still be found.

This program includes protection of natural areas, sustainable development, research, and education. Nevertheless, programs like this cannot be successful without the leadership of organizations like S.O.S. Mata Atlântica, FUNATURA, FBCN (*Fundacão Brasileira para a Conservacão da Natureza*), SPVS (*Sociedade de Pesquisa em Vida Silvestre*) or the strong participation of government agencies like IBAMA (*Instituto Brasileiro do Meio Ambiente*) with programs like *Projeto Tamar* for saving the sea-turtles.

At the government level it is important to mention the Atlantic Forest Consortium formed by the secretariats of the environment from the Atlantic forest states. Their objective is to coordinate efforts in order to increase efficiency in their programs. It will be through this project that the Atlantic forest component of a $167 million environmental program funded by the Brazilian government and the World Bank will be implemented.

As visitors fly into Rio circling the **Guanabara Bay** and flying over the **Tijuca National Park**, in the heart of the city, they must remember that they are looking at a disappearing treasure. The Atlantic forest, with all its biological diversity is hanging on to Brazil's coastline tenaciously but will disappear eventually if nothing is done to preserve and conserve it soon.

The Atlantic Forest is a sad example for a worst scenario of the Amazon's future: a contigious biogeographic region can slip toward destruction by becoming first fragmented into small parcels, which then may collapse under continuing human encroachment. It may sound like cynicism, but the nature tourist will enjoy another side of the coin: Nature reserves within this region like Tijuca, Serra dos Orgãos, Itatiaia, and Iguassu, are small and close to human settlements and provide therefore – in contrast to the Amazon – access, accommodation and hiking trails.

**Spot-billed Toucanet.**

# PARKS CLOSE TO RIO DE JANEIRO

Few international flight go directly into cities on the Amazon, and the visitor to Brazil's Amazon will often have to use **Rio de Janeiro** as gateway to Manaus, Belém, or the Pantanal. Although the tropical rain forest has been overshot at this point, a stopover in Rio provides an excellent chance to see some of the few remaining examples of southeastern Brazil's Atlantic forest, a habitat often classified as subtropical rain forest.

Like most tourist, the traveling naturalist will normally select the hotel **Copacabana**. This suburb with more than a million inhabitants provides with its long beach front, a chance to get away from the bustling city. While walking along the beach, it is worthwhile not only to try to spot human beautics, but also the numerous beautiful seabirds. Brown Boobies shoot over the waves in search of fish, and Magnificient Frigatebirds soar with Black Vultures in mixed flocks over the city.

To get away from the crowd and the traffic of Rio, it is worthwhile to visit the botanic garden, **Jardim Botanico**, about five kilometers from Copacabana. It is situated in front of an impressive backdrop of steep mountains. In the east rises the **Corcovado** to 704 m, crowned by the **Cristo Redentor statue**, a landmark of Rio. To the west are the forest covered **Tijuca mountains**.

The botanic garden is nearly 200 years old, and contains a fine selection of old tropical trees of local and exotic origin. From the surrounding forests, Common Marmosets, Grey Tree Squirrels, and several species of toucans and parakeets venture into the garden. Hummingbirds, called in Brazil "flower-kissers" (*Beija-Flores*), zoom between the trees, the large Swallow-tailed Hummingbird being the most common species. Masked Water-Tyrants catch insects between water-lilies and Rufous-bellied Thrushes sing in the trees. This garden is contiguous with the Tijuca forest, but it is not easy to identify a trail into this area.

Access to Tijuca is best by road. One of them passes between the botanic garden and Corcovado, and can be used for hiking, since there is little traffic. But it is not advisable to hike alone because of the city's high crime rate.

**Tijuca National Park**, is an anachronism of a national park, firstly since it is completely surrounded by the suburbs of Rio, and secondly, since its forests have been planted by Man. Rio was one of the continent's population centers since its colonization by the Portuguese in the 16th century, and the surrounding forests were the first to become logged. In the 18th and the beginning of the 19th century, the Tijuca mountains were exclusively used for sugarcane and coffee plantations, and virgin vegetation and wildlife survived at best in tiny pockets in inaccessible places. Around the middle of the last century, most land was acquired by the government to create a watershed for Rio's growing needs for water supply. Beginning in 1861, the plan was realized over a period of 13 years to recreate a natural landscape and to plant 100,000 native and exotic trees in the montane area of 30 square km. Over the time, the planted forest was enriched by invasion of additional native plants from remaining natural pockets, and today, 130 years later, a succession towards natural old growth forest is well-advanced.

Tijuca is an excellent example that restoration of nature is possible in the tropics, given the political will, funds, and provided, that the native species had not disappeared from earth. In Tijuca, many mammals had become locally extinct, but could be reintroduced from somewhere else. This was unfortunately not done in case of the Lion Tamarin, a native here. It is replaced by the Common Marmoset from the Brazilian northeast. This species has become one of the most common mammals of the park. Also common is the Grey Tree Squirrel. Several other mammals had either survived in the area or have been reintroduced, such as the Brown Capuchin, pacas, and agoutis.

Tijuca is excellent for birdwatching with attractive sightings of some par-

**The Margay Cat looks simply spectacular.**

ticularly colorful passerines, like Red-necked and Green-headed Tanager, and Pin-tailed, White-bearded, and Swallow-tailed Manakin. Among the larger birds are several species of toucans, seven species of parakeets and parrots like the Scaly-headed Parrot, and two species of fowl, the Rusty-margined Guan and the Spot-winged Wood-Quail.

Tijuca is split into three parts by private property, which includes some restaurants as well as a hotel. For hiking and wildlife observation, it is best to start at one of the parking lots close to the **Pico da Tijuca**, past the small but beautiful waterfall ( **Cascatinha**). It is possible to hike to the Pico da Tijuca, with 1,021 m the parks highest mountain, following a 2.8 km long trail. It is worthwhile to drive several of the other access routes to enjoy from observation points like the **Vista Chinesa**, the spectacular view of Rio and the coastline.

**Serra dos Orgãos National Park**, two hours by car to the north of Rio, invites for an attractive one-day-excursion. The park, on 50 square km,

protects the nature of the eastern flank of an impressive mountain range, that stretches between the cities of **Petrópolis** and **Teresópolis**. One entrance is at the periphery of the latter town, which also offers accommodation. With a patchwork of virgin and secondary vegetation, crystal-clear creeks and spectacular mountains like the **Dedo de Dios**, some of them higher than 2,000 m, the park caters to the naturalist, the hiker, or the mountaineer, or to people, who just want to take a cool bath in natural or artificial pools. For the serious hiker, there is a trail across the mountains from Teresópolis to Petrópolis, a hike of three to four days through a landscape of gneiss and granite rocks. These heights are in winter exposed to temperatures close to freezing point, and are hit by violent thunderstorms in summer. For small walks, start at the fringe of Teresópolis, or turn from the highway to the right at the sign "*parque nacional sub-sede*", 11 km before reaching Teresópolis.

The park's fauna is much richer than that of Tijuca. There are three species of

**Cool dip in a mountain stream.**

monkeys, the Brown Capuchin, the Buffy Tufted-ear Marmoset, and the Brown Howler Monkey, five species of cats, the coati and the Lesser Anteater. Among the numerous bird species are several tinamous and toucans.

Possibly the richest flora and fauna in this part of Brazil is found in **Itatiaia National Park**. It is located halfway between **Rio de Janeiro** and **São Paulo**, north of km 150 on the Presidente Duarte freeway. Close to the town of **Itatiaia**, turn into a 14 km long asphalted road to the park headquarters. The park includes four square kilometers of privately owned land close to the headquarters. There are a number of hotels and other facilities for visitors.

With an area of 119 square km, **Itatiaia** protects part of the slopes and the top of a mountain plateau, the **Serra da Mantiqueira**. Its **Agulhas Negra** mountains, the third highest mountain of the country at 2,787 m. Itatiaia has well-preserved examples of Atlantic forest. On its lower slopes, between 400 and 600 m, much of the vegetation is secondary with heights around 15 m. At 1,000 m, many areas are primary forest with up to 30 m height. At higher altitudes, the forest is naturally getting lower; 10 to 20 m only between 1,600 and 2,200 m and replaced by alpine meadows at above 2200 m. But impressive rock formations make up for the smaller stature of the vegetation. The park's flora is extremely rich at all elevations, and contains more than 100 endemic plants.

The park has a virtually full complement of the original fauna. Although some species are declining due to insufficient size of their habitat or to small population size – in the case for the Muriqui, the largest primate of the neotropics, for the jaguar and the Brazilian Tapir. Other mammals are faring better – the Masked Titi, the Brown Capuchin, the Brown Howler Monkey, or the Pale-throated Three-toed Sloth. There is an extremely rich birdlife of 250 species – the Red-legged Seriama, a relative of the cranes and rails, the Solitary and the Brown Tinamou, the Pileated Parrot and the Purple-bellied Parrot.

**The Emerald Tree Frog is another of the marvellous forest-dweller.**

# IGUASSU

The name **Iguassu**, which means in Guarani Indian language "big water", stands for one of the most spectacular waterfalls of the world, possibly the best known of the natural wonders of Brazil.

The Iguassu river created this waterfall as its turn from a southwesterly to a northwesterly course, by cutting into a deep layer of basalt, that originated 120 million years ago which covers a large part of southern Brazil. The waterfall forms only the central feature of a 2,200 square km large national park that bears the same name, a huge area covered with natural vegetation and rich in wildlife. Three quarters of this protected area are in Brazil, and one quarter in Argentina. The Iguassu river and the waterfall mark the border between Brazil and Argentina, and the mouth of the Iguassu into the **Paraná**, about 30 km to the northwest of the falls, is the place where the two countries meet Paraguay.

From cities in Brazil like São Paulo, the park is reached by flying into the local airport, which is halfway on the 30 km road between the falls and **Foz de Iguaçu**, a medium-sized city. From this city, one can drive to **Puerto Iguacu** on the Argentinian side, and on to the Argentinian side of the waterfall and the forests of the national park. No connection between the two countries exists directly at the falls. Accommodation is in various hotels in town or close to the airport, or – best but most often booked out – in the **Hotel das Cataratas**, close to the falls.

About 2,500 km south of the equator, the vegetation of **Iguassu National Park** can be classified as subtropical rather than tropical forest. Although there are only about 1,000 mm precipitation, half the amount falls in the Amazon or on the Atlantic coast, thus the forest is mostly evergreen rather than deciduous, since the dry season in winter is not strictly without rainfall. Much of the vegetation has small- or medium-sized

leaves, but some broad-leafed plants, palms, lianas, epiphytes and the generally quite vigorous growth of the trees, gives an impression still comparable to the tropical forests.

The forest of Iguassu had been contiguous with the Atlantic forest of southeastern Brazil until extensive clearing for agriculture of much of the state of Paraná in the middle of this century led to its isolation, an island of nature in a sea of agricultural land. Due to these previous connections, there are many relationships of flora and fauna with those of the Atlantic forest and even the Amazon, and because of today's isolation, Iguassu National Park has immense relevance for the protection of many species.

Thanks to the size of the area, viable populations persist of large mammals like Brazilian Tapir, White-lipped and Collared Peccary, capybara, jaguar, and several of the smaller cats. The locally endangered Southern River Otter persists in the park's waterways.

The park is a stronghold of large predatory birds like the Harpy and the Crested Eagle, which are seriously threatened elsewhere in forested areas of smaller dimensions. Unfortunately, one of the continents most beautiful parrots, the Glaucous Macaw, has become extinct, a fate possibly faced by several of its relatives, unless cage bird trapping is curtailed. Viable populations exist at this time of the Blue-and-Yellow Macaw, while the Vinaceous Amazon is endangered because of the decrease of its habitat, which consists of *Araucaria* groves. There is a faint possibility that one of Brazil's rarest birds, the Brazilian Merganser, still exists here.

Although, with little chance to find these rarities, the visitor should still enjoy the easy observability of the many common but beautiful and colorful tropical birds such as the Red-breasted Toucan, Red-capped Parrot, Red-rumped Cacique, Blue Dacnis, or Green-headed Tanager. One can also observe the tame and impressively large Common Iguanas while enjoying a picnic on the meadows close to the waterfall.

**Artistic impression: leaf silhouette (left) and luminescent caterpillar (below).**

# INSIGHT GUIDES
# Travel Tips

THE NOBLE TIME

# JUVENIA
## — 1860 —

*Golden Age* ®
COLLECTION
STEEL - STEEL/GOLD - 18KT GOLD AND WITH PRECIOUS STONES

# TRAVEL TIPS

# GETTING THERE

## BY AIR

Being the northernmost country in South America, Colombia is many people's first stop. There are many flights from Europe and North America to Bogotá, Cartagena and Barranquilla, with onward connections to the rest of the continent. Cheap flights are available from Costa Rica in Central America. For Australians and Asians, Aerolineas Argentinas flies from Sydney to Buenos Aires, then onwards to Bogotá.

## BY ROAD

From Ecuador, buses go to and from the border at Tulcan/Ipiales. From Venezuela, you can choose between the coastal route from Maracaibo to Santa Marta, or the highland route from Caracas and Mérida to Cucuta via San Cristobal. From Brazil, the only way to enter is via Leticia in the Amazon basin and then fly to Bogotá.

# TRAVEL ESSENTIALS

## VISAS

Citizens of United States and many European countries do not need visas, and are given a 60-day stay on arrival. Everybody else needs a visa, and the price varies per country: Australians and New Zealanders receive theirs free if they show an onward ticket from Colombia.

## MONEY MATTERS

The unit of currency is the Colombian *peso*, which is slowly being devalued every day by government regulation.

Many *casas de câmbio* change US dollars into local money. Traveler's checks are more difficult to change, although, the Banco de la Republica will do so, as well as some others, depending on the region. Otherwise, change currency in your hotel. Do not change money on the street: they are almost all unscrupulous hustlers.

## HEALTH

Tap water in Bogotá is considered by experts to be one of the purest in the world, and can be drunk without problems. Elsewhere, drink bottled mineral water. No vaccinations are necessary, although if you are spending time in the Amazon rain forest, do take malaria tablets with you.

## WHAT TO WEAR

The weather in Bogotá and the highlands stays moderate all year round: a light sweater will suffice in the evenings. The coast has a tropical climate: take light summer clothes.

## ON ARRIVAL

Bogotá's El Dorado Airport is one of the most modern in the world. Buses and taxis run to the city center. Be sure to fill out a tax exemption form when you arrive, so that you don't have to pay the Colombian citizen exit tax on departure. Even with this form, foreigners pay $15 departure tax from Bogotá when leaving by air.

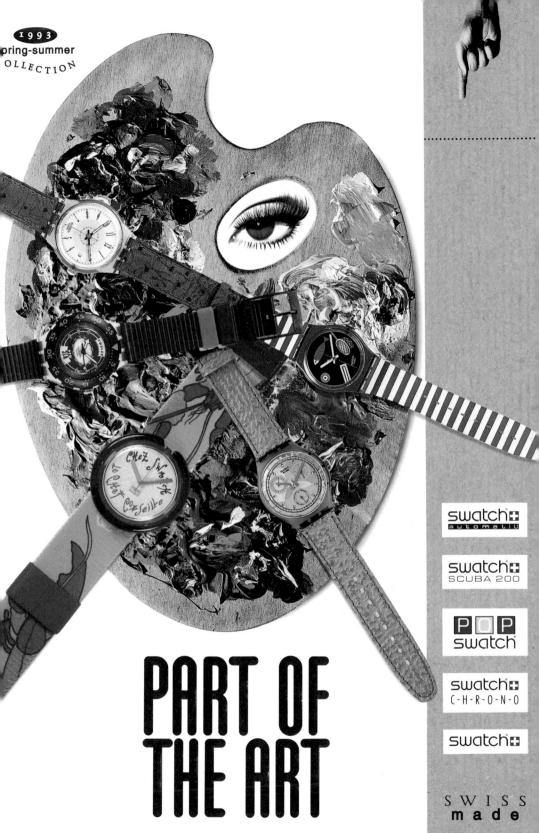

# THE WORLD IS FLAT

Its configuration may not be to Columbus' liking but to every other traveller the MCI Card is an easier, more convenient, more cost-efficient route to circle the globe.

The MCI Card offers two international services—MCI World Reach and MCI CALL USA—which let you call from country-to-country as well as back to the States, all via an English-speaking operator.

There are no delays. No hassles with foreign languages and foreign currencies. No foreign exchange rates to figure out. And no outrageous hotel surcharges.

If you don't possess the MCI Card, please call the access number of the country you're in and ask for customer service.

The MCI Card. It makes a world of difference. **MCI**

---

# GETTING ACQUAINTED

## GOVERNMENT & ECONOMY

Colombia has had a "limited democracy" where two parties have shared power for several decades. Agriculture is the most important sector of the economy, with manufacturing a close second. Colombia has one of the strongest economies on the continent.

## CLIMATE

The highlands enjoy moderate weather all year round, while the coast is tropical. The only seasons are wet and dry, varying per region: in Bogotá, the driest time is between December and March, and between July and August.

## BUSINESS HOURS

Generally Monday to Friday, 8 a.m. – noon, then 2 p.m. – 6 p.m. Banks are open from 8 a.m. – 3 p.m. Colombia is five hours behind Greenwich Meridian Time.

## MEDIA

Bogotá has several daily newspapers. Both *El Tiempo* and *El Espectador* are considered the most comprehensive in terms of news coverage.

## POSTAL SERVICES

Mail is generally very reliable in Colombia. The General Post Office in Bogotá is at the Avianca Building in the center of the city, opposite the Parque Santander.

## TELEPHONE & TELEX

The Colombian Telecom offices in all big cities have international communication facilities to enable you to dial or telex home.

## SECURITY & CRIME

The center of major cities is generally safe, but with such extreme poverty in Colombia it is worthwhile to check which outlying areas can be visited safely at night. The unsafe areas are clearly defined, usually red-light districts – ask at your hotel for details. With a little common sense, Colombia is as safe as any other country.

## LOSS

Although most Colombians are honest and friendly, theft occurs in the larger cities and the tourist should take care in looking after his valuables at all times. There are "Tourism Police" at major airports and towns to help if you have anything stolen.

## MEDICAL SERVICES

The main medical centers in Bogotá are Cruz Roja Internacional, Avenida 68, No. 66-31; or Centro Medico de Salud, Carrera 10, No. 21-36.

# GETTING AROUND

## FROM THE AIRPORT

Taxi services run from all the major airports to the city center, at fixed rates. Local buses also run to Bogotá airport.

## DOMESTIC TRAVEL

The national carrier Avianca flies to all parts of the country. Less regular services are offered by SAM and Satena.

Bus transport along the main routes is generally good, often luxurious, but deteriorates when heading into the more remote areas. Normally, the rougher the road, the poorer the bus. The alternative on main routes is a *buseta* (minibus) or *colectivo* (a shared taxi that is more expensive but much quicker).

Within the main cities, taxis are cheap and relatively reliable. A meter clocks over the price in _pesos_.

# WHERE TO STAY

## HOTELS

Hotels in Colombia range from the luxurious at $130 a night to the basic at around $2 a night.

### BOGOTÁ

Some of the five-star hotels include:

**Hilton**
Carrera 7, No. 32-16

**Tequendama**
Carrera 10, No. 26-21

**La Fontana**
Diagonal 127A, No. 21-10
Recommended by many travelers.

For a hotel with some atmosphere and still perfectly comfortable though with less conveniences – try the **Hoteria de la Candelaria**, Calle 9, No. 3-11, which is in a converted colonial mansion and furnished with antiques. (Tel: 286-1479, around US$15 a single, $20 a double).

Budget hotels are often dismal, located between Calles 13 and 17, and Avenida Caracas and Carrera 17.

### CARTAGENA

**Cartagena-Hilton**: A luxury hotel in nearby El Laguito, Tel: 50666.

**Plaza Bolívar**: In town, at the Plaza Bolívar, approximately $20 a single.

# FOOD DIGEST

## WHAT TO EAT

Colombian cooking varies by region, although in the big cities any kind of international cuisine can be enjoyed at a price.

For Colombian dishes, a few local specialties worth trying are:

_Ajico_: a soup of chicken, potatoes and vegetables, common in Bogotá.
_Arepa_: a maize pancake.
_Arroz con coco_: rice cooked in coconut oil, special to the coast.
_Bandeja paisa_: a dish of ground beef, sausages, beans, rice, plantain and avocado.
_Carne asada_: grilled meat.
_Cazuela de mariscos_: seafood stew.
_Chocolate santafareño_: hot chocolate accompanied by cheese and bread.
_Mondongo_: tripe soup.
_Puchero_: broth of chicken, beef, potatoes and pork, typical of Bogotá.
_Tamales_: chopped pork with rice and vegetables folded in a maize dough.

## WHERE TO EAT

In Bogotá, the **Casa Vieja** restaurants are considered to offer the best in regional food. They are at Avenida Jimenez, No. 3-73, Carrera 10, No. 26-50, and Carrera 3, No. 18-60. **Refugio Alpino**, Calle 23, No. 7-94 serves European food. For cheaper fare but excellent quality in a colonial setting, go to **La Pola**, Calle 19, No. 1-85.

Budget meals serving _comidas corrientes_, usually a fried piece of meat with beans and rice are found in many of the restaurants.

In Cartagena, **El Bodegon de la Candelaria** is in a beautiful colonial mansion near the San Pedro convent.

# Our history could fill this book, but we prefer to fill glasses.

When you make a great beer, you don't have to make a great fuss.

# AUTAN®

## – makes biting insects keep their distance.

*Effective for 6-8 hours indoors and outdoors.*

*Autan for protection against mosquitoes, gnats, midges, and other insects. For external use only. Do not apply to the eyes, nose or mouth.*

Bayer

# PARKS & RESERVES

## ALTO ANCHICAYA NATIONAL PARK

**Geographic Location:** Pacific slope of the Department of Valle, Colombia.

**Size:** Not defined, (protects catchment for a hydroelectric plant).

**Access:** By bus (TRANSUR company) along old Cali–Buenaventura road. Get off at Danubio (60 km). Report at gate house across river bridge and wait for a generating company bus to travel the 10 km to the work camp. If driving, leave main Cali–Buenaventura road at KM 21, turning left onto old road.

**Approximate cost of access:** Bus Cali–Danubio approximately $2.

**Tour operators:** None.

**Open:** All year, crowded during holiday periods.

**Registration:** Essential.

Book through CVC (Central Autonoma del Cauca), Carrera 56, No. 11-36, Cali.

**Accommodation:** In chalet apartments. Usually free.

**Food/restaurant:** Meals available in the camp canteen. $4 per day.

**Transport in park:** Good system of roads with side tracks. Lifts can be obtained on camp vehicles.

## AMACAYACU NATIONAL PARK

**Geographic Location:** Extreme south-east of Colombia, extending northwards from the Amazon River.

**Size:** 300,000 hectares.

**Access:** By plane to Leticia (four flights a week by Avianca). Boat upstream 60 km to the Visitors Center at Matamata.

**Approximate costs of access:** Return air ticket Bogota–Leticia $145. Boats

carrying up to 15 passengers can be hired for $80 a day. Alternatively, a public river taxi travels between Leticia and Puerto Nariño twice a week, stopping at Matamata (ask conductor) for $2. The journey takes seven hours by public transport.

**Tour operators:** Several based in Leticia. Try Amazon Explorers (Aa247, Leticia) situated in the Hotel Anaconda (Calle 8/Carrera 25, Leticia).

**Open:** All year, usually full during Easter Week.

**Registration:** Essential. Entry fee approximately 50¢.

Book through INDERENA, Diagonal 34, No. 5-84, Bogota (Tel: 232-1879) or AA006, Leticia.

**Accommodation:** Visitors Center has dormitory accommodation with hammocks or folding beds. Approximately $2.50 per night in hammock, $3.50 per night in folding bed.

**Food/restaurant:** Meals available in cafeteria. Three meals a day at $7.50 (not including drinks).

Well-marked trails enter forest. Visitors are encouraged to hire a local guide (ap-

proximately $2 per day, but ask for current standard rate at the Visitors Center). During February – early June access into forest may only be possible by dugout canoe (which can be hired from the park).

## CHINGAZA NATIONAL PARK

**Geographic Location:** On the border of Cundinamarca and Meta, east of Bogotá.
**Size:** 50,000 hectares.
**Access:** Take the road from Bogotá to La Calera. Two kilometers after La Calera, on the road to Guasca, take a right turn onto a good, but unpaved road, traveling through a checkpoint, then past a cement factory, following the route for 35 km before entering the park.
**Approximate cost of access:** No public transport available, although it may be possible to arrange transport with the water company which runs buses daily to the work camp at the reservoir in the park (Empresa de Acueductoy Alcantarillado de Bogotá, Calle 22c, No. 40-99, Bogota).
**Tour operators:** None.
**Open:** All year.

Chingaza
National Park
8 km / 5 miles

**Registration:** Essential.
Book through INDERENA, Diag. 34, No. 5-84, Bogotá (Tel: 232-1879).
**Accommodation:** New Visitors Center. Prices about $6 – $8 per room per night (accommodating up to 5 persons).
**Food/restaurant:** No restaurant facilities, so food needs to be brought. The reservoir work camp has a cafeteria and it may be possible to arrange meals there.
**Transport in park:** On foot or private vehicle.
**Trails:** Good system of paths.
**Further advice:** Be prepared for sudden changes in the weather.

## LA PLANADA RESERVE

**Geographic Location:** South-west of Colombia in the Department of Nariño, close to the town of Ricaurte.
**Size:** 3,600 hectares.
**Access:** Take the Pasto–Tumaco road. At the village of Chucunes (a few kilometers before Ricaurte) take a left turn. The reserve lies 7 km along this road. The journey time between Pasto and La Planada is approximately four hours. Buses are available for Chucunes from Pasto (try SERVICIO TRANSPORTES ESPECIALES leaving at noon from Hotel Chambu) Cost is about $3.
**Tour operators:** None.
**Open:** All year.
**Registration:** Essential.
Book through Dr Jorge Orejuela, FES, AA 5744, Cali or write directly to El Superintendente, La Planada, AA 1562, Pasto, Nariño, Colombia.
**Accommodation:** Accommodation in study-bedrooms.
**Food/restaurant:** Meals available in a small restaurant. Full board and lodging $10 a day.
**Transport in park:** On foot.
**Trails:** Well-marked system of paths and tracks.
**Further advice:** Since it can be wet, waterproofs and rubber boots are essential, although the Visitors Center does hold a stock of boots for use by visitors.

# La Planada Nature Reserve

24 km / 16 miles

Tumaco
R. Patia
R. Telembí
Barbacoas
Cajapí
R. Mira
COLOMBIA
Sotomayor
El Diviso
R. Guiza
Ancúya
Pasto
Ricaurte
Túquerres
La Planada Nature Reserve
Imués
;CUADOR
Alto Tambo
Maldonado
R. Santiago
Ipiales
↑ To Ricaurte

Parking ■
Visitor Center ■
Administration ■
Biological Station ■

LA PLANADA NATURE RESERVE

VENEZUELA
GUYANA
Atlantic Ocean
SUR.
LOMBIA
FR. GU.
JA
BRAZIL
PERU
BOLIVIA
PARAGUAY
CHILE
ARGEN-TINA

## RIO TATABRO

**Geographic Location:** On the Pacific slope of Valle, Colombia.

**Size:** Undefined.

**Access:** On the old Cali–Buenaventura road, between Agua Clara and Lianobajo.

**Approximate costs of access:** TRANSUR bus from Cali, about 4½ hours, approximate cost $3.

**Tour operators:** None.

**Open:** All year.

**Registration:** Essential.

Book through Fundación Herencia Verde, Calle 22 a norte No. 6 AN-32, Edificio

Macehy officina 302, Cali or AA 32802, Cali (Tel: 674-995).

**Accommodation:** Limited accommodation in bunk beds. $3 per night.

**Food/restaurant:** Bring food, to be prepared by caretaker's wife. About $1 per day.

**Transport in park:** Foot.

**Trails:** Two main trails; ask for directions from caretaker.

## SIERRA NEVADA DE SANTA MARTA

**Geographic Location:** The Sierra Nevada massif, situated close to the Caribbean coast on the borders of the Colombian departments Magdalena, Guajira and Cesar.

**Size:** 383,000 hectares.

**Access:** To the San Lorenzo ridge, travel to the village of Minca from Santa Marta. The road continues through the village and rises through coffee-growing areas to the INDERENA buildings at San Lorenzo at 2,200 m above sea level. The drive takes about 2½ hours. Between Minca and San Lorenzo the road is unpaved and very rough, particularly in the wet season. To visit the Ciudad Perdida, it is essential to organize a guide for the week-long trek.

**Approximate costs of access:** Regular buses (about 50¢) between Santa Marta and Minca but no public transport for the 35 km between Minca and San Lorenzo. Car hire at international rates. Treks to Ciudad Perdida average $170.

**Tour operators:** For Ciudad Perdida, Tesoro Tours (Avenida San Martin, No. 6-129, Santa Marta Tel: 654-713) and Servicios Terrestre (Edificio Mundo Marino, Loc. 1, El Rodadero, near Santa Marta Tel: 27196).

**Open:** All year.

**Registration:** Essential

Book through INDERENA, Cerrera 1, No. 22-79, Santa Marta. For Ciudad Perdida, also essential to seek permission from Instituto de Anthropologia, Carrera 7a, No. 28-66, Santa Marta Tel: 342-5925, 334-2639.

**Accommodation:** INDERENA Visitors Center at San Lorenzo, about $3.50 per night.

**Food/restaurant:** All food has to be purchased beforehand in Santa Marta.

**Transport in park:** On foot.

**Trails:** The main track along the San Lorenzo ridge provides the easiest opportunities to find Santa Marta specialities.

**Further advice:** Some armed activity occurs in the Sierra Nevada de Santa Marnta. Seek current advice from the INDERENA office in Santa Marta before venturing into the area.

# TAYRONA NATIONAL PARK

**Geographic Location:** On the Caribbean coast of the Department of Magdalena, Colombia, extends for about 35 km from just east of Santa Marta.

**Size:** 12,000 hectares.

**Access:** From Santa Marta to Canaveral (the administrative center of the park), take a bus which is heading for Rioacha from the bus terminal in Santa Marta. Ask to be dropped off at El Zaino (34 km from Santa Marta) for Canaveral. The park entrance is on the left side of the road. There is a 4-km walk along the road to the park center. It is possible to take a taxi from Santa Marta straight through to the center, or hire a vehicle. The park gate closes at 5 p.m.

**Approximate costs of access:** Entry fee approximately 50¢ ($1 for a vehicle; taxis enter free). Bus from Santa Marta to El Zaino: about $1. Taxi from Santa Marta to Canaveral about $4. Hire cars at international rates.

**Tour operators:** Check with the Santa Marta tourist office (Carrera 2, No. 16-44)

**Open:** All year.

**Registration:** Essential.
Book through INDERENA, Carrera 1, No. 22-79, Santa Marta.

**Accommodation:** At Canaveral, in "Ecohabs" (thatched-roofed cabins, with en suite bathrooms) for $17 ($15 in low season) a night for 4 persons or $26 ($22 low season) a night for 6 persons. Subject to availability, single beds for $4.50 ($4 in low season), although the ecohab may then be shared with others. Camping (maximum of 5 persons in a tent) at $4.50 per tent. The high season periods are December and January, Easter week and July. During these periods the park gets very crowded.

**Food/restaurant:** At Canaveral, two restaurants serving three meals a day. Expect to spend between $6 and $9 a day.

**Transport in park:** The only road at Canaveral is the access road from El Zaino.

**Trails:** There is well-marked system of trails, with maps available. These include a walk from Canaveral to Arrecifes (where accommodation is possible in hammocks, but food will have to be carried in) and thence to Pueblito (an archeological site).

**Further advice:** Access to Tayrona Na-

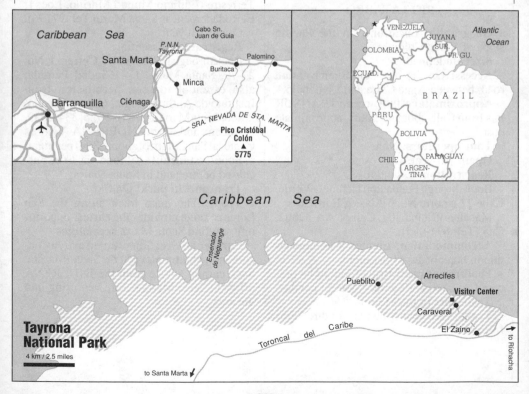

tional Park is also possible at Palangana (closer to Santa Marta) and those with hire cars may wish to explore this and other parts of the park.

Accommodation is, however, only available at Canaveral, which provides an excellent base, particularly for short-stay visitors.

# THINGS TO DO

## TOURS

Package tours as well as travel within and beyond Colombia can be arranged with El Dorado Tours in Bogotá, Tel: 234-1716 and 283-3536, Telex: TAS 45521 (Code E-01).

## MUSEUMS

Like those in most other countries, all museums in Bogotá are closed on Mondays.

**Archaeological Museum**
Carrera 6, No. 7-43

**Museo del Oro (Gold Museum)**
Parque de Santander, corner of Calle 16 and Carrera 6-A, open Tuesday to Saturday 9 a.m. – 4 p.m., Sunday 9 a.m. – noon.

**Museo Mercedes de Perez**
Carrera 7, No. 94-17, on colonial life.

**Museum of Modern Art**
Calle 24, No. 6-55

**National Museum**
Carrera 7, No. 28-66 in an old Panopticon prison.

For theaters and cinemas, see the "Espectaculos" section of daily newspapers.

## BARS & NIGHTCLUBS

In Bogotá, there are several good *salsa* bars around the intersection of Carrera 5 and Calle 27, although this is the seedy section of town. Others can be found in the Candelaria area – just wander the streets and listen for the blaring music. None gets moving until after midnight on Fridays and Saturdays.

For taped tangos from Argentina, head for **El Viejo Almacén** at Carrera 5, No. 14-23.

In Cartagena, a Caribbean trio plays on weekends at **Paco's** opposite the Santo Domingo church – a good place to nurse a drink into the small hours. The bar attached to **La Quemada** has live *salsa* on Friday and Saturday nights, with dancing. For discotheques, head for the Bocagrande district.

# SHOPPING

## WHAT TO BUY

For Colombian handicrafts, the best place is **Artesanias de Colombia**, Carrera 3, No. 18-60 next to the Iglesia de las Aguas in Bogotá. Other shops are at Carrera 10, No. 26-50 and Carrera 7, No. 23-40.

Emeralds can be bought in the *joyerias* in Bogotá's **Centro Internacional** or in **El Centro** around Calle 19/Carrera 7.

The best antique shop in Bogotá is on the **Plaza Bolívar,** next to the cathedral. Pre-Columbian pottery is sold in the **Centro Internacional**. Colombian leather goods are one of its lesser-known bargains, available in shops around the city.

# FURTHER READING

# USEFUL ADDRESSES

## BIBLIOGRAPHY

Castano U., C. (1990) *Guia del Sistema de Parques Nacionales de Colombia.* INDERENA. Bogotá.

Dix, Robert H. *The Politics of Colombia,* (Praeger, NY, 1983).

Dydynski, Krysztof. *Colombia: A Travel Survival Kit,* (1988). (This is an up-to-date guide to the country. Although aimed at the backpacker, it has information that would interest any traveler to Colombia.)

Hemming, John. *The Search for El Dorado*, (Bogotá, 1984). (This is an account of the Spanish conquest of Colombia.)

Hilty. S.L. & Brown, W.L. (1986) *A guide to the Birds of Colomnbia*. Princeton University Press. Princeton.

Tailly, Francois de. *Colombia*, (Delachaux and Niestle, 1981). (It has photographs and text in Spanish and English.)

Also worth considering is the *South American Handbook* (updated annually), a vast tome which has details on the most obscure parts of the whole continent.

Any traveler going to Colombia should read the works of national author Gabriel García Márquez. His short stories and classic work *One Hundred Years of Solitude* give an invaluable insight into Colombia's past and society.

## TOURIST INFORMATION

Can be obtained from Calle 28, No. 13A-15, ground floor (Edificio Centro de Comercio Internacional). Information is also available at the office in El Dorado airport.

There are branches of the Corporacion Nacional de Tourismo in all major towns and cities.

## EMBASSIES & CONSULATES

The following embassies and consulates are in Bogotá:

**British Embassy**
Calle 98, No. 9-03
Tel: 218-2899 or 218-1867

**United States Embassy**
Calle 37, No. 8-40
Tel: 285-1300

**West German Embassy**
Carrera 4, No. 72-36, 6th floor
Tel: 259-2501

**French Embassy**
Tel: 285-4311

**Venezuelan Consulate**
Avenida 13, Nos. 103-16
Tel: 256-3015

# GETTING THERE

## BY AIR

Venezuela has six international airports and 282 airdromes of which 250 are private. These are used mostly by small planes and helicopters. The Metropolitan area of Caracas is served by the Simon Bolívar International Airport.

## BY ROAD

There are 64,516 km of roads, about 29,032 km of which are paved and there are many freeways.

Long distance bus lines can be found operating from the Nuevo Circo bus station in the Caracas city center. Bus travel varies a lot in quality with most companies liable to run bone-shaking wrecks of buses on the same routes as luxury coaches.

# TRAVEL ESSENTIALS

## VISAS

Entry is by passport and visa or passport and tourist card.

When arriving by air, tourist cards can be issued to citizens of the United States, Canada, Japan and Western European countries (except Spain and Portugal).

When arriving by land, a visa must be obtained from a consulate which is often a lengthy process.

## MONEY MATTERS

Major credit card companies have their office in Caracas and these include:
American Express, Tel: 210-522
Visa, Tel: 575-2922
Diners, Tel: 507-1440
Mastercard, Tel: 751-3911

## HEALTH

Health conditions in Venezuela are good. Water in all major urban areas is chlorinated and safe to drink. Medical care is good. Inoculation against typhoid and yellow fever is advisable and you should have protection against malaria if you plan to visit the Orinoco basin and other swampy or forest regions in the interior. It is always good to take some form of remedy for stomach upsets and, always have a handy roll of toilet paper.

## WHAT TO WEAR

Tropical worsted clothing in normal city colors is best for Caracas while in Maracaibo and the hot, humid coastal and low-lying areas, regular washable tropical clothing is best. In western Venezuela, in the higher Andes, a light overcoat and woolen sports jacket are handy. Khaki bush clothing is a must for a visit to the oil fields but remember it is the local custom that men should wear long trousers except at the beaches. Women should wear slacks or cotton dresses with an extra wrap for cooler evenings as well as in air-conditioned restaurants and cinemas.

## ON ARRIVAL

Tourist authorities warn that taxi services at the main airports are infamous for overcharging hapless tourists and anyone else who cannot speak Spanish or handle the situation well. Corpoturismo, the Venezuelan State Tourist Authority, has made efforts to control excesses. Ask for assistance at the airport information desk if you are in doubt.

Porter services are available at all the airports and hotels – a modest tip of Bs 10 with the hint of more to come (perhaps totalling Bs 25 – Bs 30) on completion of the task will often elicit the best help available.

# GETTING ACQUAINTED

## GOVERNMENT & ECONOMY

Venezuela is a centralized federal republic formed by 20 States, a Federal District, two Federal Territories and 62 Federal Departments (corresponding to Venezuela's islands in the Caribbean Sea). It has enjoyed an uninterrupted period of democratic freedom since January 23, 1958. The government system is a representative democracy with one authority for each of the three branches of Public Power – Legislative, Executive and Judicial.

Executive power rests solely with the President designated by direct, popular and secret ballot for a five-year non-renewable term. With his Council of Ministers, he is answerable to the legislature – a two-chamber Congress of the Senate and the House of Representatives. While ex-presidents of the Republic automatically become life members of the Senate, its other members are elected, two from each State and Federal District plus 55 senators proportionally representing the minority political parties.

Seventy-seven percent of Venezuela's 17.3 million population lives in the urban areas of the country's many, middle-sized cities, making an average of 50 inhabitants per square mile.

## CLIMATE

Although Venezuela predominantly has a tropical climate with an average temperature of 27°C (80°F), four well differentiated climatic zones are represented within its boundaries: hot, mild, cool and cold.

In the Andes region the highest mountains are covered with permanent snow. Like everywhere else in the tropics, temperature depends greatly on altitude above sea level and temperate climates can be found among beautiful mountain landscapes away from the coast. Hot temperatures prevail throughout the year in the lowlands, mainly along the coast. During December, January and February there is a slight lowering of temperature and in zones like the Caracas valley and up to 1,981 m above sea level, the climate during those months is temperate and similar to France or Spain during April or the beginning of May.

The rainy season in Venezuela is, for want of a better word, called winter, generally starting in mid-May and lasting until the end of October. But showers may fall in December or January.

In the capital, Caracas, the January average temperature is 18.6°C (65°F) rising to 21°C (70°F) in July.

## CULTURE & CUSTOMS

As a result of different customs, rites, religions and musical expressions handed down through generations, Venezuelan folklore is both colorful and varied. In the warmer areas in particular, local musical instruments have African origins – the Dancing Devils of Yare and the San Juan Dance are characteristic expressions with lineage from the West African slave trade.

The "Joropo" is regarded as the national dance of Venezuela. It comes from the Llanos (Prairies) and is a lively form of music generally interpreted using the harp, "cuatro" and maracas.

## USEFUL HINTS

The metric system of measurement is used throughout Venezuela. Electric power in the country is 115 volts (60 cycles fluctuating) for domestic and personal appliances.

## BUSINESS HOURS

Business hours are 8 a.m. – noon and 2 p.m. – 6 p.m. although some stores stay open until later (8 p.m.). Like in all Latin countries, Venezuelans tend to enjoy extended lunch-hours.

## HOLIDAYS

Venezuelan time is four hours behind Greenwich Meridian Time.

The following is a list of holidays celebrated in Venezuela:

**January 1st** – New Year's Day
**April 19th** – Anniversary of the National Declaration of Independence
**May 1st** – Labour Day
**June 24th** – Anniversary of the Battle of Carabobo
**July 5th** – Anniversary of the Signing of the Venezuelan National Independence Act
**July 24th** – Anniversary of the birth of the Liberator, Simon Bolívar
**October 12th** – Columbus Day (Anniversary of the Discovery of America)
**December 25th** – Christmas Day

# COMMUNICATIONS

## MEDIA

The *Daily Journal*, founded in 1945 by Jules Waldman, is the country's only English-language daily newspaper and is favored by newcomers to Caracas as well as international businessmen and diplomats.

The main Spanish-language papers published in Caracas with a national distribution are *El Nacional, El Universal* and *El Diario*.

## POSTAL SERVICES

The Venezuelan Postal Service IPOSTEL is extremely slow and inefficient although efforts are being made to speed it up. As a result motor-cycle courier services abound in Caracas and other courier services give a very much better service to almost everywhere in the country than Ipostel.

## MEDICAL SERVICES

In time of emergencies, contact any of the following hospitals in Caracas (02):

Universitario (UCV), Tel: 662-6000
De Lidice, Tel: 811-311

Militar de las F.A.C., Tel: 462-3555
De Ninos (Children), Tel: 574-3511
Periferico de Coche, Tel: 681-1133

## SECURITY & CRIME

The crime rate in Venezuela, particularly in Caracas, is high. It is recommended that you take precautions by not wearing jewelry or carrying money in a way it can be snatched. You should not walk alone in narrow streets in downtown Caracas after dark and it is strongly recommended that you should not travel by car at night, particularly in the countryside where, should you have an accident or a breakdown, the risk of robbery and worse crimes is much greater.

## TOURIST INFORMATION

CORPOTURISMO is the Venezuelan State Tourist Authority with overall responsibility for matters relating to tourism. Although they claim to have a hotel-booking or reservation service, it is strongly recommended that you make arrangements via a local travel agency or tour organizer. For tourist information you may like to contact it at:

CORPOTURISMO
Corporacion de Turismo de Venezuela
Torre Deste, Piso 37, Parque Central
Avenida Lecuna, Caracas
Tel: (02) 507-8815 or 507-8816
Telex: 27328 TURIS VC
Telefax: (02) 574-8489

## COMPLAINTS

The Venezuelan Hotel Association, ANAHOVEN, Tel: (02) 574-3994 or 574-7172 liaises with the Venezuelan State Tourist Authority, CORPOTURISMO, to supervise hotel standards and to deal with all customer enquiries and complaints.

# WHERE TO STAY

## HOTELS

The following are some of the recommended hotels in Caracas.

**Caracas Hilton**
Avenida Libertador y Sur 25
Tel: (02) 574-1122 or 572-2122
Telex: 21171 HILTE VC.

**Grand Galaxie**
Avenida Baralt, Ministerio de Educacion
(Truco a Caja de Agua)
Tel: (02) 839-011 or 939-044
Telex: 26319 BH099 VC.

**Las Americas**
Calle Los Cerritos, Bello Monte
Tel: (02) 951-7387 or 951-7586 or 951-7798
  or 951-7985
Telex: 21497 AMERH VC.

**President Caracas**
Los Caobos
(behind La Previsora Building)
Tel: (02) 708-8111 or 782-6622 or 782-6390
Telex: 29037 PARDO VC.

**Tamanaco**
Tel: (02) 208-7111 or 208-7000 or 208-7199
Said to be the best hotel in Caracas.

# PARKS & RESERVES

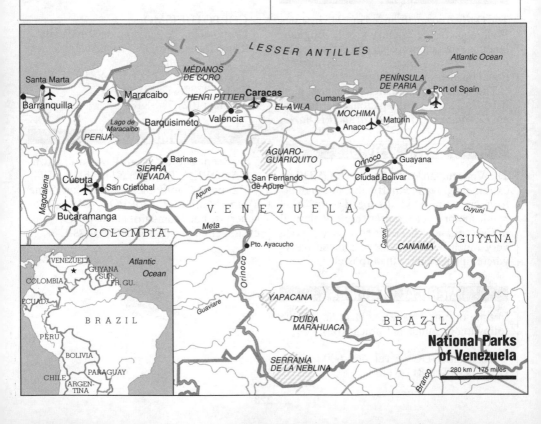

**National Parks of Venezuela**
280 km / 175 miles

**ECO-TOURISM AGENCIES**:
**VAT Venezuela Adventure Travel C.A.**
3 Av con 6 Transversal,
Quinta Mary,
Caracas
Tel: 58-2-2844712/2837984
Telefax: 58-2-2844712
Telex: 27876 CPBTH

**Manatee Venezuelan Tours**
Edif. Continental of. 10-B,
Avenida Los Jabillos con Boul. Sabana
    Grande, Caracas
Tel: 58-2-714017/726564

**G.P. Tours CA**
Avenida Miguel Angel con Calle Cervan-
    tes, Edf. Gracie Local A,
Bello Monte,
Caracas
Tel: 58-2-7511152
Telex: 27077 GP TOU
Apdo Postal 47343 Los Chaguaramos

## LODGES

**Avensa Lodge/Canaima/Angel Falls**
Served daily by Avensa Airline from Caracas. Turn to travel agent or contact Avensa
directly. Flight plus accommodation will be
from US$100 upwards. Avensa Lodge is
nicely situated at a lagoon; however, it is
normally not mentioned that visiting Angel
Falls is an additional two-day trip (six hours
by boat one way), and from December to
May often not possible, because of low water.

## Food Digest

## WHAT TO EAT

Venezuelan cuisine is very varied because
of the diverse cultural influences the country
has been subjected to over four centuries. At
Christmas and national celebrations the
*hallaca* is paramount as Venezuela's national dish – it's a stew of chicken, pork, beef
and spices used as a filling to a pie-like
dough of maize, which is then wrapped in
banana leaves and cooked in boiling water.
One of the favorites is *Pabellon* which
combines rice, black beans, shredded beef
and *tajadas* (sliced and fried ripe plantains).

In the Andes there is *pisca*, a rich and tasty
soup as well as local dishes based on trout
and sausage. Coro is famous for its *tarkari
de chivo* (made from goat) marinated fish
and goat milk preserves.

Zulia State has delicious coconut-based
specialties like *conejo en coco* (rabbit cooked
in coconut milk) and a selection of sweets
and candies. The Eastern region is widely
known for its tasty seafood specialties like
*consomé de chipichipi* (small clams broth);
cream of *guacucos* (middle-sized clams) and
*empanadas de cazón* (small shark pie). A
typical and very popular Venezuelan dish is
*mondongo* (a soup-like stew which uses
specially processed tripe as a main ingredient). The *arepa* is traditional Venezuelan
bread made from maize and served either
fried or baked.

## WHERE TO EAT

Eating out is extremely cheap by US and
European standards and there are literally
hundreds of restaurants catering to all tastes.
From the *tascas* of Candelaria where the best
Spanish cooking in the whole of Venezuela
is said to be available, to the elegant and
exorbitant French restaurants of Las
Mercedes, there is a range of gastronomic
delights to satisfy every whim.

The cheapest food is at *Fuentes de Soda*
and cafés. Food in bars may cost 50 percent
more and there is no need to feel reticent
about asking for prices before you order as
the price of a beer can often be as much as
three times higher.

A recommended list of restaurants in Caracas are:

**Arave**
Calle Mucuchies entre Avenida Rio de Ja-
    neiro y Calle Madrid, Las Mercedes
Tel: (02) 926-321 or 923-176
Arabian food.

**Boca Vante Marisqueria**
Final Avenida Venezuela, El Rosal

Tel: (02) 718-624 or 713-049
Seafood specialty.

**El Bodegon De Las Mercedes**
Edif. El Torreon, Calle Veracruz
(about 91 m from Hotel Tamanaco)
Tel: (02) 929-224
Opens midday – live music from 8 p.m.
Seafood – Tasca.

**El Cafe Naif**
Calle Madrid entre Mucuchies y Monterrey,
Urb. Las Mercedes
Tel: (02) 752-9298.

**El Gran Charolais**
Avenida Principal, La Castellana
Tel: (02) 332-723 or 326-776.

**El Palmar**
Plaza Lincoln, Colinas de Bello Monte
(at the rear of Maxy's)
Tel: (02) 751-4442 or 751-4568
Chinese food – try Peking-style duck.

**El Parque**
Edificio Anauco, Parque Central
Tel: (02) 573-7013 or 573-2976.

**El Rincon Verde**
Avenida Gamboa, San Bernadino
(at the rear of the Hospital de Clinicas)
Tel: (02) 524-898.

**El Tinajero De Los Helechos**
Avenida Rio de Janeiro, Las Mercedes
Tel: (02) 920-334
Piano Bar – live music every night.

**Gazebo**
Avenida Rio de Janeiro,
Urb. Las Mercedes
Tel: (02) 925-568 or 929-502.

**Il Ritrovo**
Torre Delta, Nivel Avenida Altamira,
Avenida Francisco de Miranda
Tel: (02) 261-1242 or 330-937
Italian pastas – Live music.

**La Brasserie Steak House**
Centro Comercial Paseo Las Mercedes, sector Trasnocho
Tel: (02) 917-646.

**La Estancia**
Avenida Principal La Castellana,
Esq. Urdaneta
Tel: (02) 331-937 or 322-419 or 338-261.

**La Fonda De Las Mercedes**
Centro Comercial Vincent, Calle Veracruz
(3 minutes walk from the Hotel Tamanaco),
Las Mercedes
Tel: (02) 913-553
Seafood specialty.

**La Cuevas Del Duque**
Edif. La Castellana II,
Avenida Principal de La Castellana con
 Avenida Francisco de Miranda
Tel: (02) 315-113 or 328-915
Spanish-style cuisine.

# THINGS TO DO

## MUSEUMS

Caracas has some fine museums and exhibitions which include: Arte Hoy, Avenida El Empalme, Urb. El Bosque. The following museums are recommended for a visit:

**Casa Natal Del Libertador**
Plaza San Jacinto A Traposos, Centro
Tel: (02) 545-7693.
 A faithful colonial re-construction on the original site where the Liberator, Simon Bolívar was born. Furnishings and memorabilia are of that period in Venezuela's history. Tuesday to Friday 9 a.m. – noon and 2.30 p.m. – 5.30 p.m., Saturday and Sunday 10 a.m. – 1 p.m. and 2 p.m. – 5 p.m. Holidays 9 a.m. – 5 p.m.

**Cuadra Bolivar**
Avenida Sr 2 entre Esq. Barcenas y Las
 Piedras
Tel: (02) 483-3971.
 A re-construction on the original site of the Bolívar family's country home "El

Palmar", with gardens and patios, colonial furniture, a restored kitchen, portraits and books of the period. Tuesday to Friday 9 a.m. – noon and 2.30 p.m. – 5.30 p.m., Saturday and Sunday 10 a.m. – 1 p.m. and 2 p.m. – 5 p.m.

### Galeria De Arte Nacional
Plaza Morelos, Urb. Los Caobos
Tel: (02) 571-3519 or 572-1070.

### Galeria Felix
3 Transversal, entre Avenida Luis Roche y San Juan Bosco, Urb. Altamira
Tel: (02) 261-4517 or 262-0204.

### Jardin Botanico
Calle Salvador Allende,
Ciudad Universitaria, UCV.

### Museo Arturo Michelena
Esq. Urapal No. 82 (a block south of La Pastora Church), Urb. La Pastora
Tel: (02) 825-853,
Former residence of the famous 19th-century Venezuelan painter Arturo Michelena. It is a stony, old-style house containing the painter's personal belongings and some unfinished canvasses. 9 a.m. – noon and 3 p.m. – 5 p.m., closed on Monday and Friday.

### Museo De Arte Colonial
Quinta Anauco, Avenida Panteon (at the Cota Mil exit), San Bernadino
Tel: (02) 518-517.
Former residence of the Marques del Toro, War of Independence hero, has been faithfully restored with a collection of colonial furniture, household implements, paintings, sculptures etc., surrounded by beautiful gardens. Tuesday to Saturday 9 a.m. – 11.30 a.m. and 2 p.m. – 4.30 p.m., Sunday and holidays 10 a.m. – 4.30 p.m.

### Museo De Caracas
Palacio Municipal (Concejo Municipal), Plaza Bolívar, Esq. Las Monjas
Tel: (02) 545-6706 or 545-8688.
Features wood-carved miniatures by Raul Santacan depicting scenes of life in Caracas from colonial times to the beginning of this century and the life works of Venezuela's internationally acclaimed impressionist painter Emilio Boggio (1857–1920). Mon-

day to Friday 9 a.m. – noon and 2.30 p.m. – 4.30 p.m., Saturday, Sunday and holidays 10.30 a.m. – 1.30 p.m.

### Museo Historico Militar
La Planicie, Urb. 23 de Enero
Tel: (02) 415-175 or 410-808.
On the site of the old Ministry of Defence buildings. It houses collections of uniforms, flags and swords. Wednesday to Sunday 10 a.m. – 5 p.m.

### Panteon Nacional
Plaza del Panteon, Avenida Panteon y Avenida Norte, Centro
Tel: (02) 821-518.
Contains the tomb of Simon Bolívar and memorials to other military heroes of the War of Independence. Tuesday to Friday 9 a.m. – noon and 2 p.m. – 5 p.m., Saturday and Sunday 10 a.m. – 1 p.m. and 2 p.m. – 5 p.m.

# SHOPPING

## WHAT TO BUY

Venezuela's craftsmen enjoy a prestigious position because of the variety and quality of their workmanship. For example, outstanding Quibor ceramics have pre-hispanic origins, molded using styles and techniques handed down through the generations. Baskets, hammocks, hats and other products made from vegetable fibers are to be found in the towns and villages along the Eastern Coast. Beautifully woven square and round ponchos, colorful blankets and caps are sold by the Andean people while craftsmen of the Llanos (Prairies) sell four-string guitar-like musical instruments called *cuatros* as well as harps and mandolins. Craftsmen of San Franscisco de Yare make Devil's masks for their Festival of the Dancing Devils while hand-carved furniture and objects made from goat skins are among representative samples from Coro.

The **Sabana Grande Boulevard** is an excellent commercial artery with hundreds of boutiques, jewelry stores, bazaars and stores. There are busy bars and coffee shops. Visit **Chacaito, C.C.C.T., Paseo Las Mercedes**, **Concresa**, **Plaza Las Americas** – all have supermarkets, department stores, cafeterias, restaurants, beauty parlors and just about every imaginable ware on display.

# USEFUL ADDRESSES

**Australia**
Quinta Yolanda, Avenida Luis Roche, entre
6 and 7 Transversal,
Altamira
Tel: (02) 261-4632.

**Austria**
Edif. Torre Las Mercedes, Piso 4,
Ofic. 408, Avenida La Estancia,
Chuao
Tel: (02) 913-888 or 913-979 or 922-956.

**Brazil**
Centro Gerencial Mohedano, Piso 6,
Avenida Los Chaguatamos con Avenida
Mohedano,
Urb. La Castellana
Tel: (02) 261-4481 or 261-5506 or 261-6529.

**Canada**
Torre Europa, Piso 7,
Avenida Francisco de Miranda,
Urg. Campo,
Alegre
Tel: (02) 951-6155 or 951-6306 or 951-6171.

**Colombia**
Calle Guiacaipuro sector Chacaito, diagonal
al Centro Comercial Chacaito,
Urb. El Rosal
Tel: (02) 951-3758 or 951-3631 or 951-5770

**France**
Edif. Los Frailes, Piso 5,
Calle La Guairita,
Urb. Chuao
Tel: (02) 910-333 or 910-324.

**Great Britain**
Torre Las Mercedes, Piso 3,
Avenida La Estancia,
Urb. Chuao
Tel: (02) 751-1022 or 751-1232.

**Italy**
Quinta Las Imeldas, Avenida Luis Roche,
Esq. 7 Transversal,
Altamira
Tel: (02) 261-0755 or 261-1779 or 261-0721.

**Japan**
Quinta Sakura, Avenida San Juan Bosco
entre 8 and 9 Transversal,
Altamira
Tel: (02) 261-8333.

**United States**
Avenida Francisco De Miranda,
Urb. La Floresta
Tel: (02) 285-2222 or 285-3111
Telefax: (02) 285-0336.

# GETTING THERE

## BY AIR

Almost touching the coast of South America, Piarco Airport in Trinidad is a good six hours from New York by plane. Tobago's Crown Point Airport is just 12 minutes flying time further, on one of the national carrier, BWIA's six daily flights.

Regular airfares to Trinidad run relatively high. There are fewer air-plus-hotel packages than for some of the more heavily advertized and touristed Caribbean countries, but several airlines do offer them occasionally.

BWIA offers flights from Miami to Piarco Airport daily, from London and New York five days a week, Baltimore–Washington three times a week, Toronto twice, Stockholm and Frankfurt once a week.

American Airlines has daily nonstop service from San Juan, linked to cities throughout the United States. There are also daily flights from London on British Airways. Air Canada's direct flights leave from Toronto; some are nonstop, depending on the day of the week. K.L.M. flies to and from Amsterdam once a week. There are also connections to other Caribbean islands and South America via BWIA, LIAT (Leeward Island Air Transport), LAV (Linea Aeropostal Venezolana), and Guyana Airways.

## BY SEA

Trinidad and Tobago is developing its assets as a cruise destination through large scale expansion of Scarborough Harbor and other improvements. This, combined with a boom in the cruise industry, means new options for travelers and more cruise lines with one or both islands on their routes. Check with a travel agent before booking.

The main problem with taking the sea route is the difficulty in extending your stay beyond the few hours the ship is docked.

Most cruise ship tickets are sold for complete voyages and you'll probably want to avoid the logistical maneuvering involved in a change.

Ships from the following cruise lines stop at Trinidad and/or Tobago:

Cunard Cruise Lines, Tel: 800-327-9501.
Epirotiki Lines Inc. (World Renaissance), Tel: 800-221-2470.
Princess Cruises, Tel: 800-421-0522.
Windjammer Cruises, Tel: 800-245-6338.

# TRAVEL ESSENTIALS

## VISAS & PASSPORTS

All visitors must have a return or ongoing ticket and a valid passport for entry into Trinidad and Tobago. Citizens of the United Kingdom, United States and Canada do not need visas unless they are doing business in Trinidad and Tobago. Visitors from certain European countries do need visas, however, check with authorities in your own country before leaving.

## MONEY MATTERS

The TT dollar (TT$) is the basic unit of currency in Trinidad and Tobago. Since many TT$ prices are the same as they were before devaluation, Trinidad and Tobago is a more reasonably priced destination than it once was, especially if you buy goods and services geared to locals or West Indian travelers.

Currently, the TT$ is classified as a "restricted currency," a result of policies aimed at curbing inflation. For visitors, this means you should reconvert all the TT$ you have not spent before you leave Trinidad and Tobago, as you probably won't be able to do so in another country. Also, be sure to get a receipt for all cash and traveler's checks you change into TT$. You must show it in order to change your money back. There is no limit to the amount of foreign currency you can bring in, but TT$1,200 is the maximum you will be able to reconvert on departure.

The National Commercial Bank of Trinidad and Tobago has a foreign exchange branch at Piarco Airport (Tel: 664-5281/5322), which changes British pounds, American and Canadian dollars and all Caribbean currencies for a small fee, but not Swiss or French francs or other European currencies. For these, you must go to a large bank in Port of Spain, so try to bring some US dollars or pounds sterling to change at the airport for initial expenses like a taxi to your hotel.

Major credit cards can be used at the larger hotels, restaurants and car rental firms throughout the islands, and at stores that cater mainly to visitors. However, many of the smaller hotels, guesthouses, restaurants and shops do not accept them, so it's wise to have an international brand of traveler's checks and a good supply of cash on hand.

## HEALTH

Vaccinations for smallpox and yellow fever are not required for entry into Trinidad and Tobago unless you have just passed through an infected area.

Throughout the Republic, the water is safe to drink and even street food is generally tasty and fresh for most of the year. However, during Carnival it's probably best to avoid the seafood and meat specials sold from numerous shanties around the Savannah. Often the food is cooked at the vendor's home in the morning, and stored without refrigeration throughout the day.

The National AIDS Program of Trinidad and Tobago also issues a timely reminder for Carnival and all year round: "AIDS is preventable; don't give it to yourself." Avoid casual sex, any procedures which pierce the skin, and remember that alcohol abuse can affect your judgement.

The sun shines with dangerous intensity all year round. Wear ample sunscreen, built tanning time slowly up from 15 minutes in the early morning and late afternoon, and bring a hat or buy one.

## WHAT TO WEAR

Casual clothes in lightweight fabrics are most comfortable for touring the countryside. In town, most men wear dark trousers and shirt-jacs, or more formal business suits. Women wear dresses or skirts and blouses for work and shopping. Note that skimpy clothing, particularly on women, will attract attention and occasionally provoke some comment.

Even at the best restaurants there's not much call for evening wear. Casual clothes in lush colors and lavish fabrics prevail: in fashionable Trinidad a sense of style is what counts. On the beaches, anything goes. But, especially in Tobago, keep in mind that this is a small-town society, so the rules of small town propriety should be observed. Wear bathing suits on the beach only.

## ON ARRIVAL

On the airplane or ship, just before you arrive, you will be asked to fill out an immigration form. Immigration officials collect one copy as you enter the country. Save the duplicate; you will need it for departure. Also be prepared to show your return ticket and give the address where you will be staying in Trinidad and Tobago. If you do not have a Trinidad and Tobago address, Tourism Development Authority staff meet all flights and can provide information on accommodations.

## RESERVATIONS

Reservations will make your travels here run smoother. But you can probably find a room without advance reservations in Port of Spain or Tobago, though you may not get your first choice. The exception to this is Carnival season when reserving a room far in advance is essential, especially if you want to stay near to the Savannah.

## ON DEPARTURE

If you are leaving by plane, plan to arrive at the airport two hours before your flight is due to depart. Last minute schedule changes are not unheard of, so reconfirm your reservation 24 hours in advance, and make sure the flight is on time before leaving for the airport. Allow extra time of at least an hour to get to Piarco if you are traveling in a taxi or private car from Port of Spain during rush hour. Between about 6.30 a.m. and 8.45 a.m., and 3.30 p.m. and 5.45 p.m., traffic is most horrendous and will defintely cause frustrating delays.

Also note that there are some strict departure rules for business people: find out what regulations apply to you at least a week before your planned departure, as they may necessitate a visit to various government offices that are not always open.

On leaving Trinidad and Tobago, you will be required to pay a TT$50 departure tax. You will also be asked to surrender the carbon copy of the Immigration Card you filled in on arrival. Beyond the customs and immigration checkpoint, the airport departure area has duty-free shops, refreshment stands, and telephones for overseas calls.

# GETTING ACQUAINTED

## GOVERNMENT & ECONOMY

Trinidad and Tobago became fully independent from Britain in 1962, and declared the Republic into being in 1976, when the President replaced the British Monarch as head of state. The government is a parliamentary democracy, headed by President, and governed by Parliament and the Prime Minister. Tobago's legislative body, the House of Assembly, sets domestic policy for the island.

Since gaining political independence, Trinidad and Tobago's leaders have also worked to achieve economic independence and stability. This goal seemed well within their grasp during the late 1970s, when prices for exported Trinidad and Tobago oil were at an all-time high. But much has changed in the last decade. Oil prices declined precipitously, unemployment claimed more of the workforce, and a populace that had grown accustomed to prosperity began to experience harder times.

Beset by economic problems, the government has turned more attention to tourism, a source of hard currency that has remained

uncultivated for many years, and encouraged light industries to diversify the country's economic base.

## SIZE & POPULATION

Of all Caribbean islands, Trinidad is furthest south, just 11 km from the coast of Venezuela across the Gulf of Paria. Roughly triangular in shape, the island is about 80 km long and almost 64 km wide, with three mountain ranges traversing the interior. These areas remain heavily forested and rich in wildlife. The east and west coasts tend to be swampy; the most popular beaches are in the north, over the mountains from Port of Spain, the capital. The south is very different, with its famous pitch lake, oilfields and several mud volcanos.

Fish shaped Tobago is 41 km long, 11 km wide and lies 32 km northeast of Trinidad. A central range of mountains forms a spine through the Tobago Forest Reserve in the north. Coral reefs surround much of the southern tip of the island. Several smaller islands lying off the coasts are wildlife sanctuaries.

As a nation, the population of Trinidad and Tobago is divided between roughly equal numbers of African and East Indian descendants, and much smaller percentages of Chinese, Europeans and Syrians. This diversity is mainly confined to Trinidad, however; in Tobago, over 90 percent of the people are of African descent.

## TIME ZONE

Trinidad and Tobago operate one hour ahead of Eastern Standard Time, or the same as Eastern Daylight Savings Time.

## CLIMATE

The weather is almost always perfect. Because of the trade winds, the sun's heat is mitigated by sea breezes. Average temperatures are 23°C (74°F) at night and 28°C (84°F) during the day.

There are two seasons: the Wet and the Dry, from December to May and June to November respectively. All this means – except in late October and November, when hurricanes further north can create stormy weather – is that from June to December it rains briefly in the afternoon. June is the wettest month; February and March the driest. It never rains during carnival.

## CULTURE & CUSTOMS

The ambience on both islands is pleasantly relaxed, but good manners are important in Trinidad and Tobago as in anywhere else. Visitors are expected to act like guests: to say "please" and "thank you," and to ask permission before taking anyone's picture. And although Trinidadians, in particular, have a disarming way of getting straight to the point in conversation, it's still inappropriate to launch right into a request for service or directions without a proper greeting – "hello, good morning," "good afternoon," "good evening."

In Trinidad and Tobago, time tends to be enlarged. If your Trini friends say they'll pick you up at eight, they'll probably arrive closer to nine. However, don't be insulted and don't worry about other appointments because the whole country is on the same schedule. "He's coming just now" may mean he'll be here any second or in the next day or two – often the latter. If you really need something done in a hurry be polite and be persistent.

## ELECTRICITY

Trinidad and Tobago operates on 110 volts and 60 cycles of electric current. If you need a transformer, most hotels can supply one.

## BUSINESS HOURS

Monday through Friday, most shops and offices are open from 8 a.m. – 4 p.m. or 4.30 p.m. Some food stores stay open until 6 p.m., except on Thursdays, when food and liquor stores close at noon. Saturday closing time is at noon or 1 p.m., except for food and liquor stores, which are open all day. Large malls like the Long Circular stay open until quite late; other malls are open late on Fridays and Saturdays.

All banks stay open from 9 a.m. – 2 p.m. Monday to Thursday, and from 9 a.m. – 1 p.m. and 3 p.m. – 5 p.m. on Fridays. Most of the major commercial banks have offices on Independence Square, as well as various branches.

## HOLIDAYS

**Good Friday**
**Easter Monday**
**Whit Monday** – eighth Monday after Easter
 Corpus Christi
**Labour Day** – June 19th
**Emancipation Day** – August 1
**Independence Day** – August 31
**Eid-Ul-Fitr** – as decreed
**Divali** – as decreed
**Republic Day** – September 24
**Christmas Day** – December 25
**Boxing Day** – December 26

In addition, Carnival, the raucous and unforgettable national party takes place on the Monday and (Shrove) Tuesday before Ash Wednesday – with the precise dates different each year. Though not a legally decreed holiday, it's safe to assume that a very relaxed attitude will prevail at any office that stays open.

## FESTIVALS

**Carnival and Steelband Music Festival** – every two years;
**Natural History Festival** – October
**Annual Indian Festivals:** Phagwah, Divali Nagar, Eid ul-Fitr and Hosay, the Best Village Folk Festival – October to December;
**Annual Flower and Orchid Shows** – March and October;
**Drama Festival** – May and June;
**Tobago's Heritage Festival** – July.

# GETTING AROUND

## ORIENTATION

For advice before your trip, consult one of the Tourism Development Authority offices in Trinidad, New York, Miami, Toronto or London. The TDA Information Office (Tel: 664-5196) at Piarco Airport is open every day except Christmas from 6 a.m. – midnight.

Here you will find friendly and efficient answers to questions about accommodations, transportation and events of interest to visitors. TDA staff also meet all cruise ships and arrange special activities like tours and calypso performances for the passengers.

The Trinidad and Tobago Hotel and Tourism Association is also happy to provide brochures on member hotels, and information about facilities and services. Write them at P.O. Box 243, Port of Spain; Tel: or Telefax: 624-3065.

## WATER TRANSPORTATION

Two ferries ply between Port of Spain and Scarborough, the *M.V. Tobago* and the *M.F. Panorama*. The trip takes five hours, and food and drink are available on board. Cabins are available for TT$80, one way double occupancy, which might be a good idea if you take the 8 p.m. sailing. Tickets are sold only at the Port of Spain and Scarborough offices; passenger ticket sales close two hours before sailing time. For more information call Tel: 625-4906 in Port of Spain and Tel: 639-2417 in Tobago.

## PUBLIC TRANSPORTATION

**Bus:** The Public Service Transport Service (PTSC) runs both buses and maxi-taxis. Buses go to every part of both islands, and are quite cheap. But because they are often hot and crowded, most tourists, and many locals, prefer the maxi taxi which costs slightly more.

**Maxi-Taxis:** These color-coded minibuses follow particular routes on both islands. In Tobago, ask directions to the nearest stop.

**Route Taxis:** These cars carry up to five passengers along set routes from fixed stands to various destinations. Sometimes drivers will digress slightly for visitors if the car isn't crowded. Unless they render special services, drivers do not expect tips. In Scarborough, Tobago, route taxi stands are located across from the bus terminal and central shopping plaza.

## PRIVATE TRANSPORTATION

**Hire Taxis:** Like route taxis, hire taxis have an H on their licence plates, but they are essentially private taxis, carrying only you and your companions exactly where you want to to. Hire taxis do not have meters and are rather expensive. Always agree on fares in advance. Also note that fares in Tobago double after 9 p.m.

Hire taxis wait near most large hotels. Below are a few taxi services:

Kapok Taxi Service, 16-18 Cotton Hill, St. Clair, Tel: 622-6995.

Himraj Taxi Rental Service, St. Helena Village, Piarco, Tel: 622-3566.

Queen's Park Taxi Stand, Queen's Park West, Tel: 625-6005.

St. Christopher Taxi Co-op Society, Ltd., Tel: 624-3560/3111 (Hilton, main office), Tel: 625-4531 ext. 1514 (Holiday Inn), or 625-3361, Tel: 625-1694.

Tobago Taxi Cab Co-op Society Ltd., Tel: 639-2659 (Carrington St.), Tel: 639-2707 (Milford Rd.).

Tobago Taxi Owners and Drivers Association, Tel: 639-2659 (Crown Reef Hotel), Tel: 639-2692 (downtown Scarborough).

**Driving Regulations:** If you have a valid driver's licence from the United States, Canada, France, United Kingdom, Germany or the Bahamas, you may drive in Trinidad and Tobago for a period of up to three months – but only the class of vehicle specified on your licence. (China, South Vietnam and South Africa are excluded from this privilege.) If your stay exceeds this limit, you must apply to the Licencing Department on Wrightston Road in Port of Spain.

While driving stay to the left, and at all times carry both your driver's licence and a document (such as a passport) certifying the date of your arrival in Trinidad and Tobago.

**Rental Cars:** Car rental in Trinidad and Tobago is handled by numerous local fleets. Many take credit cards for both rental fees and the substantial deposit (up to TT$1,000) that is often required. (*Caution:* Remember that there is a limit to the amount of TT$ you can reconvert on departure. If you expect the refund of a large TT$ deposit when you turn in your car, make sure it won't put you over the limit, or allow time to spend it.) Day rates for cars with manual transmission begin at about TT$120; weekly rates start at TT$725. The difficulty with car rental is that business clients engage many of the cars for long periods, so a short term rental can be hard to find. Plan to shop around well in advance and consult a travel agent for suggestions.

When driving in the country, you might see hitchhikers, particularly in Tobago. Everyone from schoolchildren to grandmothers hitch (though visitors shouldn't) and it's only friendly to "give a drop" to children, the elderly, and mothers with babies. Exercise more caution with young men.

# WHERE TO STAY

Trinidad has a limited number of hotels, compared to the Caribbean's resort islands, and nothing that could be called a resort, American-style. As beautiful as the beaches are here, visitors come for music, culture, business, and Carnival, then retire to Tobago for relaxation. Many business travelers stay at the Hilton or the Holiday Inn, which have facilities for business people. The Kapok Hotel and Hotel Normandie also have large business clienteles and popular restaurants. Facing the Savannah within walking distance of downtown, the Queen's Park Hotel attracts value conscious Europeans and West Indians.

In the listing below, based on winter rates and subject to change; described as "inexpensive" generally means costing below US$30 per night, "reasonable" US$30 to US$50, "moderate" US$50 to US$80, "expensive" over US$80 per night for a standard single room.

## HOTELS

### TRINIDAD

**Bel Air International**, Piarco Airport, Tel: 664-4771/3. Moderate to expensive.

**Cactus Inn**, Coconut Drive, Cross Crossing, San Fernando, Tel: 657-2657/8. Reasonable.

**Charconia Inn**, 106 Saddle Road, Maraval, Tel: 629-2101. Moderate.

**Errol J. Lau**, 66 Edward Street, Port of Spain, Tel: 625-4381. Inexpensive.

**Farrel House Hotel**, Southern Main Road, Claxton Bay, San Fernando, Tel: 659-2230/2271/2. Reasonable.

**Hilton Internationsal Trinidad**, P.O. Box 442, Port of Spain, Tel: 624-3211. Expensive.

**Holiday Inn**, P.O. Box 442, Port of Spain, Tel: 624-3211. Expensive.

**Hotel Normandie**, P.O. Box 851, Nook Avenue, St. Ann's, Port of Spain, Tel: 624-1181/4. Moderate.

**Kapok Hotel**, 16-18 Cotton Street, Port of Spain, Tel: 622-6441/4. Moderate.

**Queen's Park Hotel**, 5-5A Queen's Park West, Tel: 625-1060/1. Reasonable.

### TOBAGO

**Arnos Vale**, P.O. Box 208, Scarborough, Tel: 639-2881. Expensive.

**Blue Waters Inn**, Batteaux Bay, Speyside, Tel: 660-4341. Moderate.

**Cocrico Inn**, Plymouth, Tel: 639-2961. Moderate.

**Crown Point Beach Hotel**, Crown Point, Tel: 639-8781. Expensive.

**Grafton Hotel**, Black Road.

**Jimmy's Holiday Resort**, Crown Point, Tel: 639-8292. Reasonable to moderate.

**Kariwak Village,** P.O. Box 27, Scarborough, Tel: 639-8545. Moderate.

**Man-O-War Cottages,** Charlotte, Tel: 639-4327. Reasonable.

**Mount Irvine Bay Resort**, P.O. Box 222, Scarborough, Tel: 639-8871. Expensive.

Sandy Point Beach Club, P.O. Box 223, Crown Point, Tel: 639-8533. Moderate.

Tropikist Beach Hotel, Crown Point, Tel: 639-8512. Reasonable.

Turtle Beach Hotel, Courtland Bay, Tel: 639-2851. Expensive.

## GUEST HOUSES

### TRINIDAD

Alicia's House, 7 Coblentz Gardens, St. Ann's, Port of Spain, Tel: 623-2802.

Kestour's Sports Villa, 58 Carlos Street, Woodbrook, Port of Spain, Tel: 628-4028.

Monique's Guest House, 114 Saddle Road, Maraval, Port of Spain, Tel: 629-2233.

Mount St. Benedict Guest House, Mt. St. Benedict, Tunapuna, Tel: 662-4084.

Zollna House, 12 Ramlogan Development, LaSeiva, Maraval, Tel: 628-3731.

All are reasonable.

### TOBAGO

Coral Reef Guest House, Milford Rd. Scarborough, Tel: 639-2536.

Della Mira Guest House, Bacolet Street, Scarborough, Tel: 639-2531.

Richmond Great House, Belle Garden, Tel: 660-4467.

All are reasonable.

## BED & BREAKFAST

The Bed and Breakfast Association of Trinidad and Tobago publishes a leaflet listing a number of private homes with one or more rooms to spare. All have been inspected and many have extra features like television and access to kitchens. Breakfast is included in rates which range from US$17 to $40, slightly higher during Carnival. TDA offices have information or write to the As-sociation at Diego Martin P.O. Box 3231, Diego Martin, Republic of Trinidad and Tobago; or phone Miss Grace Steele, Tel: (809) 637-9329, or Mrs. Barbara Zollna, Tel: (809) 628-3731/660-4341.

# FOOD DIGEST

Travelers from the budget conscious student to the business executive can find food to suit their tastes and pockets, though food prices all the while are relatively high. The more expensive hotels tend to serve a kind of bland continental cuisine, but some offer special Creole meals.

Restaurants run the gamut from roti shops frequent by working people to sophisticated restaurants housed in restored Victorian mansions, serving the best French, Creole, Indian, and Asian cuisines. Roadside stands proliferate, as do American-style fast food outlets.

## DRINKING NOTES

Mixed drink specialties in Trinidad and Tobago revolve around rum, and it's not hard to find rum in a punch that will infuse any time of year with a bit of Carnival. Trinidad is also home to Angostura Bitters, a secret concoction that has been adding zest to mixed drinks for generations. Two popular brands of beer are Stag and Carib. And don't forget to sample local soft drink favorites like sorrel, ginger beer and mauby.

# PARKS & RESERVES

**ECO-TOURISM AGENCIES:**
**Peregrine Enterprises, Inc.**
P.O. Box 1003
College Park, Maryland 20740,
USA
Tel: (301) 474-1880

**ASA Wright Nature Center**
c/o Caligo Ventures, Inc.
156 Bedford Road,
Armonk, New York 10504,
USA
Tel: (800) 426-7781

**Mount St. Benedict Guest House**
Mount St. Benedict
Tunapuna, Trindad
Tel: (809) 662-4084

# THINGS TO DO

If your main stop is Trinidad, don't leave without spending some time in Tobago – or vice versa. BWIA has one-day packages to Tobago at a cost of TT$130, including transfers to the beach at Pigeon Point and a trip to Buccoo Reef. Tickets are sold only at BWIA offices and must be bought at least 24 hours in advance.

    **Nealco Air Services** (Tel; 664-5416, Piarco; Tel: 625-3426, Port of Spain) can arrange plane and helicopter sightseeing trips within Trinidad and Tobago and to other islands.

    Pan American's daily flights from New York's John F. Kennedy Airport stop first in Caracas before landing at Piarco. If you've never visited South America (or even if you have), you can easily arrange a stopover which includes a hotel, or make reservations yourself.

## CRUISES

**Jolly Roger**, Point Gourde Road, (just beyond the Small Boats Jetty), Tel: 634-4334. Wednesdays, Calypso Cruise; Fridays, Heart to Heart Cruise. Boarding time 8.30 p.m., cruise 9 p.m. to midnight.

## TOUR GUIDES

**Mr. Winston Nanan**
Tel: 645-1305

**Mr. David Ranasahai & Sons**
Tel: 663-2207 and 645-4705

## TOUR OPERATORS

**Bruce/Ying Tours**
99 Queen Street,
Port of Spain
Tel: 628-1851

**Trinidad & Tobago Sightseeing Tours**
10 Fitt Street, Woodbrook,
Port of Spain
Tel: 624-1984

**Eastman Tours**
1 Herbert Street, Newtown
Tel: 628-1851

**Tobago Travel Ltd**
Storebay
Tel: 639-8778

# GETTING THERE

## BY AIR

Direct flights from the United States into Ecuador are available with Ecuatoriana. From Europe, flights are by Avianca and Lufthansa.

Within South America, most national airlines offer services to Quito. Aerolineas Argentinas stops over at Quito to and from Buenos Aires to Bogotá. AeroPeru's flight from Lima to Quito can be included in its special "Visit South America fare".

Marisal Sucre Airport in Quito is not far from the city center. Taxis cost around US$2. There is a US$20 airport and police departure tax.

## BY LAND

Many travelers enter from Peru at the Tumbes-Huaquillas border post, changing buses at the frontier. Buses are frequent both ways. To Colombia, almost everyone takes the Quito-Ibarra, Tulcán-Pasto road.

# TRAVEL ESSENTIALS

## VISAS

Visas are not required for anyone staying up to 90 days. Stay permits (known as "T3") are issued at the border or point of arrival, usually for 30 days which can be easily extended. The immigration office in Quito is at Independencia y Amazonas 877.

For visa extensions, go to Avenida Amazonas 2639, open from 8 a.m – noon and from 3 p.m. – 6 p.m.

The main police office dealing with tourist matters is on Calle Mantúfar.

## MONEY MATTERS

The Ecuadorian currency is the *sucre*, divided into 100 *centavos*. Check the papers for the exchange rates. Travelers should bring US dollars for conversion as the dollar is highly valued and tend to fetch a higher rate of exchange at the exchange houses. Traveler's checks are also acceptable.

Diners and Visa credit cards are widely accepted for meals and purchases.

# GETTING ACQUAINTED

## GOVERNMENT & ECONOMY

There are 20 Ecuadorian provinces, including the Galápagos Islands. Ecuador has been under democratic rule since 1979. The economy is based on oil, fishing, bananas, coffee and cocoa.

## CLIMATE

Each of Ecuador's regions has a distinct climate. The coast is hottest and wettest from January to April. The mountains are driest, and therefore clearest between June and September, and in December. The Oriente is usually inundated between June and August, followed by a drier season until December. As on the coast, however, a torrential downpour is never out of the question. The best months to visit the Galápagos Islands are March, April and November.

The mountains soar and the temperature drops: Quito, at 2,850 m, is cold when the sun disappears. The other regions are invariably warm and often steaming hot.

Due to the country's equatorial climate, seasonal temperature variation is minimal.

## MEDIA

The main newspapers in Ecuador are *El Comercio*, *Hoy*, *Tiempo* and *Extra*.

## POSTAL SERVICES

Unreliable, but better in Quito than the provinces. The central post office is at Benalcázar 688 y Chile. For poste restante go to office in Toledo y Madrid.

## TELEPHONE & TELEX

Telephone center is at Benalcázar 769 y Chile. Calls to Europe are US$18 for three minutes. Telex services are offered by all major hotels.

## DOMESTIC TRAVEL

If a road exists, some form of public transport will run along it. In the Oriente, motorized dugout canoes ply the main rivers.

Flights within mainland Ecuador are remarkably cheap. Quito–Guayaquil costs $30, Quito–Cuenca $18. Flights to the Galápagos Islands is another matter: $367 return from Quito, $325 return from Guayaquil.

# WHERE TO STAY

## QUITO

For budget accommodation in Old Quito, the New Gran Casino Hotel and the Viena International are conveniently located and comfortable. In the new town, the cozy Hostal Los Alpes offers the most charming and tasteful accommodation. Residencia Lutecia is a cheaper option, while Hotel Colón provides every service and five-star comfort.

## GUAYAQUIL

The best area to stay is by the river, where the air is cleaner and cooler. The Metropolitan, Ramada and Humboldt hotels have good river views, while Hotel Moneda satisfies the luxury-conscious.

# FOOD DIGEST

## QUITO

Apart from the hilltop Panecillo, Old Quito has little to recommend by way of restaurants. Numerous places serve cheap set meals, and Chinese restaurants, known as *chifas*, offer adequate fare.

The streets surrounding Avenida Amazonas in the new town have many good "international" restaurants. These include the Columbus Steak House, the Excalibur, the El Cebiche for the national dish. The Hotel Colón offers a sumptuous buffet daily at lunchtime, and the pavement cafes on Amazonas are good for snacks, sunshine and street life.

## GUAYAQUIL

In Guayaquil, the most pleasant place to eat is aboard one of the boats moored along the Malecón. For cheaper and often more fiery seafood meals, the area just north of Parque del Centenario is good. Calle Escobedo has two popular street cafes serving breakfast.

# PARKS & RESERVES

## SANGAY NATIONAL PARK
### (CULEBRILLAS SECTION)

**Geographic Location:** South and east of the capital Quito, from the eastern *cordillera* to the Oriente. Approximate area bounded by Baños in the north to Guamote in the south and the confluence of the Rios Palora and Sangay in the east.

**Size:** 270,000 hectares.

**Access:** For Sangay volcano and culebrillas (cloud forest and *paramo*) or El Placer (cloud forest and montane forest) best access is from Alao (mountain village near Riobamba).

**Approximate costs of access:** Taxi from Riobamba to Alao (approximately $10) or get a lift on onion lorry.

Guides, porters and horses can be hired in the village at $4 each.

**Tour operators:** Expediciones Andinas at Argentinas 3860 y Zambrano, Riobamba organize all-inclusive hikes to the volcanoes.

**Open:** All year

**Registration:** Park entry permit ($10) can be obtained from the Administracion, Ministeria de Agricultura y Ganaderia in Riobamba. Check in at the park Guard Station at Alao.

**Accommodation:** Large range of hotels and hostels in Riobamba.

**Camping:** Camping may be possible at the guard station in Alao. Suitable camp site in Culebrillas valley and at La Playa, at base of volcano.

**Food/restaurant:** Purchase basic supplies in Riobamba, fresh vegetables available in Alao.

**Transport in park:** Horses, mules can be hired in Alao, but general means of access is on foot.

**Trails:** Hire guides to show the way.

## SANGAY NATIONAL PARK
### (PURSHI SECTION)

**Geographic Location:** South and east of the capital Quito, from the eastern *cordillera* to the Oriente. Approximate area bounded by Baños in the north to Guamote in the south and the confluence of the Rios Palora and Sangay in the east.

**Size:** 270,000 hectares.

**Access:** From Purshi (cloud forest and *paramo*) there is a footpath down to the Oriente at Macas.

**Approximate costs of access:** Four-wheel drive vehicle from Guamote (two hours south of Riobamba) to Laguna Negra along new road under construction. Hire guides and porters (information form Park Headquarter in Riobamba.

**Open:** All year.

**Registration:** Permit from Park Headquarter in Riobamba.

**Accommodation (Camping):** Camp at Laguna Negra on edge of the Park and along trail.

**Food/restaurant:** Purchase supplies in Riobamba or Guamote.

**Transport in Park:** On foot.

**Sangay National Park**

8 km / 5 miles

**Trails:** Guides are required. Main trail from Laguna Negra to Macas, may take several days to walk.

## PODOCARPUS NATIONAL PARK

**Geographic Location:** Southern Ecuador, between Loja and Vilcabamba and east toward Zamora.

**Size:** 146,280 hectares.

**Access:** See Western section and Northern section as follows.

**Western section** (*paramo*, Podocarpus forest, elfin forest): Car, taxi or bus from Loja, 15 km along road to Vilcabamba. Six kilometers walk from road to Guard Station. Possible to hire guides and horses in Vilcabamba.

**Northern section** (subtropical): Car, taxi from Zamora. Park guards and guides at the ranger station on the Rio Bombusgaro.

**Approximate costs of access:** Inclusive taxi fee Loja to Guard Station costs $10.

**Open:** All year.

**Registration:** Permits from Ministeria de Agricultura y Ganaderia on Riofrio (off Bolivar) in Loja.

**Accommodation:** Large range of hotels and hostels in Loja.

Alternatively, use Tourist Parador in Vilcabamba.

**Food/restaurant:** Bring your own.

**Transport in Park:** On foot.

**Trails:** Extensive well-marked trail system from Guard Stations in both sections.

## ECUADORIAN ORIENTE

**Geographic Location:** Close to Cuyabeno Reserve and Yasuni National Park. Access to the latter is difficult. Equivalent jungle experience can be acquired at jungle lodges which organize inclusive tours.

**Size:** Not applicable.

**Access:** Flight from Quito to Coca (Francisco de Orellana). River journey along Rio Napo and tributaries can be arranged independently or as part of an inclusive tour.

**Approximate costs of access:** Inclusive tours are highly variable in price depending on venue and duration.

**Tour operators:** Recommended is **La Selva** of 6 de Diciembre 2816, Quito. Acclaimed lodge for ornithologists. Company arranges inclusive tours to their La Selva Jungle Lodge close to Rio Napo. Metropolitan Touring offers luxury all-inclusive, four-day tours down the Rio Napo aboard the Flotel Orellano. Cheaper and longer excursions can be organized in Misahuallí. There are many guides and agencies to choose from, but Dayuma Lodge and Fluvial River Tours are two of the most established and reliable.

**Open:** All year.

**Registration:** Not necessary unless also intending to visit the Yasuni National Park, or the Cuyabeno Reserve Area.

**Transport in Park:** River transport and trails in vicinity of lodges.

### Podocarpus National Park

8 km /5 miles

R. Mallacatos

R. Sabanilla

Cumbaratza

R. Zamora

Zamora

Mbija

R. Campanas

Lag. di L. Compadas

R. Bombusguro

Pituca

R. Cutirinza

R. Solamaco

Romerillos

Viloabamba

PODOCARPUS

NATIONAL PARK

R. Numbala

Yangana

VENEZUELA

GUYANA

Atlantic Ocean

COLOMBIA

SUR.

FR. GU.

ECUAD.

BRAZIL

PERU

BOLIVIA

PARAGUAY

Valladolid

Tapala

Loyola

R. Loyola

CHILE

ARGEN TINA

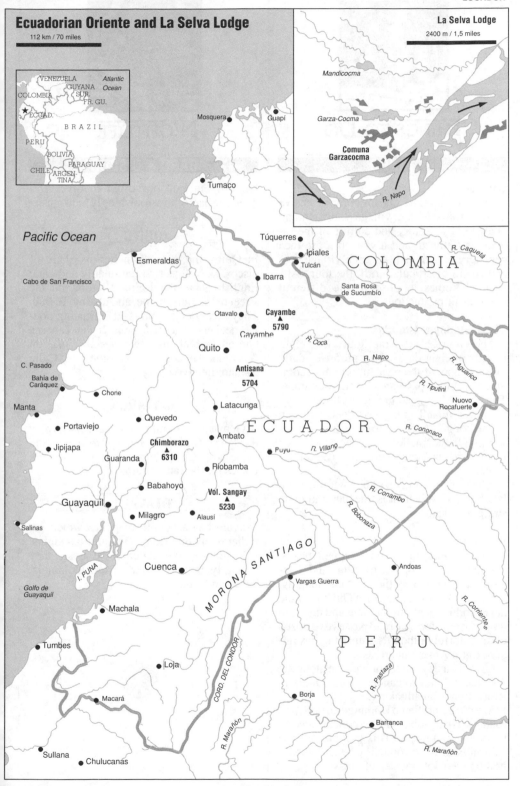

# Ecuadorian Oriente and La Selva Lodge

112 km / 70 miles

**La Selva Lodge**

2400 m / 1,5 miles

*Mandicocma*

*Garza-Cocma*

**Comuna
Garzacocma**

*R. Napo*

VENEZUELA

GUYANA

*Atlantic
Ocean*

SUR.

COLOMBIA

FR. GU.

★ ECUAD.

B R A Z I L

PERU

BOLIVIA

PARAGUAY

CHILE

ARGEN
TINA

Mosquera

Guapí

*Pacific Ocean*

Tumaco

Túquerres

Ipiales

Tulcán

COLOMBIA

*R. Caquetá*

Esmeraldas

Cabo de San Francisco

Ibarra

Santa Rosa
de Sucumbío

Otavalo

**Cayambe**

5790

Cayambe

*R. Coca*

Quito

*R. Napo*

*R. Aguarico*

**Antisana**

5704

*R. Tiputini*

C. Pasado

Bahía de
Caráquez

Chone

Latacunga

E C U A D O R

Nuovo
Rocafuerte

*R. Cononaco*

Manta

Quevedo

Portoviejo

Ambato

Puyu

*R. Villano*

Jipijapa

**Chimborazo**

6310

Guaranda

Riobamba

*R. Conambo*

Babahoyo

**Vol. Sangay**

5230

*R. Bobonaza*

Guayaquil

Milagro

Alausí

Salinas

Cuenca

M O R O N A   S A N T I A G O

Andoas

*I. PUNA*

Vargas Guerra

*Golfo de
Guayaquil*

Machala

P E R U

*R. Corrientes*

Tumbes

Loja

*CORD. DEL CONDOR*

*R. Pastaza*

Macará

Borja

Barranca

Sullana

Chulucanás

*R. Marañón*

*R. Marañón*

# THINGS TO DO

# NIGHTLIFE

## TOURS

### Otavalo
Zulaytours offers cheap, comprehensive all-day tours of the villages surrounding Otavalo. The history and culture of the Indians is outlined, and a visit to a pre-Inca cemetery is included. As each village specializes in a particular craft, the tour visits various homes to observe the different methods of production.

### The Galápagos Islands
The Gran Casino Hotel tour visits eight islands in eight days. They promise "daily lobster" aboard their yacht, which accommodates eight passengers and four crew. The cost is $430, excluding air fare.

Metropolitan Touring has a fleet of a dozen yachts for group charter for one, two or three weeks. On the smaller boats, the costs range from $350 for three nights, to $1,950 for a week.

Alternatively, organize your own tour from Isla Santa Cruz which is cheaper and quite straightforward.

### Excursions
Among Ecuador's snow-capped peaks, Chimborazo, Cotopaxi, El Altar and Tungurahua are certainly the most challenging.

Twenty kilometers north of Quito is Mitad del Mundo, where a monument and museum mark the equator. The road passes two dozen or so colorful billboards painted by prominent Latin American artists.

On a clear day, a train or bus ride along the Avenue of the volcanoes from Quito to Riobamba is breathtaking.

Shortly before Santo Domingo on the Quito road is Tinalandia, with excellent accommodation and meals. This tranquil hotel is set amidst lush subtropical vegetation inhabited by over 150 species of birds.

## BARS & DISCOS

Nightlife in Ecuador often involves frequenting bars with astonishingly loud music.

### QUITO

In Quito, the new town has its share of discos, as well as two reasonably authentic English pubs, the Reina Victoria on the street of the same name, and nearby, El Pub. They nicely complement the cruising double-deckers. The best peña is the Taberna Quiteña which has two venues, one on Avenida Amazonas, the other on Calle Manabí in the old city.

### GUAYAQUIL

In Guayaquil, countless all-night discos play the latest American and *latino* tunes, and some good Colombian *salsa*. They are not places for the faint-hearted.

### OTAVALO

Otavalo has a couple of peñas which get rather lively on Friday and Saturday nights.

In addition to excellent local music, they regularly feature groups from Colombia, Peru and Chile.

# SHOPPING

# FURTHER READING

Otavalan goods are the best buys in Ecuador. If you cannot make it to Otavalo, there are stores in Quito, Guayaquil, Cuenca and Baños where prices are only slightly higher than in the market.

Cuenca is perhaps the best place to buy quality gold and silver jewelry, and is – along with Montecristi – a center of Panama hat production.

Shops and stalls on Avenida Amazonas in Quito sell everything that Ecuador has to offer.

## HISTORY

Darwin, Charles. *The Voyage of the "Beagle"*, 1835–36.

Hassaurek, Friedrich. *Four Years Among The Ecuadorians*, 1861–65.

Michaux, Henri. *Ecuador – A Travel Journal*, 1928.

Von Hagen, Victor. W. *Ecuador and the Galápagos Islands: A History*, 1949.

Whymper, Edward. *Travel Among The Great Andes of the Ecuador*, 1891.

## GENERAL

Cuvi, Pablo. *In The Eyes Of My People*, 1985.

Hurtado, Oswaldo. *Political Power in Ecuador*, 1980.

Jackson, Michael. *Galápagos: A Natural History Market*, 1987.

Meisch, Lynn. *Otavalo – Weaving, Costume and the Market*, 1987.

Rachowiecki, Rob. *Ecuador – A Travel Survival Kit*, 1986.

# GETTING THERE

Peruvian Airlines AeroPeru and Faucett have direct flights to Lima from Miami; Aerolineas Argentinas flies from Los Angeles; Faucett also has direct Miami–Iquitos service. From Canada, Canadian Air has direct flights to Peru. From Europe: Alitalia, Air France, Iberia, KLM, Lufthansa and British Airways have flights into the country.

AeroPeru has also been offering a 45-day excursion fare, including a stopover in Lima and one other city in Peru, from Miami to anywhere on the AeroPeru route (Guayaquil, Panama City, Santiago, Buenos Aires, Caracas, Bogotá, La Paz, São Paulo and Rio de Janeiro) for US$759.

## BY LAND

There is a bus service between Lima and the major cities although in many cases the buses are not luxury vehicles. Trains run from Lima to Arequipa, then onto Puno, Cuzco and Machu Picchu, as well as from Lima to Huancayo. However, the rail service from Peru to Chile and Peru to Bolivia has been discontinued.

The Pan American Highway, once the pride of Peru and its link to Chile and Ecuador, Colombia and Venezuela to the north, has fallen into disrepair as a result of floods, earthquakes and a lack of maintenance in recent years. Car rental is available although the sluggishness of such a trip and poor conditions – especially in mountain areas – make air or bus travel preferable.

# TRAVEL ESSENTIALS

## VISAS & PASSPORT

Visitors from most European countries, the United States and Canada, do not require visas into Peru. With a valid passport, citizens of these countries receive a tourism card usually good for 90 days (or 60 days in the case of the United States). Australians and New Zealanders and Asians need visas. To extend the length of stay, contact the Dirección General de Migraciones at Paseo de la Republica 585 in Lima. A 60-day extension is available upon payment of $20 and presentation of a return travel ticket.

Vaccinations are not required for entry to Peru. On departure, a $15 exit tax must be paid in US dollars, not local currency.

## MONEY MATTERS

Since 1986, the *inti* has replaced the *sol* as the official currency of Peru. Although there are still a few old *sol* bills in circulation (1,000 *sols* equal one *inti*), most have been replaced with new bills of 10, 50, 100, 500, 1,000 and 5,000 denominations. Coins come in the denominations of five *intis*, one *inti* and 50 *centimos*.

Banks, money exchange houses, hotels and travel agencies are authorized to exchange money, either from traveler's checks, cash or, in some cases, money orders. The exchange rate fluctuates daily.

Most international credit cards are accepted, including Diners Club, Visa, Mastercard and American Express at hotels, restaurants and stores. Note, however, that the exchange rate may be calculated at the bank rate, which is generally not as favorable as the exchange house rate. American Express, Diners Club, Mastercard and Visa have offices in Lima.

## WHAT TO BRING

Tourists should bring with them any medicines or cosmetics they use; tampons and contraceptives are not available in Peru. Since Peruvians are smaller in stature than North Americans or Europeans, it may be difficult to find large-sized clothing.

## CUSTOMS

It is illegal to take archaeological artifacts out of the country. Special permits are needed to bring professional movie or video equipment into Peru. In Cuzco, a special tax is assessed for professional photographers passing their equipment through that airport.

# GETTING ACQUAINTED

## GOVERNMENT & ECONOMY

A country which has sporadically fallen into the hands of military governments, Peru since 1979 has been governed by a democratic government under a constitution. Presidential elections are held every five years and municipal elections every three years. The country is divided into 24 departments and the province of Callao – Peru's main port.

Fishing, mining and tourism play important roles in the economy, although the government of Alan García has gradually revitalized the farming sector. Petroleum is extracted from the Amazon jungle area; these supplies currently account for about half of all domestic demand.

A slump in the manufacturing sector, caused partly by the nation's refusal to pay its huge foreign debt and the corresponding cut of new aid from the World Bank and other lending agencies, has resulted in a series of problems with foreign reserves, weakening the *inti*.

## POPULATION

The nearly 20 million persons in Peru are spread across the arid coast, the jungle, Peru's fertile valleys and the Andes mountains. Because of successive immigration, Indian and *mestizo* citizens live side by side with Europeans and Chinese and Japanese.

Peru's coastal cities to the south include a cohesive black population. The official languages are Spanish and Quechua, although all school children must also study English. Near Puno and Lake Titicaca, Aymará is spoken. In the jungle, several tribal dialects are spoken.

## CLIMATE

On the coast, the temperatures average between 14°C and 27°C (58°F and 80°F), while in the highlands – or *sierra* – it is normally cold, sunny and dry for much of the year, with temperatures ranging from 9°C to 18°C (48°F to 65°F). The rainy season in the *sierra* is from December to May. The jungle is hot and humid with temperatures ranging from 25°C to 28°C (77°F to 82°F).

## BUSINESS HOURS

Lima alone suffers from the climatic condition the Peruvians call *garua* – a thick wet fog that covers the city without respite for the winter.

Lima time coincides with Eastern Standard Time in the United States (five hours behind Greenwich Mean Time). For official Peruvian time, call 652-800 in Lima.

Offices normally operate from 10 a.m. – 8 p.m. with many establishments closing from 2 p.m. – 4 p.m. Banking hours in the winter (April to December) generally stay open from 9.15 a.m. – 12.45 p.m. and from 3.30 p.m. – 5.30 p.m. or 6 p.m. from Monday to Friday.

## USEFUL HINTS

Appliances in Peru run on 220 volts, 60 cycles. The metric system is used for all units of measurements.

## SECURITY & CRIME

Pickpockets and thieves – including senior citizens and children – have become more and more common in Lima and Cuzco. It is recommended that tourists do not wear costly jewelry and that their watches, if worn, be covered by a shirt or sweater sleeve. Thieves have become amazingly adept at slitting open shoulder bags, camera cases and knapsacks; keep an eye on your belongings. Officials also warn against dealing with anyone calling your hotel room or approaching you in the hotel lobby or on the street, allegedly representing a travel agency or specialty shop. Carry your passport at all times.

A special security service for tourists has been created by the Civil Guard. These tourism police are recognizable by the white braid they wear across the shoulder of their uniforms and can be found all over Lima, especially in the downtown area. In Cuzco, all police have tourism police training. The tourist police office in Lima is Avenida Salaverry 1156, Tel: 714-313.

There has been much publicity of late about terrorism in Peru although, with only three exceptions in the past nine years, tourists have managed to avoid dangerous contact with political violence. It is recommended that tourists avoid areas where there are terrorism alerts (such as Ayacucho). For specific inquiries, contact your embassy.

## MEDICAL SERVICES

Most major hotels have a doctor on call. Three clinics in Lima and its suburbs have 24-hour emergency service and usually an English-speaking staff person on duty. They are the Clinica Anglo-Americana, Avenida Salazar in San Isidro, Tel: 403-570; the Clinica International (sic) on Washington 1475, Tel: 288-060 and the Clinica San Borja, on Avenida del Aire 333 in the suburb of San Borja, Tel: 413-141.

# GETTING AROUND

## TOURIST OFFICES

Peru's National Tourism Board offices (FOPTUR) – with two locations in Lima and offices in nearly every major city in the country – can offer the best information about how to get from one place to another.

In Lima the city tourism office in the passageway beside the city hall has information about the downtown area. FOPTUR offices are located at Jirón Belén 1066, Tel: 323-559 downtown and at Angamos Oeste 355 in the suburb of Miraflores, Tel: 453-394.

FOPTUR offices are also located at both the national and international areas of Lima's Jorge Chavez Airport.

## TOUR OPERATORS

*US Office:*
**South America Reps/Amazon Tours & Cruises**
P.O. Box 39583
Los Angeles, CA 90039
Tel: (818) 246-4816, (800) 423-2791
Telex: 362988 SAR GLND
Telefax: (818) 246-9909

*Head Office:*
**Amazon Camp**
Prospero 151
Iquitos, Peru
Tel: 094-23-39-31
Telex: 91266 PE ACATUS
Telefax: 094-23-11-25

*Lima Office:*
**Amazon Camp**
Ave. Nicolas de Pierola 677
# 105S Lima, Peru
Tel: 014-28-7813

*General Agent in Germany/Austria*
**Marion Stephan Tourisik**
Zepelinallee 33
6000 Frankfurt am Main 1
Federal Republic of Germany
Tel: (069) 706011
Telex: 411381
Telefax: (069) 7071870

## MAPS

Maps are available from FOPTUR free of charge. Larger city maps are sold at newspaper kiosks, bookstores and the Instituto Geografico National at Avenida Aramburu 1190 in the suburb of Surquillo, Tel: 451-939. One-time use of map room and files is available from the South American Explorer's Club or, for the $25 membership fee, unlimited use of the club's resources and its office (with storerooms, safe deposit boxes, library, kitchen, first aid materials) on Avenida Republica de Portugal 146, Tel: 314-480.

## FROM THE AIRPORT

There is a 24-hour bus service – TransHotel that will shuttle passengers to and from the airport and their hotels. It also will pick up passengers from other parts of town, even if they are not staying in hotels, for a fare that varies according to distance; Tel: 275-697 or 289-812.

It costs approximately US$10 to take a cab from the airport downtown or to most parts of the suburbs. Since most taxis do not have meters, make sure you agree with the driver on a price beforehand. Although cabs are inexpensive anyway you look at it, walking from the airport across the parking lot to the main roadway (Avenida Faucett) to catch a cab could mean half the price. Faucett is also where you can get city buses from the airport. Ask at the FOPTUR office at the airport for routes and fare information.

You can also rent cars at the airport, where Avis, Budget, Hertz and National have offices open 24 hours a day. An international driver's license is needed and is valid for 30 days. For additional days, it is necessary to obtain authorization from the Touring and Automobile Club of Peru, located at Cesar Vallejo 699 in the suburb of Lince; Tel: 403-270 or 221-451 or 225-975.

## DOMESTIC AIRLINES

Domestic airlines Faucett and AeroPeru serve most cities in Peru. AeroPeru's main office is at Plaza San Martin, Avenida Nicolas de Pierola (La Colmena) 914, Tel: 322-995. Faucett's principal office for domestic flights is at the corner of Avenida Garcilaso de la Vega 865 (Wilson) at Quilca, Tel: 275-000. For trips over the Nazca line or special short flights, Aerocondor has an office at the Sheraton Hotel's shopping mall in downtown Lima, Tel: 320-950, ext 117.

## PUBLIC TRANSPORT

Lima has a multitude of city bus lines, although most are overcrowded, slow and not recommended for tourists. Cab fares are generally so inexpensive by international standards that they are preferred.

For travel outside the city, cars may be rented, transportation can be arranged through travel agencies, buses are available to just about every part of the country.

There is train service east of Lima and south to Arequipa, Puno and Cuzco. Since trains are the most economical means of transportation, they are also sometimes the most crowded. Make sure you have a guaranteed seat and a tourist (first-class) ticket.

The rail line from Lima to Huancayo is the world's highest, reaching nearly 5,000 m above sea level. The train depot for all departures in Lima is the Desamparados Station behind the Presidential Palace, Tel: 289-440.

347

# WHERE TO STAY

## HOTELS

Throughout Peru, a government chain called ENTUR PERU owns tourist hotels and inns. The chain does not have hotels in Lima but it has everything from youth hostels to exclusive luxury hotels all around the rest of the country, usually going simply by the name "Hotel de Turista". Its main office and reservation center is on Avenida Javier Prado Oeste 1358, Tel: 721-928. Travel agencies can also make reservations for you.

## LIMA

**Cesar's**
At the corner of Avenida La Paz and Diez Canseco in the suburb of Miraflores.
Tel: 441-212.

**Crillon**
La Colmena 589
Tel: 283-290.

**Hotel Bolívar**
Plaza San Martin
Tel: 276-400.
The doyen of Peruvian hotels.

**Pueblo Inn**
An exclusive hotel complex on KM 11 of the Central Highway.
Tel: 350-777.

**Sans Souci Hotel**
Avenida Arequipa 2670 in suburb of San Isidro.
Tel: 226-035.

**Sheraton**
Paseo de la Republica 170
Tel: 328-676.

## CUZCO

**El Dorado Inn**
Sol 395
Tel: 232-573.

**Libertador**
San Agustin 400
Tel: 231-961; in Lima, 420-166

**Picoaga**
Santa Teresa 344
Tel: 227-691; in Lima, 186-314

## AREQUIPA

**El Portal**
Portal de Flores 116
Tel: 215-530.

**Hotel Turistas**
Plaza Bolívar on Selva Alegre
Tel: 215-110.

## NAZCA

**Hotel Turistas**
Jiron Bolognesi
Tel: 02 (no direct dialing).

**Hotel de la Borda**
Kilometer 447 of the Pan American Highway.
Tel: in Lima: 408-430.

## IQUITOS (NORTHERN AMAZON)

**Hotel Amazonas Plaza *****
Abelardo Quinonez KM 2, Iquitos
Tel: 23-5731/23-1091
Telex: 91055
Reservations in Lima, Tel: 44-1199/44-1990
Telex: HOTDIPLO
Iquitos' top class hotel, 3 km out of town.

**Acosta II**
Ricardo Palma 252
Tel: 23-2904

**Acosta I**
Esq. Huallaga & Araujo
Tel: 23-5974
Not quite as prestigious as Acosta II although includes a swimming pool and a very good restaurant.

## LODGES

### Amazon Village
Rio Mommon
Reservations in Lima: Alcanfores 285, Miraflores
Tel: 44-1199/47-8776
Telex: 21618 HOTDIPLO
Excursions to the Amazon jungle are organized here on the Rio Mommon, weather conditions permitting. Bungalows are clean and well-ventilated and good food is served. For the more comfort-conscious traveler.

### Amazon Lodge & Safaris
Putumayo 165, Iquitos
Tel: 23-3023
Reservations in Lima: Carmino Real 1106, San Isidro
Tel: 41-9194
Telex: 21371
Day trips available on the powerful "Amazon Explorer" launch as well as two-day/one-night stays.

### Explorama Lodge & Explornapo Camp
Av. La Marina 340 – Casilla 446, Iquitos
Tel: 23-5471/23-3481
Telex: 91014
Telefax: 23-4968
Reservations in Lima, Tel: 24-4764
Also offers two-day packages with charges of US$40 for each additional day. Managed by the well-organized Explorama Tours, the Explornapo Camp is better situated than the above two lodges for sighting fauna. It offers a good balance of comfort and penetration of the jungle for those who wish to experience the Amazon with a fairly structured itinerary.

For the more adventurous traveler:

### Tamshiyacu Lodge
Reservations: Wilderness Expeditions
310 Washington Ave SW
Roanoke Va. 24016 USA
Tel: (703) 342-5630
Organized trips include Yarapa River Lodge, Tambo Safari Inn Camp and wilderness expeditions.

### Selva Lodge & Yarapa Lodge
Reservations: Amazon Selva Tours
Putumayo 133
Experienced guides and great food.

### Yarapa River Camp
Reservations: Amazonia Expeditions
Putumayo 139
Strongly conservationist.

## MADRE DE DIOS REGION (SOUTHERN AMAZON BASIN)

### MANU BIOSPHERE RESERVE

### Manu Lodge
c/- Manu Nature Tours
Avenida Sol 627-B
Oficina 401, Cuzco
Tel/Telefax: (084) 23-4793
15 km from Puerto Maldonado on the Rio Madre de Dios.

### Albergue Cuzco Amazonico ***
Reservations in Lima:
Tel: 46-2775/46-9777
Telefax: 45-5598
Telex: 21475 PELATI

*Also at:*
Procuradores 48, Cuzco
Tel: 5047

## TRUJILLO

### Hotel Turistas
Plaza de Armas
Tel: 232-741

### Hotel El Golf
Urbanization El Golf, block J-1
Tel: 242-592; in Lima 317-872

## CAJAMARCA

### Hotel Turistas
Plaza de Armas
Tel: 2470

### Hostal Laguna Seca
Avenida Manco Capac just outside the city at the thermal baths, the waters of which enter the bathrooms of this hotel.
Tel: 5; in Lima, 463-270

### Hostal Cajamarca
2 de Mayo 311
Tel: 2532

# FOOD DIGEST

## WHAT TO EAT

Peru's *criolla* cuisine evolved through the blending of native and European cultures. A *la criolla* is the term used to describe slightly spiced dishes such as *sopa a la criolla*, a wholesome soup containing beef, noodles, milk and vegetables.

Throughout the extensive coastal region, seafood plays a dominant role in the Creole diet. The most famous Peruvian dish, *ceviche*, is raw fish or shrimp marinated in lemon juice and traditionally accompanied by corn and sweet potato. Other South American countries have their own version of *ceviche* but many foreigners consider Peru's to be the best.

*Corvina* is sea bass, most simply cooked *a la plancha*, while scallops, *conchitas* and muscles, *choros* might be served *a lo mancho*, in a shellfish sauce. *Chupe de Camarones* is a thick and tasty soup, of salt or freshwater shrimp.

A popular appetizer in Peru is *palta a la jardinera*, avocado stuffed with a cold vegetable salad or a *la reyna*, chicken salad. *Choclo* is corn on the cob, often sold by street vendors during lunchtime. Other Peruvian "fast food" includes *anticuchos*, shish kebabs of marinated beef heart; and *picarones*, sweet lumps of deep fried batter served with molasses. For *almuerzo* or lunch, the main meal of the day, one of four courses might be *lomo saltado*, a stir-fried beef dish, or *aji de gallina*, chicken in a creamy spiced sauce.

The most traditional of Andean foods is cuy, guinea pig, which is roasted and served with a peanut sauce. Another speciality of the Sierra is *Pachamanca*, an assortment of meats and vegetables cooked over heated stones in pits within the ground. Succulent freshwater trout is plentiful in the mountain rivers.

Peruvian sweets might be *suspiro* or *manjar blanco*, both made from sweetened condensed milk, or the ever popular ice-cream and cakes. There are many weird and wonderful fruits available in Lima, notably *chirimoya*, custard apple, *lucuma*, a nut-like fruit, delicious with ice-cream, and *tuna*, which is actually the flesh from a type of cactus.

In towns, many Peruvians drink the soft drink *chicha morada*, made with purple maize (different from the *chicha de jora*, the traditional home-made alcoholic brew known throughout the Andes). The lime green *Inca Cola* is more popular than it's northern namesake, as well as *Orange Crush*, *Sprite* and *Seven-up*. The *jugos*, juices, are a delightful alternative to sodas and there are many fruits available. Instant Nescafé is often served up even in good restaurants, although real coffee can be found at a price. Tea drinkers would be advised to order their beverage without milk, to avoid receiving some peculiar concoctions.

The inexpensive beers are of high quality. Try *Cusqueña*, *Cristal* or *Arequipena*. Peruvian wines can't compete with Chilean excellence, but for a price, *Tabernero*, *Tacama*, *Ocucaje* and *Vista Alegre* are the reliable names. The Peruvian Pisco Sour is, however, a strong contender. Try the famous, potent cathedral at the Grand Hotel Bolívar in Lima.

## WHERE TO EAT

### LIMA

Although Lima offers a huge choice of restaurants in the city area, the following are the most reliable, providing high quality food and service:

### *International Cuisine*

**Pabellon de Caza**
Alonzo de Molina 1100, Monterrico
Tel: 37-9533
Noon – 3 p.m. and 7 p.m. – 1 a.m. Monday – Saturday
Brunch: Sunday 10.30 a.m. – 4 p.m.
Located only a short stroll from the Gold Museum, containing a luxuriant garden and elegant decor.

## Carlin

La Paz 646, Miraflores
Tel: 44-4134
Noon – 4 p.m. and 7 p.m. – midnight daily
A cozy international style restaurant which is very popular with foreign residents and tourists.

## Los Condes de San Isidro

Paz Soldan 290, San Isidro
Tel: 22-2557
Noon – 3 p.m. and 7 p.m. – midnight daily
Set in a spectacular San Isidro mansion.

## El Alamo

Corner of La Paz and Diez Canseco, Miraflores
12.30 p.m. – 3 p.m. and 7 p.m. – midnight
  Monday – Saturday
A cozy spot for imported wines and cheeses. Specializing in fondues.

### Seafood Restaurants

## La Rosa Nautica

Espigon No. 4
Coasta Verde, Miraflores
Tel: 47-0057
12.30 p.m. – 2 a.m. daily
Lima's most famous seafood restaurant prestigiously located at the end of an ocean boardwalk.

## La Costa Verde

Barrangquito Beach, Barranco
Tel: 67-8218
Noon – midnight daily
Romantic at night under a thatched roof and right on the beach.

### Criolla Cuisine

## Las Trece Monedas (The Thirteen Coins)

Jr. Ancash 536, Lima
Tel: 27-6547
Noon – 4 p.m. and 7 p.m. – 11.30 p.m.
  Monday – Saturday
Formal presentation within a beautiful 18th-century colonial mansion, including courtyard and antique coach.

## Manos Morenos

Avenida Conquistadores 887, San Isidro
Tel: 42-6271
Noon – 3.30 p.m. and 6 p.m. – midnight

Tuesday – Sunday
Criollo Breakfast: Saturday and Sunday 8.30 a.m. – 11 a.m.
The best "criollo" fare Lima has to offer, specializing in *anticuchos* (shish kebab of marinated beef heart).

### French Cuisine

## L'Eau Vive

Ucayali 370, Lima
Tel: 27-5612
Noon – 2.45 p.m. and 8.15 p.m. – 10.15 p.m.
  Monday – Saturday
Fine provincial dishes prepared and served by a French order of nuns. The skylit inner courtyard is one of Lima's most pleasurable settings for luncheon – the perfect place to take a break from sightseeing. (Ave Maria is sung nightly at 10 p.m. and dinner guests are invited to join in).

### Italian

## Valentino

Manuel Banon 215, San Isidro
Tel: 41-6174
Noon – 3 p.m. and 7.30 p.m. – midnight
  Monday – Saturday
Great Italian food at Lima's best international restaurant.

## La Trattoria

Manuel Bonilla 106, Miraflores
Tel: 46-7002
1 p.m. – 3.30 p.m. and 8 p.m. – midnight
  Monday – Saturday
Authentic Italian fare.

### Japanese

## Matsuei

Canada 236, La Victoria
Tel: 72-2282
Noon – 3 p.m. and 7 p.m. – 11 p.m. Monday – Saturday; 7 p.m. – 11 p.m. Sunday
Excellent quality Japanese food.

### Steak

## La Carreta

Rivera Navarrete 740, San Isidro
Tel: 40-5424
Noon – 8.30 p.m. daily
For typically excellent Argentine beef.

**The Steak House**
La Paz 642, Mezzanine – "El Suche",
Miraflores
Tel: 44-3110
Noon – 3 p.m. and 7 p.m. – midnight
Charming setting with good steaks and salads.

### Pizza

**Las Cuatro Estaciones**
Vcayali (near corner of Jiron de la Union),
Lima. One of the very few great pizza restaurants of South America, also serving exotic local fruit juices. Wholesome and highly recommended for all travelers.

**La Pizzeria**
Benavides 322, Miraflores
Tel: 46-7793
9.30 a.m. – midnight daily
Nighttime haunt of trendy Miraflores crowd.

### "Tex-Mex"

**Villa Taxco**
Libertad 435 (F. Gerdes), Miraflores
For Reservations: 47-5264
Noon – 11.30 p.m. Monday – Thursday
The show "Mariachis" on Fridays and Saturdays starts at 9.30 p.m. Authentic Mexican dishes including *Tampiquena steak*, *burritos* and *fajitas*. Mexican decor throughout with banquet rooms available.

### Cafés

**La Tiendecita Blanca – Cafe Suisse**
Avenida Larco 111, Miraflores
Tel: 45-9797
Boasts of 30 different kinds of chocolates, excellent ice-cream and desserts. Popular with shoppers but also serves lunch and dinner.

## AREQUIPA

This attractive city has always maintained a reputation for style and affluence. There are several good restaurants around the Plaza de Armas and cafés can be found in the first block of San Francisco.

**La Rueda – Argentine Grill**
Mercaders 315
Tel: 21-9330

Barbequed meats and good wine in a cozy ranch style restaurant.

**Le Paris**
Mercaders 228
French cuisine and other international dishes.

**El Fogón**
Santa Marta 112
Tel: 21-4594
Specializes in barbequed chicken.

**Sol de Mayo**
Jerusalem 207 (Yanahuara district)
Good lunch time menu of Peruvian specialties such as *rocoto relleno*, stuffed hot peppers, or *ocopa*, potatoes in a spicy sauce with melted cheese. Include a visit to the Mirador de Yanahuara as part of your day's itinerary.

## CUZCO

While in Cuzco, sample the succulent pink trout, prepared in many of the Peruvian and international style restaurants. Cuzquenos have also mastered the art of pizza baking and there is always a new establishment to be found within the plaza's portals. Good restaurants will also provide Andean music.

**Pizzeria La Mamma**
Portal Escribanos 177, Plaza Regocijo
Just one of the many restaurants which serves mouth-watering garlic bread and fresh mixed salads as well as pizzas.

**El Trujo**
Plaz Regocijo 247
Catering to most of the large tour groups, this nightly dinner and show is the most elaborate in Cuzco and very good value.

**Trattoria Adriano**
Sol 105, corner of Mantas
Friendly service and excellent pink trout. Try it simply grilled and with a squeeze of lemon.

**Le Petit Montmarte**
Garcilaso 270
Fine French cuisine.

**Quinta Zarate**
Calle Tortera Paccha
One of the many "quintas", or inns, in the

suburbs. The specialty here is roast *cuy*, guinea pig.

## El Ayllu
Portal de Carnes 203 (beside Cathedral)
A must to visit for breakfast of ham and egg, toast and fruit juice, or homemade fruit yoghurt, cakes, teas and good coffee. Continuous opera musak and cosmopolitan clientele gives this place its unique character.

## Meson de Espaderos
Espaderos 105
A quaint wooden balcony provides perfect dining views of the Plaza. The mixed grill, *asado*, is excellent and huge.

Cuzco's best two hotels, the El Dorado Inn and the Libertador have excellent restaurants, with high prices set in US dollar.

Cafés around the plaza serve *mate de coca*, chocolate mud cake and hot milk with rum on chilly Cuzco nights.

## HUARÁZ

## Samuel's
Jr. Jose de la Mar 626
Good breakfasts.

## Chalet Suisse
Hostal Andino
Jr. Cochachin 357
Fancy international meals.

## La Familia
Avenida Luzuriaga 431
Wide selection of wholesome foods.

## Pizzeria Ticino
Avenida Luzuriaga 651

## Creperie Patrick
Avenida Luzuriaga 424

## TRUJILLO

## De Marco
Francisco Pizarro 725
Tel: 23-4251
Peruvian and international cuisine. Popular for ice-creams and desserts.

There is a concentration of restaurants near the market and a number of vegetarian places on Bolívar.

# PARKS & RESERVES

## MACHU PICCHU SANCTUARY

**Geographic Location:** Eastern *cordillera* of Peru, 100 km northwest of Cuzco.

**Size:** 32,594 hectares including Archaeological Park.

**Access:** By rail from Cuzco to Puente Ruinas, then tourist bus to the Monument.

Inclusive day-tours are available from Cuzco which include all rail travel, bus transfers and entry fee to the Monument.

Alternatively, by rail from Cuzco to rail station at KM 88 which is about 22 km beyond Ollantaytambo station, then join the Inca trail for a 33 km, three/five-day hike.

Camping overnight – equipment can be hired in Cuzco. For personal security camping is not recommended for groups of less than four persons.

**Approximate costs of access:** Inclusive day trip by train and bus $75.

Entrance ticket for independent travelers $11.

**Tour operators:** Any travel agent in Cuzco can arrange day-trip.

**Open:** All year.

**Registration:** Only required if climbing Huayna Picchu (the mountain overlooking the monument). Register at hut on the trail to the Monument.

**Accommodation:** Tourist Hotel at the Ruins at $10 per night (book at Entur Peru in Cuzco or Lima).

Cheaper rooms available 12 km back along railway towards Ollantaytambo.

**Camping:** No camping available at the Monument. Official camp site is at Puente Ruinas.

**Food/restaurant:** Food available at most railway station stops. Self-service restaurant at the Monument.

**Trails:** Various tourist trails in the vicinity of the monument, although not necessarily designed with nature observation in mind.

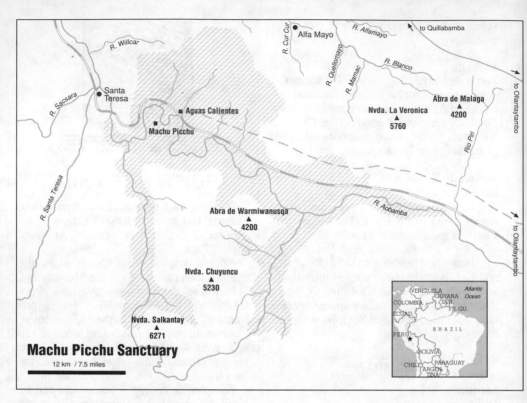

## Machu Picchu Sanctuary

R. Willcar

R. Sacsara

Santa Teresa

R. Santa Teresa

Aguas Calientes

Machu Picchu

R. Cur Cur

Alfa Mayo

R. Alfamayo

to Quillabamba

R. Quellomayo

R. Mamac

R. Blanco

Nvda. La Veronica
5760

Abra de Malaga
4200

Rio Piri

to Ollantaytambo

R. Aobamba

to Ollantaytambo

Abra de Warmiwanusqa
4200

Nvda. Chuyuncu
5230

Nvda. Salkantay
6271

12 km / 7.5 miles

VENEZUELA
COLOMBIA
GUYANA
SUR.
FR. GU.
Atlantic
Ocean
ECUAD.
BRAZIL
PERU
BOLIVIA
PARAGUAY
CHILE
ARGEN
TINA

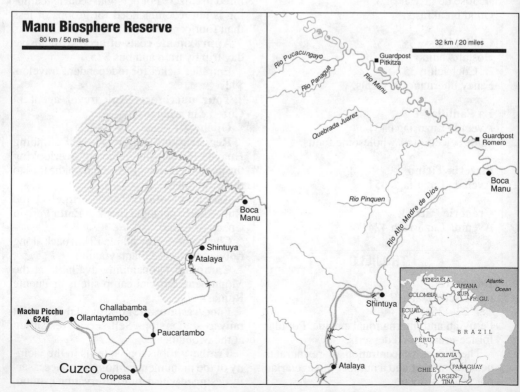

## Manu Biosphere Reserve

80 km / 50 miles

Rio Pucacungayo

Rio Panagua

Rio Manu

Guardpost Pitkitza

32 km / 20 miles

Quebrada Juarez

Guardpost Romero

Boca Manu

Rio Pinquen

Rio Alto Madre de Dios

Shintuya

Boca Manu

Shintuya

Atalaya

VENEZUELA
COLOMBIA
GUYANA
SUR.
FR. GU.
Atlantic
Ocean
ECUADOR
BRAZIL
PERU
BOLIVIA
PARAGUAY
CHILE
ARGEN
TINA

Machu Picchu
6246
Ollantaytambo
Challabamba
Paucartambo

Cuzco
Oropesa
Atalaya

**Further advice:** Ornithologists would find it worthwhile visiting the Cross Keys Bar in the Paza de Armas, Cuzco. The owner, Barry Walker, is one of Peru's top ornithologists.

---

## TAMBOPATA-CANDAMO RESERVED ZONE

**Geographic Location:** South of Puerto Maldonado – east of Cuzco. Area enclosed by the Rio Inambari (tributary of the Rio Madre de Dios) and the Rio Heath which forms the western border of Bolivia.

**Size:** 1,479,000 hectares, including 5,500 hectares of Tambopata. Adjacent to Heath Pampas National Sanctuary of 101,109 hectares.

**Access:** Flight from Lima or Cuzco to Puerto Maldonado. Hire river transport for journey up the Rio Tambopata.

**Approximate costs of access:** Hire of river transport from Puerto Maldonado to Tres Chimbadas approximately $70. Hire guides and buy food from the Esse'eja Indian community along the river.

**Tour operators:** Inclusive tours to Coco Cocha, Tres Chimbadas lake and Tambopata Macaw Lick Camp are organised by **Tambopata Nature Tours** of Avenida Sol 582, Cuzco.

Inclusive tours to the Tambopata reserve, book accommodation at the **Explorer's Inn** through **Peruvian Safaris** office in Plaza San Francisco, Cuzco at $150 for three-day/two-night tour.

**Open:** All year.

**Registration:** Not necessary.

**Accommodation:** Many hotels of all price ranges in Puerto Maldonado.

Inclusive tours as detailed above.

**Camping:** Camping at Tres Chimbasas and Macaw Lick Camp may be available through Tambopata Nature Tours as a special arrangement.

**Transport in Park:** Motorized dugout canoe and walking.

**Trails:** Extensive trail system at Explorers Inn with expert scientist guides.

Trail system in vicinity of other lodges still being developed, using local Indians as guides.

**Tambopata Candamo Reserve**
64 km / 40 miles

**Further advice:** The Tambopata Reserve is one of the oldest in the Madre de Dios region and has a high reputation amongst naturalists. The Tambopata-Candamo reserve was declared in 1990. Because of the limited tourist activity to date good opportunities may exist for nature observation.

In the UK contact Tambopata Reserve Society (TReeS), 64, Belsize Park, London NW3 4EH for conservation information.

---

## MANU BIOSPHERE RESERVE

**Geographic Location:** North of Cuzco. Eastern slopes of the Andes protecting the entire Manu watershed. Northern limit defined as high ground between the Rio De las Piedras and Manu basins, north of Fitzcarrald; western limit is the Paucartambo mountains which divides the watersheds of the Urubamba and Alto Madre de Dios.

**Size:** Biosphere Reserve of total extent 1,881,200 hectares, comprising 1,532,806 hectares which is not usually accessible.

Tourism limited to the Zona Reservada (257,000 hectares), and Zona Cultura (91,394 hectares).

**Access:** Flight from Lima to Cuzco.

Safari bus (air-conditioned and good suspension) from Cuzco to Atalaya on the Alto Madre de Dios takes 10 – 12 hours (arrange contract hire through Manu Nature Tours). Can also provide contract hire river transport on Rios Alto Madre de Dios and Manu.

Truck from Cuzco to Shintuya, 250 km can take 30 hours for $5.

The road is single track, hence vehicles should travel up or down the road on alternate days.

**Approximate costs of access:** Negotiate contract safari bus hire for return journey from Cuzco to Atalaya, around $1,250 for up to 12 persons.

Negotiate contract hire of motorized dugouts for journey up the Rio Manu, approximately $850 for up to 12 persons.

Negotiate charter plane hire from Cuzco to Boca Manu. Approximately cost $1,200 each way for up to 8 persons.

Sometimes possible to arrange private boat transport at Shintuya.

**Tour operators: Manu Nature Tours,** Avenida Sol. 582, Cuzco, organize inclusive tours visiting the Manu Biosphere Reserve and operate the only lodge in the lowland forest of the Zona Reservada. Approximately prices $800 – $1,000 (depending on number of persons) for seven-day/six-night tour involving land and boat transport in and chartered flight out from Boca Manu to Cuzco. Naturalists stay at the cloud forest lodge *en route* to Atalaya.

**Expediciones Manu** EIRL, Procuradores 372, Cuzco, arrange more adventurous trips, camping along the Rio Manu.

**Open:** All year, but rainy season from November to April may restrict access. Driest months are July to September.

**Registration:** Obtain National Park permit at Salvacion; this must be endorsed at the Park Guard station at Romero.

**Accommodation:** Tourist lodges as detailed above

**Camping:** Difficult to arrange private camping trips in this remote area, may be possible to arrange this in Shintuya.

**Food/restaurant:** Not applicable.

**Transport in Park:** River transport.

**Trails:** Trails in vicinity of lodges. Twenty-seven kilometers of trails at Manu Lodge. Take excursions to Cocha Otorongo and Salvador to view Giant Otters.

# THINGS TO DO

## TREKKING

**Preparation:** The Peruvian dry season, May through September, offers the best views and finest weather. For safety reasons and for greater enjoyment, parties of three/four or more should hike together. Groups are easily formed in Cuzco, and the Tourist office on the Plaza de Armas provides a notice board for this purpose. In Huaráz, the Casa de Guias just off the main street offers the same service.

**Equipment:** A backpack, sturdy hiking boots, sleeping bags, insulated pad, tent and stove are the major necessities for trekking. This equipment can be hired at a number of highly visible adventure travel agencies. The cost is minimal, but quality often suffers. Inspect hired equipment carefully before departure. It can get very cold at night, and by day, the Andean sun burns quickly. Bring a good sunblock, not available in Peru, and have a hat handy.

For the less adventurous, porters can be hired for the Inca Trail. In the Cordillera Blanca, *arrieros*, or mule-drivers, are quite inexpensive and readily available.

**Food:** All food should be brought from a major town. Very little is available in the smaller villages. Freeze-dried food is not available, but a variety of dried fruits, cheeses, fruits, packaged soups, and tinned fish can be easily acquired in a number of "supermercados", *bodegas*, and open-air markets. Drinking water should be treated with iodine, and instant drink mixes like *Tang* can be added to offset the unpleasant taste.

**High Altitude:** The effects of high altitude can be significantly diminished by following a few simple precautions. Alcohol, overeating and physical exertion should be avoided for the first day or two. Drinking ample liquids helps the system adjust quickly.

The sugar in hard candy stimulates the metabolism, and aspirin eases the headache. *Mate de coca*, tea brewed from coca leaves, is said to be the best overall remedy.

# MUSEUMS

## LIMA

### Museo de la Nación
Javier Prado Este 2466, San Borja
Tel: 37-7822/37-7776
9 a.m. – 6 p.m. Tuesday – Friday; 10 a.m. – 6 p.m. Saturday and Sunday
Entrance US$3.00
Lima is justly proud of its newest museum, a large modern structure containing many floors of meticulously prepared exhibitions. They include impressive models of new and established archaeological sites as well as a spacious mannequin showroom displaying Peru's magnificent folkloric costumes. Worth at least one long visit.

### Museo de Oro del Peru (Gold Museum)
Alonso de Molina 100, Monterrico
Tel: 35-2917
Noon – 7 p.m. daily
Entrance US$5.00
There are actually two private museums at this outer suburb address. Apart from Peru's unique collection of Pre-Inca and Inca gold, there is an astonishingly complete Arms Museum which includes a number of highly decorative uniforms.

### Museo Rafael Larco Herrera
Bolívar 1515, Pueblo Libre
Tel: 61-1312
9 a.m. – 1 p.m. and 3 p.m. – 6 p.m. Monday – Saturday; 9 a.m. – 1 p.m. Sunday
Entrance US$4.00
Private collection of over 400,000 well-preserved ceramics, pre-Columbian art and artifacts. Famous erotic *"huacos"* ceramics from the Moche culture are housed in a separate room beside the gift shop.

### Museo Nacional de Antropologia y Arqueologia
Plaza Bolivar, Pueblo Libre
Tel: 63-5070
10 a.m. – 6 p.m. daily
Entrance US$4.00
Highly comprehensive exhibitions depicting the prehistory of Peru from the earliest archaeological sites to the arrival of the Spaniards. All the major Peruvian cultures are well represented.

### Museo Nacional de la Republica
(the National Museum of History)
Plaza Bolívar (next door to the Archaeological Museum), Pueblo Libre
Tel: 63-2009
9 a.m. – 6 p.m. daily
Entrance US$4.00
Both San Martín and Bolívar had lived in this building, now containing the furnishings and artifacts from colonial, republican and independent Peru. A must for those interested in the Peruvian revolution.

### Museo Pedro de Osma
Pedro de Osma 501, Barranco
Tel: 67-0915/670019
11 a.m. – 1 p.m. and 4 p.m. – 6 p.m.
Admission by appointment only.
Treasured collection of Viceregal painting, sculpture and silver.

### Museo de Arte
Paseo de Colon 125, Lima
Tel: 23-4732
9 a.m. – 5 p.m. Tuesday – Sunday
Entrance US$1.00
Grand building containing four centuries of Peruvian artworks, including modern painting. Also houses pre-Columbian artifacts and colonial furniture.

### Amano Museum
Retiro 160 (off 11th of Angamos), Miraflores
Tel: 41-2909
Admission by appointment only, guided tours in small groups at: 2 p.m., 3 p.m., 4 p.m. and 5 p.m. Monday – Friday
Particularly beautiful is the textile collection from the lesser known Chancay culture.

### Museo del Banco Central de Reserva
Ucayali 291, Lima
Tel: 27-6250
10 a.m. – 5 p.m. Monday – Saturday; 10 a.m. – 1 p.m. Sunday
A welcome relief from the hoards of money changers in this part of town, the bank offers a collection of looted pre-Columbian artifacts recently returned to Peru, paintings from the Viceroyalty to the present and numismatics.

## Museum of the Iquisition

Junin 548 (right side of the Plaza Bolívar as you face the Congress), Lima
9 a.m. – 7 p.m. Monday – Friday; 9 a.m. – 4.30 p.m. Saturday
Admission free
Explicit representations of torture methods in the Dungeon of this, the headquarters of the Inquisition for all Spanish America from 1570–1820.

## Museo de Historia Natural

Arenales 1256, Jesus Maria
Tel: 71-0117
9 a.m. – 3.30 p.m. Monday – Friday; 9 a.m. – 1 p.m. Saturday
Collection of Peruvian flora and fauna.

## Museo de Ciencias de la Salud

(Health Sciences Museum)
Junin 270, Lima
Tel: 27-0190
10 a.m. – 4 p.m. Wednesday – Sunday
Entrance US$1.00
Details pre-Columbian medical practices with English translations. The museum also arranges pre-Columbian banquets for organized groups.

## Museo Etnografico de la Selva

Avenida Tacna (first block)
Within the grounds of the Santuario de Santa Rosa de Lima
9.30 a.m. – 1 p.m. and 3.30 p.m. – 7 p.m. daily
Dominican missionaries have collected jungle Indian artifacts from south-eastern Peru.

## Numismatic Museum

Banco Wiese, 2nd Floor
Cuzco 245, Lima
Tel: 27-5060 ext. 553
9 a.m. – 1 p.m. Monday – Friday
Admission free
Peruvian coin collection from Colonial times to the present.

## CUZCO

A bewildering number of museums and historic sights can be found in Cuzco. Independent travelers can visit them separately with a US$10 Visitor's Ticket which is available from several of the major attractions. Hours are listed on the ticket and attractions explained in the Cuzco section of this book.

## IQUITOS

### The Iron House

Putumayo & Raymondi
(corner of the Plaza de Armas)
Supposedly designed by Eiffel and imported piece-by-piece during rubber boom days.

# NIGHTLIFE

Lima is full of colorful *peñas* where folk music and dance performances go non-stop until 2 a.m. or 3 a.m. Go about 9 p.m. to get a good seat.

### La Casa de Edith Barr

Ignacio Merino 250 in the suburb of Miraflores.
Tel: 410-612.

### Hatuchay

Trujillo 228
Tel: 247-779.

### La Palizada

Avenida del Ejercito 800 in the suburb of Miraflores.
Tel: 410-552.

# GETTING THERE

## BY AIR

The national Bolivian airline, Lloyd Aero Boliviano flies direct from the United States to La Paz. From Europe, Lufthansa has a service via Lima. Aerolineas Argentinas flies passengers from the United States, Europe and Australia to La Paz via Buenos Aires. Aero-Peru flies regularly from Lima to La Paz (which can be included in their 45-day round South America excursion fares from the States). Lan Chile flies connections from Santiago.

**AeroPeru**
Calle Colón 157
Tel: 370-002.

**Lloyd Aero Boliviano**
Avenida Camacho 1460
Tel: 353-606.

## BY LAND

The great majority of travelers enter Bolivia via Lake Titicaca in Peru. Many companies offer minibus connections between Puno and La Paz. Crillon Tours offers a luxury hydrofoil service both ways across the lake, allowing a stopover on the Island of the Sun, a meal in Copacabana and drinks on the water as well as a visit to their new Museum of the Altiplano with audio-visual displays, which gives an excellent introduction to Bolivia. The charge is US$160 each way. Crillon Tours is at Avenida Camacho 1223 in La Paz (Tel: 350-363, 374-566), with an agent in Puno.

It is also possible to travel from Argentina by land, crossing the frontier at La Quiaca and continuing by bus or train – a slow but fascinating journey. From Brazil, the jokingly-named "Train of Death" comes up from Corumba in the jungle to Santa Cruz.

It's not particularly comfortable but favored by more adventurous travelers. To and from Chile, there are train and bus links.

# TRAVEL ESSENTIALS

## VISAS

Citizens of the United States and most European countries do not need visas; Australians, New Zealanders and most Asians do. At the border, the guards will grant entry for between 30 and 90 days. The Immigration Office for visa extensions is at Avenida Gonzalvez 240. On departure, a US$10 tax is levied.

## MONEY MATTERS

Bolivia's currency is the *boliviano*, divided into *centavos*. The exchange rate has been relatively constant since 1986 at about 2.4 *bolivianos* to the US dollar, although this could change.

Banks, hotels and street money changers all offer similar rates of exchange for foreign currency – the black market is more or less dead. Take money in US dollars and change in larger cities.

Credit cards are not always accepted outside of major hotels and restaurants. American Express is the most widely recognized. Visa cash advances are possible from the Banco de La Paz in the capital.

## HEALTH

Travelers of all ages arriving by air in La Paz are likely to need a half day's rest to become accustomed to the altitude. Take it easy – walking at a relaxing pace for the first few days, drink plenty of *mate de coca* and you will adjust. No inoculations are required to enter Bolivia. If going to the Amazon basin, take malaria tablets and a yellow fever shot may be advisable.

The Bolivian highlands can be bitterly cold at night all year round. Bring warm clothes and take a jumper out with you even when the day is warm – the temperature is likely to drop dramatically when the sun sets.

Note that winter is the dry season and it rarely snows in La Paz or Potosí. Winter nights do not go much below freezing in the capital, but can be bitterly cold higher up in the Andes mountains.

# GETTING ACQUAINTED

## GOVERNMENT & ECONOMY

Bolivia may have suffered an average of one coup per year since independence from Spain in 1825, however, it has had civilian rule since 1982. The President is elected by popular vote every four years under the 1967 constitution.

The country is divided into nine departments, each controlled by a Delegate appointed by the President. Bolivia has La Paz for an effective capital, as the seat of the Government and Congress, with the legal capital being the small city of Sucre, where the Supreme Court sits.

Bolivia is the poorest Southern American republic. Over half the population subsist on agriculture. Until the collapse of prices in 1985, tin was the major export earner for the country.

Now Bolivia still mines quantities of gold, silver and zinc. Natural gas is exported. The biggest money-spinner in Bolivia is coca growing, which the Government hopes to curb – it employs thousands of peasants and contributes an undetermined but significant sum to the official economy.

## POPULATION

The 6.5 million Bolivians are mostly crowded onto the bleak highlands – some 70 percent of the population lives here, mostly around La Paz, Oruro and Lake Titicaca. About two-thirds of the total population are pure Indian stock, many speaking only Quechua or Aymará. Most of the rest are *mestizos*, or mixed Spanish and local descent, locally referred to as *cholos*. About one percent is of African heritage, and a fraction of pure European and Japanese descent.

## CLIMATE

The average temperatures in La Paz are between 6°C and 21°C (42°F and 70°F) during summer and 0°C and 17°C (32°F and 63°F) in winter. Naturally, the higher mountains are much colder while the Amazon basin is hot and wet all year around.

## USEFUL HINTS

For visitors who bring along their electrical appliances, the electricity in Bolivia runs at 220 volts, 50 cycles.

## BUSINESS HOURS

Normally 9 a.m. – noon and 2 p.m. – 6 p.m. Banks are only open until 4.30 p.m. Government offices close on Saturday.

## NEWSPAPERS

In La Paz, you can choose from *El Diario, Hoy* and *Ultima Hora*. The other cities each have their own local papers.

## POSTAL SERVICES

The main post office in La Paz is at Ayacucho, just above Potosí, open from 8 a.m. – 8 p.m. daily except Sunday, when it closes at 6 p.m. Poste Restante keeps letters for three months. The outbound mail seems quite efficiently posted to the rest of the world.

## TELEPHONE & TELEX

Telephone calls can now be made by satellite to the United States and Europe, although an operator is required. You can either call from your hotel or the ENTEL office in La Paz, Edificio Libertad, Calle Potosí.

## MEDICAL SERVICES

Major hotels have doctors on call. The Clinica Americana in La Paz, or the Clinica Santa Isabel (opposite the Hotel Crillon) are competent and well-run to handle any problems that you may encounter.

# GETTING AROUND

## FROM THE AIRPORT

To get from El Alto to the above-mentioned city, La Paz, you can choose a taxi for US$10 or a minibus for about $2 which runs along the main street of the city and will fetch you to your hotel.

## PUBLIC TRANSPORT

The bus network in Bolivia is well-developed, although the quality of service depends on the conditions of roads (of which only some 2 percent are paved). Luxurious services run between La Paz, Cochabamba and Santa Cruz; quite rough services to Potosí.

A wide choice of services are available to Coroico and Copacabana. Trains run from La Paz to Oruro and Potosí, only for the hardy and those with plenty of time: worth paying for a pullman seat if it's available. To get off the beaten track, trucks are the standard transport in the Andes.

Most tour companies also offer services to major cities and attractions.

Within La Paz, taxis are probably the best way to get around. They can be hired privately to go anywhere in the city for US$2.50 or caught as *colectivos*, sharing with passengers along a fixed route for about 30 *centavos*.

## DOMESTIC TRAVEL

The domestic airline Lloyd Aero Boliviano covers routes between all major cities. Fares are cheap by foreign standards. It can be worthwhile to avoid a back-breaking bus journey, especially from La Paz to Potosí (on this route, it is necessary to fly to Sucre then take the four-hour bus journey to Potosí. Crillon Tours offers a comfortable minibus connection that can be arranged in La Paz – their office is at Avenida Camacho 1223).

# WHERE TO STAY

### LA PAZ

**Sheraton**
Avenida Arce, US$117 to US$147 a double.
Tel: 356-950.

**Plaza**
Avenida 16 de Julio, US$93 a double.
Tel: 378-311.

**Crillon**
Plaza Isabel La Catolica, US$30 a double.

**Residencial Rosario**
Illampu 730, at about US$5 a double (one of the best budget hotel).

**Andes**
Avenida Manco Capac 364, basic. (Recommended by budget travelers as the cheapest in La Paz.)

### POTOSÍ

**Hostal Colonial**
Hoyos 8
Tel: 24265.

**Bolivar**
Near corner of Bolivar and Oruro, new hotel in colonial house, about US$20 a double.

# Food Digest

Bolivian food varies by region. Lunch is the main meal of the day and many restaurants offer an *almuerzo completo* or three-course fixed menu. Traditional foods of the highlands are starchy, with potatoes, bread and rice. Their meats are highly spiced. Trout from Lake Titicaca is excellent.

Many restaurants in La Paz serve international cuisine – from Japanese to Swiss. These are mostly along Avenida 16 de Julio (the Prado) and Avenida 6 de Agosto. Below are some of the recommended restaurants.

**Los Escudos**
Edif. Club de la Paz
Tel: 322-028.
(Set lunch, at only $2.50, is excellent)

**Casa del Corregidor**
Calle Murillo 1040
Tel: 353-633.
(International and local dishes)

**Club de la Prensa**
Calle Campero.

**Hoyo 19**
Avenida 16 de Agosto
Tel: 322-320.

## DRINKING TIPS

The usual tea, coffee and soft drinks are readily available in restaurants and cafés, as is the *mate de coca* in tea bags. For a coffee, head for the Cafe La Paz, corner of Camancho and the Prado.

Bolivian beer is surprisingly good, brewed under German supervision. The favorite drink of everyday Bolivians is *chicha*, the potent maize liquor produced near Cochabamba. Keep in mind that altitude intensifies the effects of alcoholic beverages – both the intoxication and the hangover! Also remember that most Bolivians drink to get drunk.

# Things to Do

## MUSEUMS

**Museum of Ethnography and Folklore**
Exhibits relating to Indian culture, corner Ingavi and Sanjines.

**Museo Tiahuanaco**,
Relics from the site, on Calle Tiwanaku just below the Prado; 50 *centavos* admission.

**National Art Museum**
In restored colonial building, corner of Comercio and Socabaya near the Plaza Murillo.

## EXCURSIONS

Most tour companies offer services to major cities. For information on trekking and adventure tours, call Club Andino Boliviano, Calle Mexico, Tel: 324-582.

## NIGHTLIFE

La Paz has several *peñas*. Try the most popular, Peña Naira on Sagarnaga just above Plaza San Francisco, open every night except Sunday with a US$5 cover charge. Others are at the Casa del Corregidor and Los Escudos restaurants.

Bars are not the most pleasant places to hang out in La Paz, usually being reserved for inebriation rather than relaxation.

## SHOPPING

The **artisan's market** is along Calle Sagarnaga, between Mariscal Santa Cruz and Isaac Tamayo. You can pick up ponchos, vests, jackets and mufflers of llama and alpaca wool, as well as some extraordinary tapestries. Many also sell various types of Andean musical instruments, as well as various small sculptures. A little haggling is expected, but don't overdo it – after all, many of these handicrafts take months to make. Prices vary depending on the quality, but are a fraction of what they'd be if bought from the United States or Europe.

# PARKS & RESERVES

# USEFUL ADDRESSES

## CONSULATES & EMBASSIES

**Peruvian Consulate**
Avenida Mariscal Santa Cruz 1285
Tel: 352-031
Open 8.30 a.m. – 1.30 p.m.

**US Embassy**
Calle Colon and Mercado, Consulate on Avenida Potosí.
Tel. 350-251.

**UK Consulate**
Also representing Australian and New Zealand, Avenida Arce casi Campos 2732.
Tel: 329-401.

**National Parks of Bolivia**

384 km / 240 miles

# GETTING THERE

A total of 28 airlines offer international service to and from Brazil with a variety of routes. Although most incoming flights head for Rio de Janeiro, depending on where you are coming from, there are also direct flights to São Paulo and Brasilia, Salvador and Recife on the northeastern coast, and to the northern cities of Belém and Manaus on the Amazon River.

## BY ROAD

There are bus services between a few of the larger Brazilian cities and major cities in neighboring South American countries, including Asunción (Paraguay), Buenos Aires (Argentina), Montevideo (Uruguay) and Santiago (Chile). While undoubtedly a good way to see a lot of the countryside, where the distances are great, you sit in a bus for several days and nights.

# TRAVEL ESSENTIALS

## VISAS

Most Asian, Australian, American and French citizens are required to arrive with a visa; Britons and Germans are not. Your passport will normally be stamped for a 90-day stay.

## MONEY MATTERS

Banks and hotels will exchange your foreign currency into new *cruzados* (NCz) at the official rate (fewer *cruzados* per dollar) but do not usually exchange traveler's checks and cannot change any leftover *cruzados* back into your currency at the end of your stay. Money exchangers at special shops (*casa de câmbio*) or at tourist agencies will give you the parallel exchange rate for both buying and selling currency.

Unless you don't have the time, it's obviously best to exchange money at the parallel rate. Ask at your hotel where the nearest money exchange is located. Try to calculate so as not to have too many *cruzados* left at the end of your stay or you will be "buying" your foreign currency back at the highest rate (higher than you exchanged them for).

## HEALTH

Brazil does not normally require any health or inoculation certificates for entry, nor will you be required to have one to enter another country from Brazil. If you plan to travel in areas outside of cities in the Amazon region or in the Pantanal in Mato Grosso, however, it is recommended for your own comfort and safety, that you have a yellow fever shot (protects you for 10 years, but is effective only after 10 days, so plan ahead). It is also a good idea to protect yourself against malaria in these jungle areas and although there is no vaccine against malaria, there are drugs that will provide immunity while you are taking them. Consult your local public health service and be sure to get a certificate for any vaccination.

## WHAT TO WEAR

Brazilians are very fashion-conscious but actually quite casual dressers. What you bring along, of course, will depend on where you will be visiting and your holiday schedule. São Paulo tends to be more dressy; small inland towns are more conservative. If you are going to a jungle lodge, you will want sturdy clothing and perhaps boots. However if you come on business, a suit and tie for men, and suits, skirts or dresses for women are the office standard.

If you come during Carnival, remember

that it will be very hot to begin with and you will probably be in a crowd and dancing nonstop. Anything colorful is appropriate.

## CUSTOMS

You will be given a declaration form to fill out in the airplane before arrival.

# GETTING ACQUAINTED

## GOVERNMENT

Brazil is a federal republic with 27 states, each with its own state legislature. Since the federal government exercises enormous control over the economy, the political autonomy of the states is restricted. The overwhelming majority of government tax receipts are collected by the federal government and then distributed to the states and cities. The head of government is the president with executive powers and, in fact, exercises more control over the nation than the American president does over the United States. The legislative branch of the federal government is composed of a Congress divided into a lower house, the Chamber of Deputies, and an upper house, the Senate.

In February 1987, however, the Federal Congress was sworn in as a National Constitutional Assembly to draft a new federal constitution for Brazil. The Constituent Assembly completely revised Brazil's constitution in 1987–88. The new constitution, approved in 1988, opened the way for direct presidential elections held in November 1989, with a new president taking office in March 1990.

## TIME ZONES

Despite the fact that Brazil covers such a vast area, over 50 percent of the country is in the same time zone and it is in this area, which includes the entire coastline, that most of the major cities are located. The western extension of this zone is a north-south line from the mouth of the Amazon River, going west to include the northern state of Amapá, east around the states of Mato Grosso and Mato Grosso do Sul and back west to include the south. This time zone, where Rio de Janeiro, São Paulo, Belém and Brasilia are located, is three hours behind Greenwich Mean Time (GMT). Another large zone encompassing the Pantanal states of Mato Grosso and Mato Grosso do Sul, and most of Brazil's north is four hours behind GMT. The far western state of Acre and the westernmost part of Amazonas state are in a time zone five hours behind GMT.

## CLIMATE

In jungle region, the climate is humid equatorial, characterized by high temperatures and humidity, with heavy rainfall all year round.

The eastern Atlantic coast from Rio Grande do Norte to the state of São Paulo has a humid tropical climate, also hot, but with slightly less rainfall than in the north and with summer and winter seasons.

Most of Brazil's interior has a semi-humid tropical climate, with a hot, rainy summer from December through March and a drier, cooler winter (June – August).

Mountainous areas in the Southeast have a high-altitude tropical climate, similar to the semi-humid tropical climate, but rainy and dry seasons are more pronounced and temperatures are cooler, averaging from 18°C to 23°C (64°F to 73°F).

Part of the interior of the Northeast has a tropical semi-arid climate – hot with sparse rainfall. Most of the rain falls during three months, usually March – May, but sometimes the season is shorter and in some years there is no rainfall at all. Average temperature is 24°C to 27°C (75°F to 80°F).

Brazil's South, below the Tropic of Capricorn, has a humid subtropical climate. Rainfall is distributed regularly throughout the year and temperatures vary from 0°C to 10°C (30°F to 40°F) in winter, with occasional frosts and snowfall (but the latter is rare) to 21°C to 32°C (70°F to 80°F) in summer.

## CULTURE & CUSTOMS

While handshaking is a common practice, here it is customary to greet not only friends and relatives but also complete strangers to whom you are being introduced with hugs and kisses. The "social" form of kissing consists usually of a kiss on each cheek.

## ADDRESSES

To understand the addresses, here's what the Portuguese words mean:
*Alameda* (Al.) = lane;
*Andar* = floor, story;
*Avenida* (Av.) = avenue;
*Casa* = house;
*Centro* = the central downtown business district also frequently referred to as a *cidade* or "the city";
*Conjunto* (Cj.) = a suite of rooms or a group of buildings;
*Estrada* (Estr.) = road or highway;
*Fazenda* = ranch, also a lodge;
*Largo* (Lgo.) = square of plaza;
*Lote* = Lot;
*Praça* (Pça.) = square or plaza;
*Praia* = beach;
*Rio* = river;
*Rodovia* (Rod.) = highway;
*Rua* (R.) street;
*Sala* = room.

Ordinal numbers are written with ° or a degree sign after the numeral, so that 3° andar means 3rd floor. BR followed by a number refers to one of the federal interstate highways, for example BR 101.

Telex/telephone numbers are given with the area code for long-distance dialing in parentheses. *Ramal* = telephone extension.

## TIPPING

Most restaurants will usually add a 10 percent service charge onto your bill. If you are in doubt as to whether it has been included, it's best to ask (*O serviço está incluido?*). Give the waiter a bigger tip if you feel the service was special. Although many waiters will don a sour face if you don't tip above the 10 percent included in the bill, you have no obligation to do so.

Hotels will also add a 10 percent service charge to your bill, but this doesn't necessarily go to the persons who were helpful to you.

## ELECTRICITY

Electric voltage is not standardized throughout Brazil, but most cities have a 127-volt current, as is the case of Rio de Janeiro and São Paulo, Belém, Belo Horizonte, Corumbá and Cuiabá, Curitiba, Foz do Iguaçu, Porto Alegre and Salvador. The electric current usage is 220 volts in Brasilia, Florianópolis, Fortaleza, Recife and São Luis. Manaus uses 110-volt electricity.

## BUSINESS HOURS

Business hours for offices in most cities are 9 a.m. – 6 p.m. Monday through Friday. Lunch "hours" may last literally hours.

Banks open from 10 a.m. – 4.30 p.m. Monday through Friday. The *casas de câmbio* currency exchanges operate usually from 9 a.m. – 5 p.m. or 5.30 p.m.

## HOLIDAYS

**January 1**
– **New Year's Day** (national holiday)
– **Good Lord Jesus of the Seafarers** (four-day celebration in Salvador; starts off with a boat parade)

**January 6**
– **Epiphany** (regional celebrations, mostly in the Northeast)

**January (3rd Sunday)**
– **Festa do Bonfim** (one of the largest celebrations in Salvador)

**February 2**
– **Iemanjá Festival** in Salvador (the Afro-Brazilian goddess of the sea in syncretism with Catholism corresponds with Virgin Mary)

**February/March (movable)**
– **Carnival** (national holiday; celebrated all over Brazil on the four days leading up to Ash Wednesday. Most spectacular in Rio, Salvador and Recife/Olinda)

**March/April (movable)**
– **Easter** (Good Friday is a national holiday; Colonial Ouro Preto puts on a colorful procession; passion play staged at Nova Jerusalem)

**April 21**
- **Tiradentes Day** (national holiday in honor of the martyred hero of Brazil's independence – celebrations in his native Minas Gerais, especially Ouro Preto)

**May 1**
- **Labor Day** (national holiday)

**May/June (movable)**
- **Corpus Christi** (national holiday)

**June/July**
- **Festas Juninas** (a series of street festivals held in June and early July in honor of Saints John, Peter and Anthony, featuring bonfires, dancing and mock marriages)

**June 15 – 30**
- **Amazon Folk Festival** (held in Manaus)

**June/July**
- **Bumba-Meu-Boi** (processions and street dancing in Maranhao are held in the second half of June and beginning of July)

**September 7**
- **Independence Day** (national holiday)

**October**
- **Oktoberfest** in Blumenau (put on by descendants of German immigrants)

**October 12**
- **Nossa Senhora de Aparecida** (national holiday honoring Brazil's patron saint)

**November 2**
- **All Souls Day** (national holiday)

**November 15**
- **Proclamation of the Republic** (national holiday, also election day)

**December 25**
- **Christmas** (national holiday)

**December 31**
- **New Year's Eve** (on Rio de Janeiro beaches, gifts are offered to Iemanjá).

## MEDIA

A daily English-language newspaper, the *Latin America Daily Post,* circulates in Rio de Janeiro and São Paulo, carrying international news from wire services, including sports and financial news, as well as domestic Brazilian news. The *Miami Herald*, the Latin America edition of the *International Herald Tribune* and the *Wall Street Journal* are available on many newsstands in the big cities, as are such news magazines as *Time* and *Newsweek*.

## POSTAL SERVICES

Post offices generally are open from 8 a.m. – 6 p.m. Monday – Friday, 8 a.m. – noon on Saturday and are closed on Sunday and holidays. In large cities, some branch offices stay open until later. (The post office in the Rio de Janeiro International Airport is open 24 hours a day.)

Post offices are usually designated with a sign reading *correios* or sometimes "ECT" (for *Empresa de Correios e Telégrafos* = Postal and Telegraph Company).

## TELEPHONE & TELEX

Pay phones in Brazil use tokens which are sold at newsstands, bars or shops, usually located near the phones. Ask for *fichas de telefone* (the "i" is pronounced like a long "e" and the "ch" has an "sh" sound). Each *ficha* is good for three minutes, after which your call will be cut off. To avoid being cut off in the middle of a call, insert several tokens into the slot – unused tokens will be returned when you hang up. The sidewalk *telefone público* is also called an *orelhão* (big ear) because of the protective shell which takes the place of a booth – yellow for local or collect calls, blue for direct-dial long-distance calls within Brazil. The latter requires a special, more expensive token. You can also call from a *posto telefônico*, a telephone company station (most bus stations and airports have such facilities), where you can either buy tokens, use a phone and pay the cashier afterward, or make a credit card or collect call.

Domestic long-distance rates go down 75 percent every day between 11 p.m. and 6 a.m. and are 50 percent less expensive between 6 a.m. and 8 a.m. and 8 and 11 p.m. on weekdays, between 2 p.m. and 11 p.m. on Saturday and between 6 a.m. and 11 p.m. on Sunday and holidays.

## MEDICAL SERVICES

Should you need a doctor while in Brazil, the hotel you are staying at will be able to recommend reliable professionals who often speak several languages. Many of the better hotels even have a doctor on duty. Your consulate will also be able to supply you with a list of physicians who speak your language. In Rio de Janeiro, the Rio Health Collective (English-speaking) runs a 24-hour referral service, Tel: (021) 325-9300 ramal or extension 44 for the Rio area only.

Check with your health insurance company before traveling – some insurance plans cover any medical service that you may require while abroad.

## DRINKING WATER

Don't drink tap water in Brazil. Although water in the cities is treated and is sometimes quite heavily chlorinated, people filter water in their homes. Any hotel or restaurant will have inexpensive bottled mineral water, both carbonated (*com gas* or "with gas") and uncarbonated (*sem gas* or "without gas"). If you are out in the hot sun, make an effort to drink extra fluids.

# GETTING AROUND

## FROM THE AIRPORT

Until you get your bearing, you are best off taking a special airport taxi for which you pay in advance at the airport at a fixed rate set according to your destination. There will be less of a communication problem, no misunderstanding about the fare and even if the driver should take you around by the "scenic route", you won't be charged extra for it. However, If you should decide to take a regular taxi, check out the fares posted for the official taxis so that you will have an idea of what is a normal rate.

## PUBLIC TRANSPORT

Taxis are the best way for visitors to get around. It's easy to get "taken for a ride" in a strange city. Whenever possible, take a taxi from your hotel where someone can inform the driver where you want to go.

## BUSES

Comfortable, on-schedule bus service is available between all major cities, and even to several other South American countries. Remember that distances are far and bus rides can be long, i.e., a few days. But you could break the long journey with a stop along the way.

## BOATS

Local boat tours and excursions are available in coastal and riverside cities. There are also options for longer trips.

There are Amazon River boat trips lasting a day or two to up to a week or more. These range from luxury floating hotels to more rustic accommodations. Boat trips can be taken on the Rio São Francisco in the Northeast and in the Pantanal marshlands of Mato Grosso, where many visitors go for the fishing. The *Blue Star Line* will take passengers on its freighters which call at several Atlantic coast ports. *Linea C* and *Oremar* have cruises out of Rio which stop along the Brazilian coast on the way down to Buenos Aires or up to the Caribbean. Book well in advance for the longer trips .

Cities along the coast (and along major rivers) offer short sightseeing or day-long boat tours. Many towns have local ferry service across bays and rivers and to islands. And schooners and yachts, complete with crew, may be rented for an outing.

## TRAINS

Except for crowded urban commuter railways, trains are not a major form of transportation in Brazil and rail links are not extensive. There are a few train trips, which are tourist attractions in themselves, either because they are so scenic or because they run on antique steam-powered equipment.

In the southern state of **Paraná**, the 110-km Curitiba–Paranaguá railroad is famous for spectacular mountain scenery.

The train to Corumbá, in the state of Mato Grosso do Sul, near the Bolivian border, crosses the southern tip of the **Pantanal** marshlands. There are train links all the way to São Paulo, over 1,400 km away – a long ride. The most scenic part is the 400-km stretch between Campo Grande and Corumbá.

In the **Amazon** region, you can ride on what is left of the historic Madeira–Mamoré Railway which a 27-km of track between Porto Velho and Cachoeira de Teotônio in the state of Rondônia. The Madeira–Mamoré runs on Sundays only and strictly as a tourist attraction.

## DOMESTIC TRAVEL

For travel within Brazil, the major airlines are Transbrasil, Varig, Cruzeiro and Vasp with several other regional carriers which service the smaller cities.

Different lines have similar prices for the same routes. To get the best value call up for several quotes. There is a 30 percent discount for night flights (*vôo econômico or vôo noturno*) with departures between midnight and 6 a.m.

Transbrasil and Varig also offer air passes which must be bought outside Brazil. There are two types: one costing US$250 is valid for 14 days and allows you to visit four cities; the other costs US$330, is valid for 21 days and permits unlimited travel. Ask your travel agent about these – they are a good deal if you plan to travel extensively within Brazil.

The large airlines also cooperate in a shuttle service between Rio and São Paulo (with flights every half hour), Rio and Brasilia (flights every hour) and Rio and Belo Horizonte (usually about 10 flights per day). Although you may be lucky, a reservation is a good idea.

# WHERE TO STAY

## HOTELS

### RIO DE JANEIRO (CITY)

**Caesar Park**
Avenida Vieira Souto, 460
Ipanema
Tel: (021) 287-3122
Telex: (021) 21204

**Copacabana Palace**
Avenida Atlântica, 1702
Copacabana
Tel: (021) 255-7070
Telex: (021) 21482

**Inter-Continental Rio**
Rua Prefeito Mendes de Morais, 222
São Conrado
Tel: (021) 322-2200
Telex: (021) 21790

**Leme Palace**
Avenida Atlântica, 656
Leme
Tel: (021) 275-8080
Telex: (021) 23265

**Luxor Copacabana**
Avenida Atlântica, 2554
Copacabana
Tel: (021) 257-1940
Telex: (021) 23971

**Meridien-Rio**
Avenida Atlântica, 1020
Leme
Tel: (021) 275-9922
Telex: (021) 23183

**Praia Ipanema**
Avenida Vieira Souto, 706
Ipanema
Tel: (021) 239-9932
Telex: (021) 31280

**Trocadero**
Avenida Atlântica, 2064
Copacabana
Tel: (021) 257-1834
Telex: (021) 22655

## SÃO PAULO (CITY)

**Bourdon Hotel**
Avenida Vieira de Carvalho, 99
Centro
Tel: (011) 223-224
Telex: (011) 32781

**Caesar Park**
Rua Augusta, 1508/20
Cerqueira Cesar
Tel: (011) 285-6622
Telex: (011) 22539

**Maksoud Plaza**
Alameda Campinas, 150
Bela Vista
Tel: (011) 251-2233
Telex: (011) 30026

**Moferrej Sheraton**
Alameda Santos, 1437
Cerqueira Cesar
Tel: (011) 284-5544
Telex: (011) 34170

**São Paulo Hilton**
Avenida Ipiranga, 165
Centro
Tel: (011) 256-0033
Telex: (011) 21981

## BRASILIA

**Brasilia Carlton**
Setor Hoteleiro Sul
Quadra 5, Bloco G
Tel: (061) 224-8819
Telex: (061) 1981

**Eron Brasilia**
Setor Hoteleiro Norte
Quadra 5, Lote A
Tel: (061) 226-2125
Telex: (061) 1422.

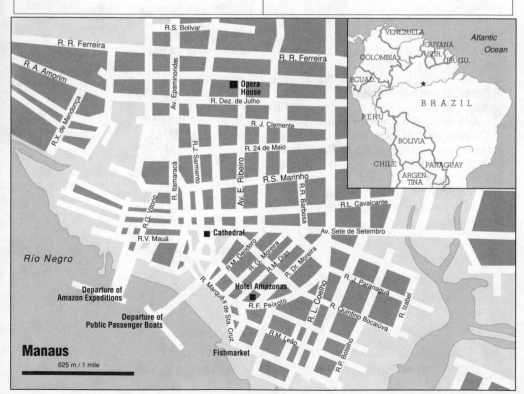

Manaus
625 m / 1 mile

# MANAUS, AMAZONAS

**Note:** Manaus is a major industrial and commerical city with free-port status. Prices tend to be higher than in other parts of the Amazon, but lucily the forest begins just a few kilometers away from town. Road trips to Porto Velho and Boa Vista are best attempted in the dry season. Roads to Itacoatiara and Manicaparu are passable all year round, but local commuters seem to prefer the river ferry. Telephone prefix 092.

**Amazonas**
Praça Adalberto vale (Center)
Tel: 234-7679
Telefax: (092) 2277

**Ana Cássia**
Rua dos Andradas 14 (Center)
Tel: 232-6201
Telex: (092) 2713

**Anaconda**
Rua Quintino Bocaiúva 791 (Center)
Tel: 233-1110
Telex: (092) 1359

**Aquarius**
Rua Guilherme Moreira 116 (Center)
Tel: 232-5620, 232-5470
Telex: (092) 2420

**Central Manaus**
Rua Dr. Moreira 202 (Center)
Tel: 232-7887
Telex: (092) 2765

**DaVince**
Rua Belo Horizonte 240-A (Adrianópolis)
Tel: 611-1213
Telex: (092) 1024

**Flamboyant**
Avenida Eduardo Ribeiro 926 (Center)
Tel: 234-0296

**Imperial**
Avenida Getúlio Vargas 227 (Center)
Tel: 233-8711
Telex: (092) 2231

**International**
Rua Dr. Moreira 168 (Center)
Tel: 234-1315

**Lider**
Avenida 7 de Setembro 827 (Center)
Tel: 234-4965
Telex: (092) 2679

**Lord**
Rua Marcílio Dias 217 (Center)
Tel: 234-9741
Telex: (092) 2278

**Manaus**
Rua Lobo D'Almada 48
Tel: 234-3626

**Mônaco**
Address: Rua Silva Ramos 20 (Center)
Tel: 232-5211
Telex: (092) 2802

**National**
Address: Rua Dr. Moreira 59
Tel: 232-3514

**Novotel Manaus**
Avenida Mandii 4 (Industrial District)
Tel: 237-1211
Telex: (092) 2429

**Panair Palace**
Avenida Getúlio 678 (Center)
Tel: 234-0518
Telex: (092) 2770

**Plaza**
Avenida Getúlio Vargas 215 (Center)
Tel: 233-8900
Telex: (092) 2563

**Regente**
Rua Cel. Sérgio Pessoa 189 (Center)
Tel: 234-6864
Telex: (092) 2420

**Rei Salomão**
Rua Dr. Moreira 119 (Center)
Tel: 234-7374

**Rio Mar**
Rua Guilherme Moreira 329 (Center)
Tel: 234-7409

**São Pedro**
Rua Rui Barbosa 166 (Center)
Tel: 232-8664

**Solimões**
Rua Saldanha Marinho 435 (Center)
Tel: 234-0123

**Sombra Palace**
Avenida 7 de Setembro 1325 (Center)
Tel: 234-8777

**Tropical Manaus**
Ponta Negro (beach)
Tel: 238-5757
Telex: (092) 2173
This is the largest luxury resort hotel in
Northern South America.

## SALVADOR

**Bahia Othon Palace**
Avenida Presidente Vargas, 2456
Ondina
Tel: (071) 247-1044
Telex: (071) 1217

**Club Mediterranee**
Estrada Itaparica-Nazaré
KM 13 Itaparica
Tel: (071) 833-1141

Telex: (071) 2143
Reservations in Salvador
Tel: (071) 247-3488

## OLINDA

**Marolinda**
Avenida Beira-Mar, 1615
Tel: (081) 429-1699
Telex: (081) 3249

## RECIFE

**Boa Viagem**
Avenida Boa Viagem, 5000
Boa Viagem
Tel: (081) 341-4144
Telex: (081) 2072

**Recife Palace**
Avenida Boa Viagem, 4070
Boa Viagem
Tel: (081) 325-4044
Telex: (081) 4258

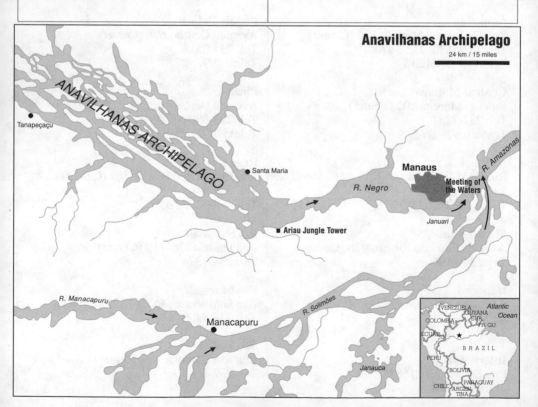

### Anavilhanas Archipelago
24 km / 15 miles

# LODGES

## NEAR MANAUS, AMAZONAS

**Note:** These forest lodges can be booked in Manaus. Stays of two or more days are common. Be sure to enquire what is included in the package deals offered (transport, meals, tours, canoe rental, etc.)

**Adventure Camp**
(40 km from Manaus, on Rio Tarumã)
Information: Regiatur
Avenida 7 de Setembro 788, shop 1
Tel: 234-6900
Telex: (092) 2579

**Amazon Lodge**
(100 km from Manaus, on Juma Lake)
Information: Transamazonas
Rua Leonardo Malcher 734 (Center)
Tel: 232-1454
Telex: (092) 2641

**Amazon Village**
(60 km from Manaus, on Puraquequara Lake)
Information: Transamazonas

Rua Leonardo Malcher 734 (Center)
Tel: 232-1454
Telex: (092) 2641

**Ariaú Jungle Tower**
(60 km from Manaus, near the Avalihanas Archipelago on the Rio Negro)
Information: Rio Amazonas Tours
Rua Silva Ramos 42 (Center)
Tel: 234-7308
Telex: (092) 2730
Telefax: 233-5615
The tall tower is a good point for watching wildlife. Nice trail system. Walkway out to Rio Negro.

**Janauacá Jungle Lodge**
(120 km from Manaus, on Janauacá Lake)
Information: Concyatur
Rua Dr. Almínio 36 (Center)
Tel: 234-2348
Telex: (092) 1147

**Lago Salvador Lodge**
(30 km from Manaus, up the Rio Negro)
Information: Tropical Turismo
Lobby of Hotel Tropical Manaus,

Pico da Neblina
National Park
80 km / 50 miles

Ponte Negra
Tel: 238-5757
Telex: (092) 2173

**Parque Pousadas Rio Negro**
(1 hour by boat from Manaus, up the Rio
  Negro)
Information: ATA Turismo
Avenida Eduardo Ribeiro 639 (Center)
Tel: 232-2657
Telex: (092) 2459

**Pousada dos Guanavenas**
(on Silvas Island, 4½ hours by car from
  Manaus)
Information: Pousada dos Guanavenas
Rua Ferreira Pena 755 (Center)
Tel: 236-2352
Telex: (092) 1101

**Tapiri Pousada na Selva**
(camp at January Lake, 1½ hours by boat from
  Manaus)
Information: Selvatur
Praça Adalberto Vale (Center)
Tel: 234-8639
Telex: (092) 1500

**Tapiri Lodge Puraquequara**
(60 km, 2½ hours by boat from Manaus, on.
  Puraquequara Lake)
Information: Luciatur
Rua Barroso 307 (Center)
Tel: 234-0733
Telex: (092) 2457

## BELÉM

**Note:** Belém, the capital of Pará, is a major
river and ocean port, the main commercial
center for the lower Amazon. Its mercantile
on the been broken by the new highways to
Manaus, Santerém, and Porto Velho, but old
habits (like 100% profit margins) die hard. If
you are outfitting Amazon expedition in
Belém, a good rule of thumb is to shop
around. The same holds true for accommodations and services, as well. Telephone
prefix 091.

**Akemi**
Avenida Ceará 81
Tel: 226-5961

**Avenida**
Avenida Pres. Vergas 404 (Center)
Tel: 222-9953, 223-8893

**Hilton International Belém**
Avenida Pres. Vargas 882 (Praça da
  República)
Tel: 223-6500

**Cambará**
Rua 16 de Novembro 300 (Cidade Velha)
Tel: 224-2422

**Central**
Avenida Pres. Vargas 290 (Center)
Tel: 222-3011

**Diplomata**
Trav. 1 de Queluz 29 (Canudos)
Tel: 231-3321

**Executivo**
Avenida Alcindo Cacla 855
Tel: 223-0767

**Gentil**
Avenida Gentil Bittencourt 918 (Nazaré)
Tel: 224-9022

**Parque Nacional
de Amazonia (Tapajós)
and Trombetas River**

160 km / 100 miles

**Milano**
Avenida Pres. Vargas 636
Tel: 223-7722

**Novotel**
Avenida Bernardo Sayão 4804 (Guamá)
Tel: 229-8011
Telex: (091) 1241

**Verde Oliva**
Rua Boaventura da Silva 1179 (São Brás)
Tel: 224-7682

**Equatorial Palace**
Avenida Braz de Aguiar 612 (Nazaré)
Tel:224-8855
Telex: (091) 1605

**Excelsior Grão Pará**
Avenida Pres. Vargas 718 (Center)
Tel: 222-3255
Telex: (091) 1171

**Zoghbi Park**
Rua Padre Prudêncio 220 (Center)
Tel: 241-1800

**Plaza**
Praça da Bandeira 130 (Cidade Velha)
Tel: 224-2800

**Regente**
Avenida Gov. José Malcher 485 (Center)
Tel: 224-0755
Telex: (091) 1796

**Vila Rica**
Avenida Júlio César 1777 (nr. airport)
Tel: 233-4222
Telex: (091) 1585

**Sagres**
Avenida Gov. José Malcher 2927 (Sãn Brás)
Tel: 228-3999
Telex: (091) 1662

**Torino**
Rua 28 de Setembro 956
Tel: 241-2412, 224-5944, 224-5913

**Transbrasil**
Avenida Cipriano Santos 243 (São Brãs)
Tel: 228-2500

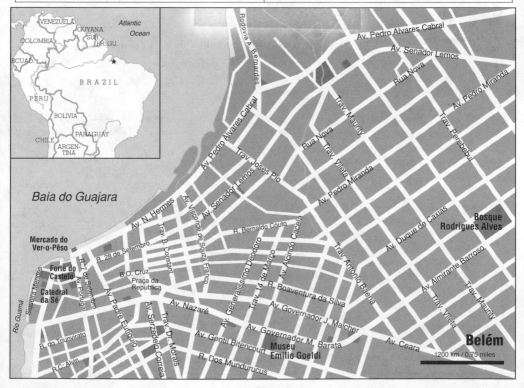

**Vanja**
Rua Benjamin Constante 1164
Tel: 222-6457
Telex: (091) 1343

**Ver-O-Peso**
Avenida Boulevard Castilhos França 208/
214 (Center)
Tel: 224-2267, 241-4236
With restaurant, overlooking port.

**Vidonho's**
Rua Ó de Almeida 476 (Center)
Tel: 225-1444

## BELO HORIZONTE

**Belo Horizonte Othon Palace**
Avenida Afonso Pena, 1050
Centro
Tel: (031) 226-7844
Telex: (031) 2052

## SANTARÉM

**Tropical Santarém**
Avenida Mendonça Furtado 4120

Tel: 522-5285
Telex: (091) 5505

**Santarém Palace**
Avenida Rui Barbosa 726
Tel: 522-5285
Telex: (091) 2029

**Brasil Grande Hotel**
Trav. 15 de Agosto 213
Tel: 522-5660

**City**
Trav. Francisco Correa 200
Tel: 522-4719

**Uirapuru**
Avenida Adriano Pimentel 140
Tel: 522-1531

**Camino Hotel**
Praça Dr. Rodrigues dos Santos 887
Tel: 522-2981

**Modelo**
Rua Sen. Lameira Bittencourt 344
Tel: 522-1689

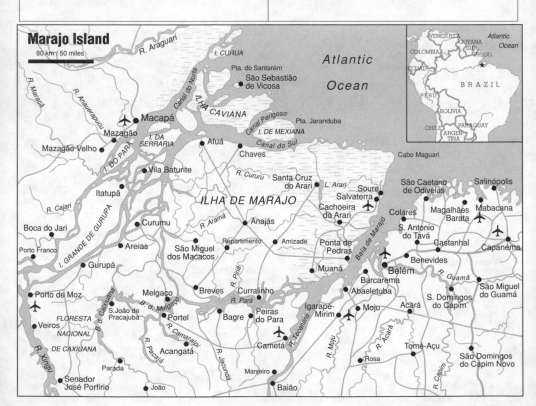

**Horizonte**
Trav. Sen. Lemos 737
Tel: 522-3435

**São Luiz**
Trav. Sen Lemos 118
Tel: 522-1043

**São Francisco**
Avenida Maglhães Barata 734
Tel: 522-1957

**Nossa Estrela**
Avenida Cuiabá 1210
Tel: 522-1449

**Equatorial**
Rua Silvino Pinto/Rui Barbosa
Tel: 522-1135

# FOOD DIGEST

## WHAT TO EAT

A country as large and diverse as Brazil naturally has regional specialties when it comes to food. Immigrants, too, influence Brazilian cuisine. In some parts of the south, the cuisine reflects a German influence; Italian and Japanese immigrants brought their cooking skills to São Paulo. Some of the most traditional Brazilian dishes are adaptations of Portuguese or African foods. But the staples for many Brazilians are rice, beans and manioc.

Lunch is the heaviest meal of the day and you might find it very heavy indeed for the hot climate. Breakfast is most commonly *café com leite* (hot milk with coffee) with bread and sometimes fruit. Supper is often taken quite late.

Although not a great variety of herbs is used, Brazilian food is tastily seasoned, not usually peppery – with the exception of some very spicy dishes from Bahia. Many Brazilians do enjoy hot pepper (*pimenta*) and the local *malagueta* chilis can be infernally fiery or pleasantly nippy, depending on how they're prepared. But the pepper sauce (most restaurants prepare their own, sometimes jealously guarding the recipe) is almost always served separately so the option is yours.

Considered Brazil's national dish (although not found in all parts of the country), *feijoada* consists of black beans simmered with a variety of dried, salted and smoked meats. Originally made out of odds and ends to feed the slaves, nowadays the tail, ears, feet, etc. of a pig are thrown in. *Feijoada* for lunch on Saturday has become somewhat of an institution in Rio de Janeiro, where it is served *completa* with white rice, finely shredded kale (*couve*), *farofa* (manioc root meal toasted with butter) and sliced oranges.

The most unusual Brazilian food is found in Bahia, where a distinct African influence can be tasted in the *dendê* palm oil and coconut milk. The Bahianos are fond of pepper and many dishes call for ground raw peanuts or cashew nuts and dried shrimp. Some of the most famous Bahian dishes are *Vatapá* (fresh and dried shrimp, fish, ground raw peanuts, coconut milk, *dendê* oil and seasonings thickened with bread into a creamy mush); *moqueca* (fish, shrimp, crab or a mixture of seafood in a *dendê* oil and coconut milk sauce); *xinxim de galinha* (a chicken *fricasse* with *dendê* oil, dried shrimp and ground raw peanuts); *caruru* (a shrimp-okra gumbo with *dendê* oil); *bobó de camarão* (cooked and mashed manioc root with shrimp, *dendê* oil and coconut milk); and *acarajé* (a patty made of ground beans fried in *dendê* oil and filled with *vatapá*, dried shrimp and *pimenta*). Although it is delicious, note that the palm oil as well as the coconut milk can be too rich for some delicate digestive tracts.

Seafood is plentiful all along the coast, but the Northeast is particularly famed for its fish, shrimp, crabs and lobster. Sometimes cooked with coconut milk, other ingredients that add a nice touch to Brazilian seafood dishes are coriander, lemon juice and garlic. Try *peixe a Brasileiro*, a fish stew served with *prião* (manioc root meal cooked with broth from the stew to the consistency of porridge) and a traditional dish served along the coast. One of the tastiest varieties of fish is *badejo*, a sea bass with firm white meat.

A favorite with foreign visitors and very

popular all over Brazil is the *churrasco* or barbecue, which originated with the southern gaucho cowboys who roasted meat over an open fire. Some of the finest *churrasco* can be eaten in the South. Most *churrascarias* offer a *rodizio* option: for a set price diners eat all they can of a variety of meats.

A few exotic dishes found in the Amazon include those prepared with *tucupi* (made from manioc leaves and having a slightly numbing effect on the tongue), especially *pato no tucupi* (duck) and *tacacá* broth with manioc starch. There are also many varieties of fruit that are found nowhere else. The rivers produce a great variety of fish, including piranha. River fish is also the staple in the Pantanal.

Two Portuguese dishes that are popular in Brazil are *bacalhau* (imported dried salted codfish) and *cosido*, a glorified "boiled dinner" of meats and vegetables (usually several root vegetables, squash and cabbage and/or kale) served with *pirão* made out of broth. Also try delicate palmito palm heart, served as a salad, soup or pastry filling.

*Salgadinhos* are a Brazilian style of finger food, served as appetizers, canapés, ordered with a round of beer or as a quick snack at a lunch counter – a native alternative to US-style fast food chains. *Salgadinhos* are usually small pastries stuffed with cheese, ham, shrimp, chicken, ground beef, palmito, etc. There are also fish balls and meat croquettes, breaded shrimp and miniature quiches. Some of the bakeries have excellent *salgadinhos* which you can either take home or eat at the counter with a fruit juice or soft drink. Other tasty snack foods include *pão de queijo* (a cheesy quick bread), and *pastel* (two layers of a thinly rolled pasta-like dough with a filling sealed between, deep-fried). Instead of French-fried potatoes, try *aipim frito* (deep-fried manioc root).

## DRINKING NOTES

Brazilians are great social drinkers and love to sit for hours talking and often singing with friends over drinks. During the hottest months, this will usually be in open air restaurants where most of the people will be ordering *chope*, cold draft beer, perfect for the hot weather. Brazilian beers are really very good. Take note that although *cerveja* means beer, it is usually used to refer to bottled beer only.

Brazil's own unique brew is *cachaça*, a strong liquor distilled from sugar cane, a type of rum, but with its own distinct flavor. Usually colorless, it can also be amber. Each region boasts of its locally produced *cachaça*, also called *pinga*, *cana* or *aguardente*, but traditional producers include the states of Minas Gerais, Rio de Janeiro, São Paulo and in the northeastern states where the sugar cane has long been a cash crop.

Out of *cachaça*, some of the most delightful mixed drinks are concocted. Tops is the popular *caipirinha*, also considered the national drink, It's really a simple concoction of crushed lime – peel included – and sugar topped with plenty of ice. Variations on this drink are made using vodka or rum, but you should try the real thing. Some bars and restaurants mix their *caipirinhas* sweeter than you may want – order yours *com pouco açucar* (with a small amount of sugar) or even *sem açucar* (without sugar). *Batidas* are beaten in the blender or shaken and come in as many varieties as there are types of fruit in the tropics. Basically fruit juice with *cachaça*, some are also prepared with sweetened condensed milk. Favorites are *batida de maracujá* (passion fruit) and *batida de coco* (coconut milk), exotic flavors for visitors from cooler climates. When sipping *ba-tidas*, don't forget that the *cachaça* makes them a potent drink, even though they taste like fruit juice.

Finally there is wonderful Brazilian coffee. *Café* is roasted dark, ground fine, prepared strong and taken with plenty of sugar. Coffee mixed with hot milk (*café com leite*) is the traditional breakfast beverage throughout Brazil. Other than at breakfast, it is served black in tiny demitasse cups, never with a meal. (And decaffeinated is not in the Brazilian vocabulary.) These *cafezinhos* or "little coffees", offered the visitor to any home or office, are served piping hot at any *botequim* (there are even little stand-up bars that serve only *cafezinho*). However you like it, Brazilian coffee makes the perfect ending to every meal.

# THINGS TO DO

## CONCERTS

Musical forms have developed in different parts of the country, many with accompanying forms of dance. While the Brazilian influence (especially in jazz) is heard around the world, what little is known of Brazilian music outside the country is just the tip of an iceberg.

Take in a concert by a popular singer or ask your hotel to recommend a nightclub with live Brazilian music: *bossa nova*, *samba*, *choro* and *seresta* are popular in Rio and São Paulo – each region has something different to offer. If you are visiting at Carnival, you'll see and hear plenty of music and dancing in the streets, mostly *samba* in Rio and *frevo* in the Northeast. There are also shows all year long designed to give tourists a taste of Brazilian folk music and dance. If you like what you hear, get some records or tapes to bring back with you.

The classical music and dance season runs from Carnival through mid-December. Besides presentations by local talents, major Brazilian cities (mainly Rio, São Paulo and Brasilia) are included in world concert tours by international performers. One of the most important classical music festivals in South America takes place in July each year in Campos do Jordáo in the state of São Paulo.

## MUSEUMS

Brazil's historical museums are unlikely to be the highlight of your visit. With rare exceptions, there are just not enough resources available for proper upkeep and acquisitions. What follows is a partial listing. Temporary exhibits are announced in the newspapers under *Exposicoes*.

## RIO DE JANEIRO

**City Museum**
(Museu da Cidade)
Estrada de Santa Marinha
Parque da Cidade, Gávea
Tel: (021) 322-1328
Tuesday – Sunday noon – 4.30 p.m.

**Folk Art Museum**
(Museu do Folclore Edison Carneiro)
Rua do Catete
Tel: (021) 285-0891
Tuesday – Friday 11 a.m. – 6 p.m.; Saturday, Sunday and holidays 3 p.m. – 6 p.m.

**Indian Museum**
(Museu do Indio)
Rua das Palmeiras, 55
Botafogo
Tel: (021) 286-8799
Tuesday – Friday 10 a.m. – 5 p.m.; Saturday and Sunday 1 p.m. – 5 p.m.

**Museum of Modern Art**
(Museu de Arte Moderna)
Avenida Infante D. Henrique, 85
Parque do Flamengo
Tel: (021) 210 – 2188
Tuesday – Sunday noon – 6 p.m.

**National History Museum**
(Museu Histórico Nacional)
Praça Marechal Ancora
(near Praça 15 de Novembro)
Centro
Tel: (021) 240 – 7978/ 220 – 2628
Tuesday – Friday 10 a.m. – 5.30 p.m.; Saturday, Sunday and holidays 2.30 p.m. – 5.30 p.m.

**National Museum**
(Museu Nacional)
Quinta da Boa Vista
São Cristóvão
Tel: (021) 264 – 8262
Tuesday – Sunday 10 a.m. – 4.45 p.m.

**National Museum of Fine Arts**
(Museu Nacional de Belad Artes)
Avenida Rio Branco, 199
Centro
Tel: (021) 200 – 160/240 – 0068
Tuesday and Thursday 10 a.m. – 6.30 p.m.; Wednesday and Friday noon – 6.30 p.m.; Saturday, Sunday and holidays 3 p.m. – 6 p.m.

## SÃO PAULO

**Museum of Brazilian Art**
(Museu de Arte Brasileira)
Rua Alagoas, 903
Higienópolis
Tel: (011) 826 – 4233
Tuesday – Friday 2 p.m. – 10 p.m.; Saturday, Sunday and holidays 1 p.m. – 6 p.m.

**Museum of Contemporary Art**
(Museu de Arte Contemporanea)
Parque do Ibirapuera
Pavilhão da Bienal
3° andar
Tel: (011) 571 – 9610
Tuesday – Sunday 1 p.m. – 6 p.m.

**Museum of Image and Sound – (Cinema)**
(Museu da Imagem e do Som)
Avenida Europa, 158
Jardim Europa
Tel: (011) 852 – 9197
Tuesday, Sunday and holidays 2 p.m. – 10 p.m.

**Museum of Modern Art**
(Museu de Arte Moderna)
Parque do Ibirapuera
Grande Marquise
Tel: (011) 549 – 9688
Tuesday – Friday 1 p.m. – 7 p.m.; Saturday and Sunday 11 a.m. – 7 p.m.

## MANAUS

**Indian Museum**
(Museu do Indio)
Rua Duque de Caxias
Avenida 7 de Setembro
Tel: (092) 234 – 1422
Monday – Saturday 8 – 11 a.m. and 2 – 5 p.m.

**Museum of the Port of Manaus**
(Museu do Porto de Manaus)
Boulevard Vivaldo Lima
Centro
Tel: (092) 232 – 4250
Tuesday – Sunday 8 – 11 a.m. and 2 – 5 p.m.

## OURO PRETO

**Inconfidencia Historial Museum**
(Museu da Inconfidência)
Praça Tiradentes
Tuesday – Sunday noon – 5.30 p.m.

## SALVADOR

**Afro – Brazilian Museum**
(Museu Afro – Brasileiro)
(old medical school/Faculdade de Medicina building)
Terreiro de Jesus
Tel: (071) 243 – 0384
Tuesday – Saturday 9 – 11.30 a.m. and 2 – 5.30 p.m.

# SHOPPING

## WHAT TO BUY

Most visitors to Brazil just can't resist the stones. One of the major attractions of shopping for **gemstones** in Brazil, besides the price, is the tremendous variety not found anywhere else. Brazil produces amethysts, aquamarines, opals, topazes, the many – colored tourmaline – to name just a few of the most popular buys – as well as diamonds, emeralds, rubies and sapphires.

Although you may find some tempting offers, unless you are an expert gemologist, it's wiser to buy from a reliable jeweler, where you will get what you pay for and can trust their advice, whether you are selecting a gift for someone (or treating yourself) or whether you have an investment in mind. The three leading jewelers operating nationwide are H. Stern, Amsterdam Sauer and Roditi, but there are other reliable smaller chains. The top jewelers have shops in the airports and shopping centers and in most hotels.

Another good buy in Brazil is **leather** goods, especially shoes, sandals, bags, wallets and belts. Although found everywhere, some of the finest leather comes from the south of Brazil. Shoes are plentiful and handmade leather items can be found at handicraft street fairs.

# SPECIAL INFORMATION

## TOURIST INFORMATION

Brazil's national tourism board Embratur, headquartered in Rio de Janeiro, will send information abroad. Write to: Embratur, Rua Mariz e Barros, 13, 9° andar, Praça da Bandeira, 20000 Rio de Janeiro, RJ, Brazil.

In Brazil, each state has its own tourism bureau.

## AMAZON BOATS

The government-owned ENASA service for tourists ply the Belém–Manaus–Santarem–Belém run over 12 days. The modern boats have air-conditioned cabins, swimming pool, etc. Cost is US$370 per person in a 4-person cabin. Offices in Avenida Pres Vargas 41, Belém, and Rua Marechal Deoforo, Manaus.

Hammock space on the somewhat irregular ENASA passenger boats cost US$30 from Belém to Manaus and vice versa. Other services make the run, often in better conditions than ENASA boats. They take five to six days from Manaus to Belém, as well as working along the more remote tributaries to Colombia and Peru.

Arranging a passage is a matter of patience and luck in most cases as there are no fixed schedules for bookings.

## LOWER AMAZON BOAT TRIP

River craft: *Catamaran Pará*
Company: ENASA
Address: Avenida Presidente Vargas 41 (near docks)
Tel: 223-3011
Telex: (091) 2064
Telefax: 223-3234
This is a luxury trip with a pool on board, but don't expect to see much of the forest on the banks of the lower Amazon, with the exception of the Breves Channel. Belém–Manaus

seven-day/six-night; Manaus–Belém six-day/five-night. The route taken is the Narrows of the Breves Channel, Santarém, Partintins and Manaus. Downstream includes Soure.

## TOUR BOATS OUT OF MANAUS

**Note:** These boats can be booked in Manaus, but advance reservations are suggested for longer excursions.

Boats: *Dinossauro*, *Manati*, and *Uairana*
Information: Adventure Tours
Address: Rua Lima Bacuri 204
Tel: 233-6910
Seven/six-day packages for minumum of 8 passengers, US$840/ person.

Boat: *Chicla Ocellaris*
Information: Amazon Expeditions
Address: Caixa Postal 703
Tel: 232-7492
Telefax: 232-7492 (keeping trying)
Moacir Fortes will take you anywhere you want to go and show you everything on the Rio Negro. Recommended.

Boat: *Dona Carlota*
Information: Onzenave
Address: Rua Isabel 183 (Center)
Tel: 232-2481
60 cabins hold up to 100 passengers.

Agency: Pernoite na Selva (with three medium-sized boats)
Information: Amazon Explorers
Address: Rua Quintino Bocaiúva 189, suite 11 & 14
Tel: 233-9339
Telex: (092) 2859
Two-day/one-night for US$292 per person.

Boat: *Tunã*
Information: Safari Ecológico
Address: Rua Lima Bacuri 204
Tel: 233-6910
Naturalist-author John Harwood is often the guide aboard the *Tunã*. Three-day/three-night for US$499 per person.

Boats: (8 person sleepers)
Information: Amazon Nut Safaris Adventure Expeditions
Address: Rua 6 de Setembro, 43
Tel: 233-0154

## BELÉM

**Note:** Belém's travel agencies are mostly directed toward the sale of air passages and package tours, but can help to get travelers out of the city and into the "interior".

**Adetur**
Avenida Nazaré 121
Tel: 225-0155, 225-0056, 225-0267
Telex: (091) 1728

**Amazon Travel Service**
Rua do Mundurucus 1826 (Jurunas)
Tel: 223-1099, 223-1259
Telex: (091) 1260
Branch in Manaus.

**Beltur**
Avenida A. Vasconcelos 207
Tel: 241-1506
Telefax: 241-5011

**Carajás**
Avenida Presidente Vargas 762, shop 12
Tel: 225-1550
Telex: (091) 1625
Telefax: 223-0937

**Castur**
Avenida Gov. José Malcher 2463 (São Brás)
Tel: 226-5626
Telex: (091) 2353
Branches in Pará interior.

**Ciatur**
Avenida Pres. Vargas 645;
Rua dos Mundurcus 1688
Tel: 228-0011
Telex: (091) 2018, (091) 1390
Telefax: 226-4166
Daily river excursions. Booking for Marajó ranches.

**Curuá-Una**
Branch of Santarém agency, specializing in ecological tourism.

**Expresso Mercantil**
Rua Gaspar Viana 488
Tel: 223-8155, 223-8224
Telex: (091) 2070
Telefax: 225-3607
Bookings for Marajó.

**Gran-Pará**
Avenida Pres. Vargas 882 (Hilton Hotel shop 8)
Tel: 224-2111, 224-3233
Telex: (091) 1534, (091) 1565
Telefax: 224-2030
Patrick Barbier speaks French. Tours to Mosqueiro Island beach and Village of Icoaraci. Bookings for Marajó ranches.

**Lusotur**
Avenida Braz de Aguiar 612 (Equatiorial Hotel)
Tel: 224-2013 and 224-8855 ext. 7508
Telex: (091) 2999

**Marco Polo**
Avenida Piedade 539
Tel: 223-8682
Telex: (091) 2608
Telefax: 222-4434

**Metur - Marajó Turismo**
Rua Sen. Manoel Barata 727
Tel: 223-2128, 223-3100
Telex: (091) 1240
Marajó Island tours.

**Monopolo**
Avenida Presidente Vargas 325
Tel: 223-4099, 223-3177
Telex: (091) 2455
Telefax: 224-1384

**Muiraquitá**
Trav São Pedro 566, suite 101
Tel: 224-9066
Telex: (091) 0993
Telefax: 224-9494

**Mundial** (Tágide)
Avenida Presidente Vargas 780
Tel: 222-8108, 222-7414
Telex: (091) 0996
Telefax: 223-1981
Bookings for river tour, Mosqueiro, and Marajó.

**Mururé**
Rua Boaventura da Silva 562-A
Tel: 241-0891
Telex: (091) 2613
Excursions to Marajó Island. Same owner as Pousada Marajoada in Soure.

**Native Amazônia Turismo**
Address: Rua Carlos Gomes 305
Tel: 222-7720
Telex: (091) 2697
Telefax: 241-5238

**Neytur**
Rua Carlos Gomes 300
Tel: 224-2469, 224-4552, 223-9943
Telex: (091) 1595
Daily four-hour river excursion.

**Uirapuru**
Rua Sezedelo Correa 958
Tel: 225-3092
Telex: (091) 2936

## SANTARÉM, PARÁ

**Amazônia**
Trav. 15 de Novembro 185-A
Tel: 522-3325

**Lago Verde**
Rua Galdino Veloso 664
Tel: 522-1645
Fishing trips

**Tapan**
Trav. 15 de Agosto 127-A
Tel: 552-1946

**Amazon Tours**
Trav. Turiano Meira 1084
Tel: 522-1098
Steve Alexander, owner and operator, outfits river excursions for fishing and exploring. He operates Santa Lúcia Woods, near Santarém, with local forest guides.

**Turisan – Santarém Turismo**
Trav. São Francisco Correa 168, shop 3
Tel: 522-4543, 522-4706
Telex: (091) 5515

**Tarcísio Lopes** (Marine)
Trav. Francisco Corrêa 9
Tel: 522-2034
Books river transport.

**Agênsias Tropicais de Thurismo**
Avenida Mendinça Furtado 4120 (Hotel Tropical Santarém)
Tel: 522-1533

**Agência de Navegaçáo Marítima Tapajós**
Avenida Tapajós 905, Second Floor, Room 2
Tel: 522-1138
Bookings for ENASA vessels to Belém and Manaus.

## MANAUS, AMAZONAS

**EMAMTUR – Empresa Amazonense de Turismo**
Avenida Taraumã (Praça 24 de Outubro)
Tel: 234-9417, 234-5900
Telex: (092) 2279
This is the state tourist department.

**Acram**
Rua Monsenhor Coutinho 175
Tel: 234-2474
Telex: (092) 2578
Telefax: 234-2647

**Amazon Explorers**
Quintino Bocaiúva 189, suites 11/13
Tel: 234-9741
Telex: (092) 2859

**Amazonas**
Avenida Tarumã 379
Tel: 234-5983
Telex: (092) 2279

**BIC Turismo**
Rua Marcílio Dias 305
Tel: 233-7872
Telex: (092) 2503

**Itapemirim**
Avenida Getúlio Vargas 1169
Tel: 234-9877

**Selvatur**
Praça Adelberto Valle (Amazonas Hotel)
Tel: 234-8984

**Tarumã**
Avenida Darcy Vargas 520
Tel: 642-2100
Telex: (092) 1048
Telefax: 642-2255

**Tarusa**
Rua Monsenhor Coutinho 527
Tel: 233-1573
Telex: (092) 2792
Telefax: 232-0059

**Uatumã**
Rua Henrique Martins 116
Tel: 234-5071
Telex: (092) 1230
Telefax: 232-8539

**Vianatur**
Rua Salanha Marinho 606
Tel: 234-1427
Telex: (092) 3027

## GOVERNMENT TOURISM OFFICE

**Manaus**
Emamtur
Avenida Tarumã, 379
Tel: (092) 234-2252
Information Center: Airport.

**Rio de Janeiro**
Riotur – Rua da Assembleia, 10
8-9° andares
Tel: (021) 242-1947/242-8000;
Flumitur – Rua da Assembleia, 10
8° andar
Tel: (021) 398-4077
Information Centers: International Airport,
Bus Station, Corcovado, Sugar Loaf,
Cinelândia Subway Station, Marina da
Gloria.

**Salvador**
Bahiatursa – Praça
Municipal
Palácio do Rio Branco
Tel: (071) 241-4333
Information Centers: Airport, Bus Station,
Mercado Modelo, Porto da Barra.

**São Paulo**
Anhembi Centro de Feiras e Congressos
Avenida Olvavo Fontoura, 1209
Tel: (011) 267-2122
Information Centers: Praça da República,
Praça da Liberdade, Sé, Praça Ramos de
Azevedo, Avenida Paulista in front of Top
Center and at the corner of Rua Augusta,
Shopping Morumbi, Shopping Ibirapuera.

# FURTHER READING

Bruce, G. *Brazil and The Brazilians*. New
York, NY: Gordon Press Pubs., 1976.
Cooke M. *Brazil on the March*. New York,
NY: Gordon Press Pubs., 1976.
Denis, Pierre. *Brazil*. New York, NY: Gor-
don Press Pubs., 1977.

# PARKS & RESERVES

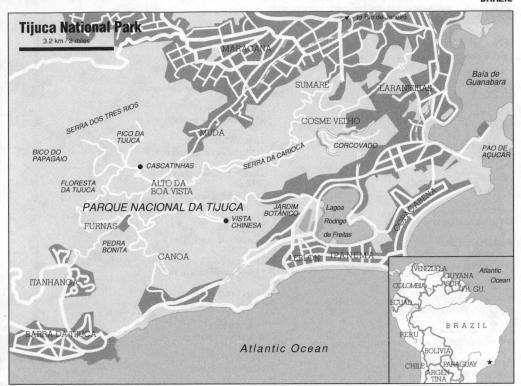

# Tijuca National Park

3.2 km / 2 miles

to Rio de Janeiro

MARACANA

SUMARE

LARANJEIRAS

Baía de Guanabara

SERRA DOS TRES RIOS

PICO DA TIJUCA

MUDA

COSME VELHO

BICO DO PAPAGAIO

CASCATINHAS

SERRA DA CARIOCA

CORCOVADO

PAO DE AÇUCAR

FLORESTA DA TIJUCA

ALTO DA BOA VISTA

PARQUE NACIONAL DA TIJUCA

JARDIM BOTÂNICO

Lagoa Rodrigo de Freitas

COPACABANA

FURNAS

VISTA CHINESA

PEDRA BONITA

CANOA

LEBLON

IPANEMA

ITANHANGA

BARRA DA TIJUCA

Atlantic Ocean

VENEZUELA
GUYANA
COLOMBIA
SUR.
FR. GU.
ECUAD.
Atlantic Ocean
PERU
B R A Z I L
BOLIVIA
CHILE
PARAGUAY
ARGEN-TINA

# Iguassu Waterfalls

16 km / 10 miles

Itaipu Lake

Itaipu Dam

Paraná R.

VENEZUELA
GUYANA
COLOMBIA
SUR.
FR. GU.
ECUAD.
Atlantic Ocean
B R A Z I L
PERU
BOLIVIA
CHILE
PARAGUAY
ARGEN-TINA

BR 277

to Curitiba

to Asunción

BRAZIL

Bridge

Foz de Iguaçu

Airport

Cataratas Road

Iguasu R.

PARAGUAY

Bridge

IGUASSU NAT'L

Puerto Iguazu

PARK

ARGENTINA

Iguassu Falls

A list of the scientific names of Amazon wildlife mentioned in this book.

## MAMMALS

Acouchi
(*Myoprocta exilis*)

Agouti, Red-rumped
(*Dasyprocta agouti*)

Alpaca
(*Lama pacos*)

Anteater, Giant
(*Myrmecophaga tridactyla*)

Anteater, Lesser
= Anteater, Tamandua
(*Tamandua tetradactyla*)

Anteater, Silky
= Anteater, Pigmy
(*Cyclopes didactylus*)

Armadillo, Giant
(*Priodontes maximus*)

Armadillo, Nine-banded
(*Dasypus novemcinctus*)

Armadillo, Three-banded
(*Tolypeutes matacus*)

Armadillo, Yellow
(*Euphractes sexcinctus*)

Bat, Bulldog
= Bat, Fishing
(*Noctilio leporinus*)

Bat, Common Vampire
(*Desmodus rotundus*)

Bat, Proboscis
(*Rhynconucteris naso*)

Bear, Spectacled
(*Tremarctos ornatus*)

Capuchin, Brown
(*Cebus apella*)

Capuchin, Weeping
(*Cebus olivaceus*)

Capuchin, White fronted
(*Cebus, albifrons*)

Capybara
(*Hydrochaeris hydrochaeris*)

Cat, Margay
(*Felis wiedii*)

Coati, South American
(*Nasua nasua*)

Cuy, Sacha
(*Cavia spec.*)

Deer, Grey Brocket
(*Mazama gouazoubira*)

Deer, Marsh
(*Blastocerus dichotomus*)

Deer, Red Brocket
(*Mazama americana*)

Deer, White-tailed
(*Odocoileus virginianus*)

Dolphin, Amazonian River
= Boto
(*Inia geoffrensis*)

Dolphin, Grey
= Tucuxi
(*Sotalia fluviatilis*)

Dog, Bush
(*Speothos venaticus*)

Dog, Small-eared
(*Atelocynus microtis*)

Fox, Andean
(*Dusicyon griseus*)

Grison
(*Galicitis vittata*)

Jaguar
(*Panthera onca*)

Jaguarundi
(*Felis jagouarundi*)

Kinkajou
(*Potos flavus*)

Manatee, Amazonian
(*Trichechus inunguis*)

Marmoset, Bare-ear
= Marmoset, silvery
(*Callithrix, argentata*)

Marmoset, Buffy Tufted-ear
(*Callithrix aurita*)

Marmoset, Common
(*Callithrix jacchus*)

Marmoset, Pigmy
(*Cebuella pygmaea*)

Marmoset, Tassel-eared
(*Callithrix humeralifer*)

Marmoset, Yellow-handed
(*Saguinus midas*)

Monkey, Black Howler
(*Alouatta caraya*)

Monkey, Brown Howler
(*Alouatta fusca*)

Monkey, Common Woolly
= Monkey, Common Humbold's
(*Lagothrix lagothricha*)

Monkey, Goeldi's
(*Callimico goeldii*)

Monkey, Mantled Howler
(*Alouatta palliata*)

Monkey, Night
(*Aotus vociferans*)

Monkey, Red Howler
(*Alouatta seniculus*)

Monkey, Red-handed Howler
(*Alouatta belcebul*)

Monkey, Spider
= Monkey, Black Spider
(*Ateles paniscus*)

Monkey, Squirrel
(*Saimiri scicureus*)

Monkey, Woolly Spider
(*Brachyteles arachnoides*)

Monkey, Yellow-tailed Woolly
(*lagothrix flavicauda*)

Mouse, House
(*Mus musculus*)

Muriqui
(*Brachyteles arachnoides*)

Ocelot
(*Felis pardalis*)

Olingo
(*Bassaricyon gabbii*)

Oncilla
(*Felis tigrina*)

Opossum, Bare-tailed
(*Caluromys philander*)

Opossum, Common
(*Didelphis marsupialis*)

Opossum, Grey four-eyed
(*Philander opossum*)

Opossum, Water
(*Chironectes minimus*)

Opossum, Western Woolly
(*Caluromys lanatus*)

Otter, Giant
(*Pteronura brasiliensis*)

Otter, Southern River
(*Lutra longicaudus*)

Paca
(*Agouti paca*)

Peccary, Collared
(*Tayassu tajacu*)

Peccary, White-lipped
(*Tayassu pecari*)

Porcupine, Prehensile-tailed
(*Coendon prehensilensis*)

Pudu
(*Pudu pudu*)

Puma
(*Felis concolor*)

Raccoon, Crab-eating
(*Procyon cancrivorous*)

Rat, Bamboo
(*Dactylomus dactylinus*)

Rat, Black
(*Rattus rattus*)

Rat, Spiny
(*Proechymis spec.*)

Rat, Tree
(*Echymis spec.*)

Saki, Monk
(*Pithecia monachus*)

Saki, Red-backed
(*Chiropotes satanas*)

Saki, White-faced
(*Pithecia pithecia*)

Skunk, Hog-nosed
(*Conepatus semistriatus*)

Sloth, Brown-throated Three-toed
(*Bradypus variegatus*)

Sloth, Maned
(*Bradypus torquatus*)

Sloth, Pale-throated Three-toed
(*Bradypus tridactylus*)

Sloth, Southern Two-toed
(*Choloepus didactylus*)

Squirrel, Fire-vented Tree
(*Sciurus igniventris*)

Squirrel, Grey Tree
(*Sciurus aestuans*)

Tamarin, Black-mantle
(*Saguinus negricollis*)

Tamarin, Cotton-top
(*Saguinus oedipus*)

Tamarin, Emperor
(*Saguinus imperator*)

Tamarin, Lion
(*Leontopithecus rosalia*)

Tamarin, Red-handed
(*Saguinus midas*)

Tamarin, Saddle-back
(*Saguinus fuscicollis*)

Tapir, Baird's
(*Tapirus bairdii*)

Tapir, Brazilean
(*Tapirus terrestris*)

Tapir, Mountain
(*Tapirus pinchaque*)

Tayra
(*Eira barbara*)

Titi, Dusky
(*Callicebus moloch*)

Titi, Masked
(*Callicebus personatus*)

Uakari, Black
(*Cacajao melanocephalus*)

Uakari, Red
(*Cacajao calvus*)

Vicuna
(*Vicugna vicugna*)

Wolf, Maned
(*Chrysocyon brachyurus*)

# BIRDS

Amazon, Vinaceous
= Parrot, Vinaceous
(*Amazona vinacea*)

Anhinga
(*Anhinga anhinga*)

Ani, Greater
(*Crotophaga major*)

Antbird, White-bellied
(*Myrmeciza longipes*)

Antpitta, Fulvous-bellied
(*Hylopezus fulviventris*)

Antshrike, Variable
(*Thamnophilus caerulescens*)

Ant-Tanager
(*Habia spec.*)

Bare-eye, Black-spotted
(*Phlegopsis nigromaculata*)

Booby, Brown
(*Sula leucogaster*)

Booby, Masked
(*Sula dactylatra*)

Buzzard-Eagle, Black-chested
(*Geranoaetus melanoleucus*)

Cacique, Red-rumped
(*Cacicus haemorrhous*)

Cacique, Yellow-rumped
(*Cacicus cela*)

Canastero, Many-striped
(*Asthenes flammulata*)

Caracara, Carunculated
(*Phalcoboenus carunculatus*)

Caracara, Crested
(*Polyborus plancus*)

Caracara, Mountain
(*Phalcoboenus megalopterus*)

Caracara, Red-throated
(*Daptrius americanus*)

Caracara, Yellow-headed
(*Milvago chimachima*)

Cardinal, Red-capped
(*Paroaria gularis*)

Chachalaca
(*Ortalis spec.*)

Chlorophonia, Blue-naped
(*Chlorophonia cyanea*)

Cock-of-the-Rock, Andean
(*Rupicola peruviana*)

Cock-of-the-Rock, Guianan
(*Rupicola rupicola*)

Condor, Andean
(*Vultur gryphus*)

Conebill, Rufous-browed
(*Canirostrum rufum*)

Conure, Golden
= Conure, Golden-plumed
= Parrot, Golden-plumed
(*Leptositaca branickii*)

Conure, Mitred
= Parakeet, Mitred
(*Aratinga mitrata*)

Conure, Pearly
= Parakeet, Pearly
(*Pyrrhura perlata*)

Cormorant, Neotropical
(*Phalacrocorax olivaceus*)

Cotinga, Black-faced
(*Conioptilon mcilhenny*)

Cotinga, Plum-throated
(*Cotinga maynana*)

Cotinga, Black-necked Red
(*Phoenicircus nigricollis*)

Cuckoo, Black-bellied
(*Piaya melanogaster*)

Cuckoo, Guira
(*Guira guira*)

Cuckoo, Squirrel
(*Piaya cayana*)
Cuckoo, Yellow-billed
(*Coccyzus erythropthalmus*)

Curassow, Black
(*Crax alector*)

Curassow, Nocturnal
(*Nothocrax urumutum*)

Curassow, Razor-billed
(*Mitu mitu*)

Curassow, Southern Horned
(*Pauxi unicornis*)

Dacnis, Blue
(*Dacnis cayana*)

Dipper, White-capped
(*Cinclus leucocephalus*)

Donacobius, Black-capped
= Mocking-Thrush, Black-capped
(*Donacobius atricapillus*)

Duck, Muscovy
(*Cairina moschata*)

Duck, Torrent
(*Merganetta armata*)

Eagle, Crested
(*Morphnus guianensis*)

Eagle, Harpy
(*Harpia harpyja*)

Egret, Great
(*Casmerodius albus*)

Emerald, Glittering-throated
(*Amazilia fimbriata*)

Fairy, Black-eared
(*Heliothrix aurita*)

Falcon, Aplomado
(*Falco femoralis*)

Falcon, Bat
(*Falco rufigularis*)

Flowerpiercer, Masked
(*Diglossa cyanea*)

Flycatcher, Boat-billed
(*Megarhynchus pitangua*)

Flycatcher, Fork-tailed
(*Muscivora tyrannus*)

Foliage-gleaner
(*Philydor etc. spec.*)

Frigatebird, Magnificent
(*Fregata magnificens*)

Fruitcrow, Bare-necked
(*Gymnoderus foetidus*)

Gallinule, Azure
(*Porphyrula flavirostri*)

Gallinule, Purple
(*Porphyrula martinica*)

Gnateater, Slate-throated
(*Conopophaga spec.*)

Goose, Orinoco
(*Neochen ubata*)

Greenlet, Lesser
(*Hylophilus minor*)

Grosbeak, Slaty
(*Pitylus grossus*)

Ground-Cuckoo, Banded
(*Neomorphus radiolosus*)

Ground-Cuckoo, Red-billed
(*Neomorphus pucheranii*)

Ground-Dove, Ruddy
(*Columbina talpacoti*)

Guan, Andean
(*Penelope montagnii*)

Guan, Black-fronted Piping
(*Pipile jacutinga*)

Guan, Marail
(*Penelope marail*)

Guan, Piping
(*Aburria pipile*)

Guan, Rusty-margined
(*Penelope superciliaris*)

Hawk, Puna
(*Buteo poecilochrous*)

Hawk, Red-backed
(*Buteo polysoma*)

Hawk, Roadside
(*Buteo magnirostris*)

Helmetcrest, Bearded
(*Oxypogon guerinii*)

Hermit, Straight-billed
(*Phaethornis bourcieri*)

Heron, Agami
= Heron, Chestnut-bellied
(*Agamia agami*)

Heron, Boat-billed
(*Cochlearius cochlearius*)

Heron, Cocoi
= Heron, White-necked
(*Ardea cocoi*)

Heron, Great Blue
(*Ardea herodias*)

Heron, Little Blue
(*Florida caerulea*)

Heron, Zigzag
(*Zebrilus undulatus*)

Hillstar, Andean
(*Oreotrochilus estella*)

Hoatzin
(*Opisthocomus hoazin*)

Honeycreeper, Green
(*Chlorophanes spiza*)

Honeycreeper, Red-legged
(*Cyanerpes cyaneus*)

Hornero, Pale-legged
(*Furnarius leucopus*)

Hummingbird, Giant
(*Patagona gigas*)

Hummingbird, Swallow-tailed
(*Eupetomena macroura*)

Hummingbird, Sword-billed
(*Ensifera ensifera*)

Ibis, Green
(*Mesembrinibis cayennensis*)

Ibis, Scarlet
(*Eudocimus ruber*)

Ibis, Wood
= Wood-Stork, American
(*Mycteria americana*)

Inca, Collared
(*Coeligena torquata*)

Jabiru
(*Jabiru mycteria*)

Jacamar
(*Galbula spec.*)

Jacana, Wattled
(*Jacana jacana*)

Jay, Azure
(*Cyanocorax caeruleus*)

Jay, Green
(*Cyanocorax yncas*)

Jewelfront
(*Polyplancta aurescens*)

Kingfisher, Amazon
(*Chloroceryle amazona*)

Kingfisher, Green
(*Chloroceryle americana*)

Kingfisher, Green-and-Rufous
(*Chloroceryle inda*)

Kingfisher, Pigmy
(*Chloroceryle aenea*)

Kingfisher, Ringed
(*Ceryle torquata*)

Kiskadee, Great
(*Pitangus sulfuratus*)

Kite, Snail
= Kite, Everglade
(*Rostrhamus sociabilis*)

Lapwing, Andean
(*Vanellus resplendens*)

Lapwing, Southern
(*Vanellus chilensis*)

Leaftosser
(*Sclerurus rufigularis*)

Macaw, Blue-and-Yellow
(*Ara ararauna*)

Macaw, Glaucous
(*Anodorhynchus glaucus*)

Macaw, Golden-naped
= Macaw, Golden-collared
(*Ara auricollis*)

Macaw, Hyacinthine
(*Anodorhynchus hyacinthinus*)

Macaw, Red-and-Green
(*Ara chloroptera*)

Macaw, Red-fronted
(*Ara rubrogenys*)

Macaw, Scarlet
(*Ara macao*)

Macaw, Spix's
(*Cyanopsitta spixii*)

Manakin, Blue-backed
(*Chiroxiphia pareola*)

Manakin, Golden-headed
(*Pipra erythrocephala*)

Manakin, Pin-tailed
(*Ilicura militaris*)

Manakin, Swallow-tailed
(*Chiroxiphia caudata*)

Manakin, White-bearded
(*Manacus manacus*)

Manakin, Wire-tailed
(*Teleonema filicauda*)

Mango, Black-throated
(*Anthracocorax nigricollis*)

Martin, Bank
(*Riparia riparia*)

Martin, Southern
(*Procne modesta*)

Merganser, Brazilian
(*Mergus octosetaceus*)

Metaltail, Tyrian
(*Metallura tyrianthina*)

Motmot, Highland
(*Momotus aequatorialis*)

Mountain-Tanager, Scarlet-bellied
(*Anisognathus igniventris*)

Mountaineer
(*Oreonympha nobilis*)

Nighthawk, Band-tailed
(*Nyctiprogne leucopyga*)

Nighthawk, Short-tailed
(*Lurocalis semitorquatus*)

Nightjar, Blackish
(*Caprimulgus nigrescens*)

Nothura, Spotted
(*Nothura maculosa*)

Nunbird, Black-fronted
(*Monasa nigrifrons*)

Nunlet, Rusty-breasted
(*Nonnula rubecula*)

Oropendola, Russet-backed
(*Psarocolius angustifrons*)

Osprey
(*Pandion haliaetus*)

Owl, Tropical Screech
(*Otus choliba*)

Palmcreeper
(*Berlepschia spec.*)

Parakeet, Andean
(*Bolborhynchus orbygnesius*)

Parakeet, Canary-winged
(*Brotogeris versicolurus*)

Parakeet, Cobalt-winged
(*Brotogeris cyanoptera*)

Parakeet, Dusky-headed
(*Aratinga weddellii*)

Parakeet, Golden-winged
(*Brotogeris chrysopterus*)

Parakeet, Tui
(*Brotogeris sanctithomae*)

Parrot, Festive
(*Amazona festiva*)

Parrot, Mealy
(*Amazona farinosa*)

Parrot, Orange-winged
(*Amazona amazonica*)

Parrot, Pileated
(*Pionopsitta pileata*)

Parrot, Purple-bellied
(*Trichlaria malachitacea*)

Parrot, Red Fan
(*Deroptyus accipitrinus*)

Parrot, Red-spectacled
(*Amazona pretrei*)

Parrot, Red-tailed
=Parrot, Scaly-headed
(*Pionus maximiliani*)

Parrot, Vinaceous
= Parrot, Vinaceous-breasted
(*Amazona vinacea*)

Pauraque
(*Nyctidromus albicollis*)

Pelican, Brown
(*Pelecanus occidentalis*)

Pigeon, Ruddy
(*Columba subvinacea*)

Pigmy-Tyrant
(*Euscarthmus spec.*)

Piha, Screaming
(*Lipaugus vociferans*)

Plover, Collared
(*Charadrius collaris*)

Potoo, Common
(*Nyctibius griseus*)

Potoo, Great
(*Nyctibius grandis*)

Potoo, Long-tailed
(*Nyctibus grandis*)

Puffbird, Black-breasted
(*Notharchus pectoralis*)

Puffbird, Rufous-necked
(*Malacophila rufa*)

Puffbird, Spotted
(*Bucco tamiata*)

Quetzal, Golden-headed
(*Pharomachrus auriceps*)

Quetzal, Pavonine
(*Pharomachrus pavoninus*)

Redstart, Golden-fronted
(*Myiborus ornatus*)

Rhea
(*Rhea americana*)

Sapphirewing, Great
(*Pterophanes cyanopterus*)

Screamer, Horned
(*Anhima cornuta*)

Scythebill, Red-billed
(*Campylorhamphus trochilirostris*)

Seedeater, Chestnut-bellied
(*Sporophila castaneiventris*)

Seriema, Red-legged
(*Cariama cristata*)

Shrike-Vireo, Slaty-capped
(*Smaragdolanius leucotis*)

Sierra-Finch, Plumbeous
(*Phrygilus unicolor*)

Skimmer, Black
(*Rynchops nigra*)

Snipe, Andean
(*Gallinago jamesoni*)

Snipe, Noble
(*Gallinago nobilis*)

Spadebill
(*Platyrinchus spec.*)

Spinetail
(*Synallaxis spec.*)

Spoonbill, Roseate
(*Ajaja ajaja*)

Stork, Jabiru
see Jabiru

Stork, Wood
= Wood-Ibis, American
(*Mycteria americana*)

Sunbittern
(*Eurypyga helias*)

Sungrebe
= Finfoot
(*Heliornis fulicula*)

Swallow, Barn
(*Hirundo rustica*)

Swallow-Wing
(*Chelidoptera tenebrosa*)

Swift, Short-tailed
(*Chaetura brachyura*)

Swift, White-collared
(*Aeronautes andecolus*)

Tanager, Blue-grey
(*Thraupis episcopus*)

Tanager, Blue-whiskered
(*Tangara johannae*)

Tanager, Dusky-faced
(*Mitrospingus cassinii*)

Tanager, Gold-and-green
(*Tangara schrankii*)

Tanager, Golden-crowned
(*Iridosornis rufivertex*)

Tanager, Golden-hooded
(*Tangara larvata*)

Tanager, Grass-green
(*Chlorornis riefferii*)

Tanager, Green-headed
(*Tangara seledon*)

Tanager, Grey-and-Gold
(*Tangara palmeri*)

Tanager, Opal-rumped
(*Tangara velia*)

Tanager, Palm
(*Thraupis palmarus*)

Tanager, Paradise
(*Tangara chilensis*)

Tanager, Red-necked
(*Tangara cyanocephala*)

Tanager, Scarlet-and-White
(*Erythrothlypis salmoni*)

Tanager, Silver-beaked
(*Ramphocelus carbo*)

Tanager, Tawny-crested
(*Tachyphonus cristatus*)

Teal, Puna
(*Anas puna*)

Tern, Large-billed
(*Phaetusa simplex*)

Tern, Yellow-billed
(*Sterna superciliaris*)

Thistletail, White-chinned
(*Schizoeaca fuliginosa*)

Thornbill, Rainbow-bearded
(*Chalcostigma herrani*)

Thrush, Cocoa
(*Turdus fumigatus*)

Thrush, Glossy Black
(*Turdus serranus*)

Thrush, Great
(*Turdus fuscater*)

Thrush, Lawrence's
(*Turdus lawrencii*)

Thrush, Pale-breasted
(*Turdus leucomelas*)

Thrush, Rufous-bellied
(*Turdus rufiventris*)

Tiger-Heron, Rufescent
(*Tigrisoma lineatum*)

Tinamou, Brown
(*Crypturellus obsoletus*)

Tinamou, Great
(*Tinamus major*)

Tinamou, Red-winged
(*Rhynchotus rufescens*)

Tinamou, Solitary
(*Tinamus solitarius*)

Toucan Grey-breasted Mountain
(*Andigena hypoleuca*)

Toucan, Red-breasted
(*Ramphastos dicolorus*)

Toucan, Yellow-ridged
(*Ramphastos dicolorus*)

Toucanet, Spot-billed
(*Selenidera maculirostris*)

Trainbearer, Black-tailed
(*Lesbia victoriae*)

Trumpeter, Grey-winged
(*Psophia crepitans*)

Trumpeter, Pale-winged
(*Psophia leucoptera*)

Tyrannulet, White-throated
(*Mecocerculus leucophrys*)

Tyrant, White-winged Black
(*Lessonia oreas*)

Umbrellabird, Amazonian
(*Cephalopterus ornatus*)

Umbrellabird, Long-wattled
(*Cephalopterus penduliger*)

Vulture, Black
(*Coragyps atratus*)

Vulture, Greater Yellow-headed
(*Cathartes melambrotus*)

Vulture, Lesser Yellow-headed
(*Cathartes burrovianus*)

Vulture, King
(*Sarcoramphus papa*)

Vulture, Turkey
(*Cathartes aura*)

Warbler, Black-and-White
(*Mniotilta varia*)

Warbler, Blackburnian
(*Dendroica fusca*)

Water-Tyrant, Masked
(*Fluvicola nengeta*)

Woodcreeper
(*various genera like Xiphorhynchus, Lepidocolaptes etc.*)

Woodpecker, Cream-colored
(*Celeus flavus*)

Woodpecker, Crimson-crested
(*Campephilus melanoleucos*)

Woodpecker, Red-necked
(*Campephilus rubricollis*)

Woodpecker, Yellow-tufted
(*Melanerpes cruentatus*)

Wood-Quail, Spot-winged
(*Odontophorus capueira*)

Wood-Rail, Grey-necked
(*Aramides cajanea*)

Wren, Musician
(*Cyphorhinus arada*)

Wren, Nightingale
(*Microcerculus marginatus*)

Wren, Thrush-like
(*Campylorhynchus turdinus*)

## BIBLIOGRAPHY

Anhalzer, J.J. 1990. *National Parks of Ecuador*. Imprenta Mariscal, Quito.

Bradbury, A. 1990. *Backcountry Brazil, the Pantanal, Amazon, and the North-east Coast*. Hunter Publ., Edson, N.J.

Campbell, D, and Hammond, H.D. (eds). 1989. *Floristic Inventory of Tropical Countries*. New York Botanical Garden.

Cockburn, A., and Hecht, S.B. 1989. *The Fate of the Forest: Developers, Destroyers, and Defenders of the Amazon*. Verso, London and New York.

Diamond, A.W., and Lovejoy, T.E. 1984. *Conservation of Tropical Forest Birds*. International Council for Bird Preservation Technical Publication No. 4. ICBP, 219, Huntingdon Road, Cambridge, England.

Dixon, J.R., and Soini, P. 1986. *The Reptiles of the Upper Amazon Basil, Iquitos Region, Peru*. Milwaukee Public Museum, Milwaukee, Wisconsin.

Dunning, J.S. 1982. *South American Land Birds, A Photographic Aid to Identification*. Harrowood Books, Newton Square, Pennsylvania. ISBN 0-915-180-22-7.

Eisenberg, J.F. 1989. *Mammals of the Neotropics. Volume I: The Northern Neotropics*. University of Chicago Press. ISBN 0-226-19540-6.

Emmons, L.H., and Feer, F. 1990. *Neotropical Rainforest Animals, A Field Guide*. University of Chicago Press. ISBN 0-226-20716-1.

French, R. 1976. *A Guide to the Birds of Trinidad and Tobago*. Harrowood, Valley Forge, Pennsylvania, USA.

Fjeldsa, J., and Krabbe, N. 1989. *Birds of the High Andes*. Apollo Books, Svendborg, Denmark.

Goulding, M. 1989. *Amazon, the Flooded Forest*. BBC Books. ISBN 0-563-2-7043.

Guia Quatro Rodas 1991. *Guia Brasil*. Editora Abril, Sao Paulo.

Hemming, J. 1987. *Amazon Frontier, The Defeat of the Brazilian Indians*. Macmillan, London.

Hilty, S.L. and Brown, W.L. 1986. *A Guide to the Birds of Colombia*. Princeton University Press, Princeton.

Jacobs, M. 1988. *The tropical rain forest, A First Encounter*. Springer Verlag. ISBN 3-540-17996-8 (Berlin), ISBN 0-38717996-8 (New York).

Kricher, J. 1989. *A Neotropical Companion: A Guide to the Animals, Plants*. Princeton, New Jersey, USA.

Meyer de Schauensee, R., and Phelps, W.H. 1978. *A Guide to the Birds of Venezuela*. Princeton University Press.

Murphy, W.L. 1987. *A Birder's Guide to Trinidad and Tobago*. Peregrine Enterprises, College Park, Maryland, USA.

Orlog, C.C. 1984. *Las Aves Argentinas*. Administracion de Parques Nacionales. ISBN 84-499-5802-4.

Padua, M.T.J., and Filho, A.F.C. *Os Parques Nacionais do Brasil*. Instituto de Cooperacao Iberoamerica. ISBN 84-85389-19-0.

Penny, N.D., and Arias, J.R. 1982. *Insects of an Amazon Forest*. Columbia University Press, New York.

Ridgely, R.S., and Tudor, G. 1989. *The Birds of South America. Vol. I. The Oscine Passerines*. University of Texas Press. ISBN 0-292-70756-8.

Smith, A. 1990. *Explorers of the Amazon*. Viking, London.

Smith, N.J.H. 1982. *Rainforest Corridors: The Transamazon Colonization Scheme*. University of California Press, Berkeley.

Trupp, F. 1983. *Amazonas*. Verlag Schroll, Wien. ISBN 3-7031-0582-9.

# ART/PHOTO CREDITS

*Photography by*

# INDEX

# G

# H

# I